ENVIRONMENTAL SCIENCE, ENGINEERING AND TECHNOLOGY

AUTOMOTIVE EXHAUST EMISSIONS AND ENERGY RECOVERY

ENVIRONMENTAL SCIENCE, ENGINEERING AND TECHNOLOGY

Additional books in this series can be found on Nova's website under the Series tab.

Additional e-books in this series can be found on Nova's website under the e-book tab.

ENVIRONMENTAL SCIENCE, ENGINEERING AND TECHNOLOGY

AUTOMOTIVE EXHAUST EMISSIONS AND ENERGY RECOVERY

APOSTOLOS PESIRIDIS
EDITOR

nova publishers
New York

For permission to use material from this book please contact us:
Telephone 631-231-7269; Fax 631-231-8175
Web Site: http://www.novapublishers.com

Library of Congress Cataloging-in-Publication Data

ISBN: 978-1-63321-493-4

Published by Nova Science Publishers, Inc. † New York

CONTENTS

PREFACE

Since the invention of the first commercially successful internal combustion (IC) engine by Nikolaus Otto and Eugen Langen in 1866, the ICE has remained the most significant and widely used form of energy conversion technology in the transportation sector. Throughout the years, IC engines have been principally diversified by type of fuel, type of fuel injection and combustion mixing process as well as through the type of air handling and exhaust energy recovery technology used and improvements have made it more efficient and reliable. However, years of air pollution as a result of emissions from IC engines have made the development and integration of exhaust emissions mitigation technologies and systems increasingly significant in the continued effort to provide engines to the market which conform to increasingly stringent emissions regulations.

ICE development has to take into account air pollutants such as Nitrogen Oxides (NO_x), Carbon Monoxides (CO) and Hydrocarbons (HC), particulate matter (PM) as well as greenhouse gases (GHG) such as Carbon Dioxide (CO_2) and Nitrous Oxide (N_2O). In Europe, for example, Euro1 introduced in 1992 is the first of many subsequent EU regulations which regulates air pollutants. In 2009, the European Commission brought about mandatory CO_2 emission targets to regulate the new passenger car fleet CO2 emissions at 130 g/km by 2015 and 95 g/km by 2020. Emissions regulations require the mitigation of certain, previously unregulated emissions posing a health risk such as Nitrogen Dioxide (NO_2), ammonia as well as the formation of GHGs. New legislation is now including limits for NH_3 and certain specialist applications have been required to cap NO_2 and N_2O.

Another immediate concern for improved engine efficiency and fuel economy includes customer demand to own and drive more fuel efficient vehicles. This market-driven demand places additional pressure towards the development of more efficient engines in addition to emissions mitigation requirements and has resulted in a proliferation of systems and technologies in the exhaust system of the IC engine to recover the significant levels of exhaust gas energy expended after the combustion/power stroke. These include new forms of turbocharging, turbocompounding and waste heat recovery technologies.

The above driving concerns for fuel economy and reduced emissions make the topic of emissions control and exhaust energy recovery a timely one for both gasoline and diesel engines. Whereas diesel engines have been predominantly turbocharged only a relatively small percentage of gasoline engines is similarly equipped (especially in the US and large Asian markets) which has led towards significant efforts by engine manufacturers in recent years to downsize and downspeed these engines. On the other hand, the relative focus in diesel engine development in terms of emissions and exhaust energy recovery has shifted towards devices other than the turbocharger for enhanced energy recovery and in emissions control technologies to allow the diesel engines of the future to keep up with the twin demand for very low emissions and increasing levels of fuel economy.

The present volume on "Automotive Exhaust Emissions and Energy Recovery" focusses, therefore, on the exhaust system and on the technologies and methods used to reduce emissions and increase fuel economy by capitalising upon the exhaust gas energy availability (either in the form of gas kinetic energy or as waste heat extracted from the exhaust gas). It is projected that in the short to medium term, advances in exhaust emissions and energy recovery technologies will lead the way in IC engine development and pave the way towards increasing levels of engine hybridisation until full electric vehicle technology can claim a level of maturity and corresponding market share to turn the bulk of this focus away from the ICE.

The book is comprised of ten chapters which in most cases provide a review of recent developments as well as future directions for both gasoline and diesel four-stroke engines. As such the present volume is aimed at engine research professionals in the industry and academia in the first place but also towards students of powertrain engineering. The collection of articles in this book aims to review both fundamentals of relevant, recent exhaust system technologies but to also detail recent or on-going projects and to uncover future research directions and potentials where relevant. The content is not divided in sections but individual chapters follow the approximate route of the exhaust gas from in-cylinder formation in the initial chapters to waste heat recovery technologies at the end with discussion on bio-fuels included where relevant.

The initial chapter run of six chapters is principally dedicated to the emissions (mitigation and control) part of the book. Chapter 1 starts off with a description and review of emissions mitigation and control systems for both gasoline and diesel engines. The systems covered are three-way-catalysts, exhaust gas re-circulation (EGR), oxidation catalysts, particulate filters, selective catalytic reduction (SCRs) and leanNO_x trap designs as well as water injection systems. Diesel in-cylinder NO_x and soot formation by means of optical techniques is investigated in Chapter 2 for rapeseed methyl ester (RME) combustion. Further to the topic of NO_x, the influence of different EGR rates to diesel engine emissions is experimentally investigated and presented in Chapter 3. Chapter 4 is a review of literature on the effects of biofuel/diesel blends on particulate matter (PM) emissions from diesel engines operating under transient conditions and a statistical analysis allows comparisons to be drawn for the different types of fuel. Chapter 5 is a review of aftertreatment technologies with the provision of not only the physico-chemical phenomena and the respective mathematical modeling equations describing the transport and reaction processes but moves beyond the discussion of Chapter 1, also, in that it focusses on system design challenges from the control and optimisation points of view. Chapter 6, concludes the initial chapter run on in-cylinder measurements and aftertreatment technologies by focussing exclusively on Diesel Particulate Filters (DPFs); the chapter is a review of DPFs with a focus on filter material choices and the wall flow DPF design considerations.

The final chapter run of four chapters focusses on exhaust (both mechanical and thermal) energy recovery technologies and its impact on fuel economy (as well as emissions). Chapter 7 is a review of turbocharging technology, covering fundamentals as well as engine-turbocharger matching and applications of such systems in modern use. Chapter 8, focusses on small, high power density, directly injected (DI), turbocharged engines which are of wide interest given the industry's focus of today on downsized, turbocharged, SI engines. This chapter reports on the trends in the turbo-gasoline DI technology and includes the implications from the use of three way catalytic converter aftertreatment for energy recovery and fuel economy while complying with pollutant emissions standards. The final two chapters review waste heat recovery technologies most recently introduced in the product range of several manufacturers. Chapter 9 focusses on automotive Organic Rankine Cycle (ORC) applications from the point of view of the design challenge associated with process component design (mainly of the expander in small scale systems) due to the low IC engine waste heat power availability and further challenges associated with the restricted available space for the process heat exchangers. The concluding chapter (no.10) is a review of technologies associated with mechanical as well as hybrid (electric) exhaust energy recovery systems, as well as of most waste heat recovery technologies currently in development including the development of systems based on Bottoming (including Rankine) Cycles as well as Thermoelectric generator systems

This book has been made possible by the dedication of contributing authors to agree to and to then proceed to complete their works within the agreed, final publication schedule, for which I am grateful. I would also like to thank Carra Feagaiga and the staff of NOVA Science Publishers for their professional support in preparing this book.

Dr Apostolos Pesiridis
College of Engineering, Design and Physical Sciences
Brunel University London
UK

December 2013

In: Automotive Exhaust Emissions and Energy Recovery ISBN: 978-1-63321-493-4
Editor: Apostolos Pesiridis

Chapter 1

EMISSIONS MITIGATION AND CONTROL SYSTEMS

David Lemon*, BTech CEng FIMechE
David Lemon Consultants

ABSTRACT

This chapter describes the various systems that have been used for the mitigation of regulated air quality exhaust emissions in both Gasoline and Diesel engines; i.e. oxides of nitrogen (NO_x), particulate matter (PM), particulate number (PN), hydrocarbons (HC) and carbon monoxide (CO). Potential "horizon" technologies are also investigated.

The systems covered in this chapter are: Three-Way-Catalyst (Gasoline NO_x, HC and CO), Exhaust Gas Re-Circulation (Diesel NO_x), Oxidation Catalyst (Diesel PM, HC and CO), Particulate Filter (Diesel PM plus HC and CO if catalysed), Selective Catalytic Reduction (Diesel NO_x) and LeanNO_x Trap (also known as NO_x Storage Catalyst or NO_x Adsorber Catalyst) designs (Diesel and Gasoline NO_x). Water Injection (Diesel or Gasoline NO_x) is also discussed as this has seen some limited use in larger Diesel engines and may be adopted in greater numbers in the future. Where there are significant variants in design approach these are also highlighted.

Reference is made to the timeline whereby each system, either by itself or in combination, was introduced to meet legislation requirements. For clarity, European OEMs' practice has been followed as driven by the European legislation. This is for both Light Duty and Heavy Duty vehicles where a different approach was undertaken for Diesel engines (as there was between Europe and the USA for instance). Against this, the port injected Gasoline engines (PISI) have used the same technology from Euro 1 of Three-Way-Catalyst with the variation being in the development of the chemistry, control and warm-up techniques across the same timeline.

Each device has its method of operation described together with typical performance indications. The demands the system places on the vehicle are discussed, both in the installation and, if relevant, their operation including On-Board-Diagnostics (OBD), together with any maintenance and infrastructure requirements. Also included are any desirable characteristics of fuels and / or lubricants to enable robust performance where appropriate. Where there are special issues when running combinations of devices – these are highlighted.

* Email: david.lemon@virgin.net.

If there is an associated control system requirement specific to the device its characteristics are described with an indication of calibration strategies. This includes the minimisation of certain emissions that were previously unregulated where there exists a risk to health such as ammonia (NH_3) and Nitrogen Dioxide (NO_2) or the formation of greenhouse gas; e.g. Nitrous Oxide (N_2O). New legislation is now including limits for NH_3 and certain specialist applications have been required to cap NO_2 and N_2O and an example is included.

For catalytic devices the main chemical reactions are stated together with an indication of their special materials requirements. If there is a beneficial effect on unregulated emissions then this is also included.

Keywords: Three-Way-Catalyst, Exhaust Gas Re-Circulation, Oxidation Catalyst, Diesel Particulate Filter , Selective Catalytic Reduction, LeanNO$_x$ Trap, NO$_x$ Storage Catalyst, NO$_x$ Adsorber Catalyst, Water Injection

INTRODUCTION

Design of the engine combustion, fuel injection system and valve operation will modify engine-out emissions. Each engine generation would have had significant changes in these areas in parallel to the addition of emissions mitigation systems. This chapter addresses the other means to obtain lower levels of regulated emissions from the tailpipe. As an example of the development of emission control systems the following tables are based on European Passenger Car Diesel, Heavy Duty Diesel and Passenger Car Gasoline legislation and the typical technology approaches taken for each level. There was divergence elsewhere globally according to the local legislation but for clarity these three examples were chosen to show how technologies were adopted over a 22 year timeline.

The following two tables list the abbreviations in typical use by the industry which will appear throughout the chapter:

Table 1. Engine and fuel systems abbreviations

Abbreviation	Application	Full name
DI	Diesel	Direct Injection
DISI	Gasoline	Direct Injection Spark Ignition
EUI	Diesel	Electronic Unit Injector
EUP	Diesel	Electronic Unit Pump
HPCR	Diesel	High Pressure Common Rail
IDI	Diesel	Indirect Injection
PISI	Gasoline	Port Injected Spark Ignition

Table 2. Emissions mitigation systems abbreviations

Abbreviation	Application	Full name	Mitigated emissions
DOC	Diesel	Diesel Oxidation Catalyst	HC, CO, PM
DPF	Diesel	Diesel Particulate Filter	PM (plus HC, CO if catalysed)
EGR	Diesel and Gasoline	Exhaust Gas Recirculation	NO_x
LNT	Diesel and Gasoline (DISI)	Lean NO_x Trap	NO_x
NAC	Diesel and Gasoline (DISI)	NO_x Adsorber Catalyst	NO_x
SCR	Diesel	Selective Catalytic Reduction	NO_x
NSC	Diesel and Gasoline (DISI)	NO_x Storage Catalyst	NO_x
TWC	Gasoline	Three-Way-Catalyst	NO_x, HC, CO

[1] LNT, NAC and NSC are essentially similar and these abbreviations are used by different suppliers.

Table 3. European Passenger Car Diesel: Development of emissions regulations and typical technologies

Legislation	Engine	Cycle	CO g/km	HC g/km	HC+NO_x g/km	NO_x g/km	PM g/km	PN n/km	FIE	NO_x	CO+HC	PM PN
Euro1 Jul-1992	IDI	ECE15+ EUDC	2.72		0.97		0.14		Mechanical Rotary 400 bar	EGR Open loop + On/Off	None	None
Euro2 Jan-1996	IDI DI	ECE15+ EUDC	1.00 1.00		0.70 0.90		0.08 0.10		Electronic Rotary or EUI 1000+ bar (DI)	EGR Closed loop + Variable lift valve	DOC	DOC
Euro3 Jan-2000	DI	NEDC	0.64		0.56	0.50	0.05		Electronic HPCR or EUI 1200+ bar	EGR Closed loop + Variable lift valve + Cooler	DOC	DOC or DPF
Euro4 Jan-2005	DI	NEDC	0.50		0.30	0.25	0.025		Electronic HPCR 1400+ bar	EGR Closed loop + Variable lift valve + Cooler	DOC	DPF

Table 3. (Continued)

Legislation	Engine	Cycle	CO g/km	HC g/km	HC+NO$_x$ g/km	NO$_x$ g/km	PM g/km	PN n/km	FIE	NO$_x$	CO+HC	PM PN
Euro5a Sep-2009	DI	NEDC	0.50		0.23	0.18	0.005		Electronic HPCR 1600+ bar	EGR Closed loop + Variable lift valve + Cooler	DOC	DPF
Euro5b Sep-2011	DI	NEDC	0.50		0.23	0.18	0.005	6x 10^{-11}	Electronic HPCR 1600+ bar	EGR Closed loop + Variable lift valve + Cooler	DOC	DPF
Euro6 Sep-2014	DI	NEDC	0.50		0.17	0.08	0.005	6x 10^{-11}	Electronic HPCR 1800+ bar	EGR Closed loop + Variable lift valve + Cooler	DOC	DPF

[1] The ECE15+EUDC cycle measured emissions from time = 40 seconds whereas the NEDC measured from time = zero (otherwise the cycles used the same speed / load versus time).

[2] DI engine technology superseded IDI for Euro 3.

[3] European nomenclature used PI for Diesel; i.e. "Positive Ignition".

Table 4. European Heavy Duty Diesel - Development of emissions regulations and typical technology

Legislation	Cycle	CO g/kWh	HC g/kWh	NO$_x$ g/kWh	PM g/kWh	PN m^{-1}	FIE	NO$_x$	CO+HC	PM PN
Euro1 Oct-1992	ECE R49	4.5 4.5	1.1 1.1	8.0 8.0	0.612 0.36		Mechanical In-Line 1000 bar	None	None	None
Euro2a Oct-1996	ECE R49	4.0	1.1	7.0	0.25		Mechanical In-Line 1000 bar	None	None	None
Euro2b Oct-1998	ECE R49	4.0	1.1	7.0	0.15		Mechanical In-Line 1000 bar	None	None	None
Euro3 Oct-2000	ESC ETC	2.1 5.45	0.66 0.78	5.0 5.0	0.10 0.16		Electronic In-line or HPCR or EUI / EUP 1200 bar to 1600 bar	None	None	None

Legislation	Cycle	CO g/kWh	HC g/kWh	NO_x g/kWh	PM g/kWh	PN m^{-1}	FIE	NO_x	CO+HC	PM PN
Euro4 Oct-2005	ESC ETC	1.5 4.0	0.46 0.55	3.5 3.5	0.02 0.03		HPCR or EUI / EUP 1400 bar to 1800 bar	SCR or EGR Closed loop + Cooler	None	None with SCR DPF with EGR
Euro5 Oct-2008	ESC ETC	1.5 4.0	0.46 0.55	2.0 2.0	0.02 0.03		HPCR or EUI / EUP 1600 bar to 2000 bar	SCR or EGR Closed loop + Cooler	None	None with SCR DPF with EGR
Euro6 Jan-2013	WHSC WHTC	1.5 4.0	0.13 0.16	0.4 0.46	0.01 0.01	$8x10^{-11}$ $8x10^{-11}$	HPCR or EUI / EUP 1800 bar to 2200 bar	SCR and EGR Closed loop + Cooler	DOC	DPF

[1] From Euro 3 the ELR test has also been run and this with its smoke opacity targets have been omitted for reasons of space.

[2] For R49 and ESC the regulations are for THC (total hydrocarbons). For ETC they are expressed as NMHC (non-methane HC for diesel) and CH_4 (methane hydrocarbons) for gas engines.

[3] For Euro 6 there is an additional requirement of a maximum of 10 ppm NH_3 for both WHSC and WHTC cycles.

[4] NO_2 proportion of total NO_x limitation may follow as NO_2 is the parameter that is legislated for air quality.

Light Commercial Vehicle (LCV) legislation also commenced from Euro 1. To save space the full list of regulations has been omitted as they are tiered for vehicle reference weights. The trend was for less stringent measures compared to Passenger Car as the legislation and thus the applied technology lagged that of Passenger Car by one level of legislation; i.e. Euro 3 emission control technologies were similar to Passenger Car Euro 2. Similarly Off-Highway legislation was effectively a further level behind and is now driving considerable development exercises to meet future, far more stringent requirements.

The European legislation also included a durability requirement from Euro 4.

Table 5. European emissions durability legislation

Vehicle Category	Euro 4 and Euro 5	Euro 6
N1 & M2	100,000 km / 5 years	160,000 / 5 years
N2 N3<= 16 tonne M3 Class I, Class II, Class A, Class B	200,000 km / 6 years	300,000 km / 6 years
N3 > 16 tonne M3 Class III, Class B > 7.5 tonne	500,000 km / 7 years	700,000 km / 7 years

The European definitions of vehicle categories define the emissions legislation group that a particular vehicle type must be homologated.

Table 6. European vehicle categories

Category	Description
M	Motor vehicles with at least four wheels designed and constructed for the carriage of passengers
M_1	Vehicles designed and constructed for the carriage of passengers and comprising no more than 8 seats in addition to the driver's seat
M_2	Vehicles designed and constructed for the carriage of passengers comprising of more than 8 seats in addition to the driver's seat and having a maximum mass ("technically permissible maximum laden mass") not exceeding 5 tonnes
M_3	Vehicles designed and constructed for the carriage of passengers comprising of more than 8 seats in addition to the driver's seat and having a maximum mass exceeding 5 tonnes
N	Motor vehicles with at least four wheels designed and constructed for the carriage of goods
N_1	Vehicles designed and constructed for the carriage of goods and having a maximum mass not exceeding 3.5 tonnes
N_2	Vehicles designed and constructed for the carriage of goods and having a maximum mass exceeding 3.5 tonnes but not exceeding 12 tonnes
N_3	Vehicles designed and constructed for the carriage of goods and having a maximum mass exceeding 12 tonnes
O	Trailers (including semi-trailers)
G	Off-Road vehicles

Symbol G shall be combined with either symbol M or N. For example a vehicle of category N_1 which is suited for off-road use shall be designated as N_1G.

Table 7. European Passenger Car Gasoline: Development of emissions regulations and technologies

Legislation	Engine	Cycle	CO g/km	HC g/km	$HC+NO_x$ g/km	NO_x g/km	PM g/km	PN n/km	FIE	NO_x	CO+HC	PM PN
Euro1 Jul-1992	PISI	ECE15+ EUDC	2.72		0.97				EFI 2b-4 bar	TWC	TWC	
Euro2 Jan-1996	PISI	ECE15+ EUDC	2.20		0.50				EFI 2b-4 bar	TWC	TWC	
Euro3 Jan-2000	PISI DISI	NEDC	2.30	0.20		0.15			EFI 2b -4 b (DISI) 150b - 200b (DISI)	TWC + LNT (DISI only)	TWC	
Euro4 Jan-2005	PISI DISI	NEDC	1.0	0.10		0.08			EFI 2b -4 b (DISI) 150b - 200b (DISI)	TWC + LNT (DISI only)	TWC	
Euro5 Sep-2009	PISI DISI	NEDC	1.0	0.10		0.06	0.005		EFI 2b -4 b (DISI) 150b - 200b (DISI)	TWC + LNT (DISI only)	TWC	none
Euro6 Sep-2014	PISI DISI	NEDC	1.0	0.10		0.06	0.005	$6x\ 10^{-11}$	EFI 2b -4 b (DISI) 150b - 200b (DISI)	TWC + LNT (DISI only)	TWC	none

[1] The ECE15+EUDC cycle measured emissions from time = 40 seconds whereas the NEDC measured from time = zero (otherwise the cycles used the same speed / load versus time).

[2] European nomenclature used SI for Gasoline; i.e. "Spark Ignition".

[3] EFI=Electronic fuel injection.

For completeness European Passenger Car Gasoline legislation is also included. The emissions mitigation system has effectively remained the same for Port Injected engines (PISI) as the TWC has been developed over the years and rapid light up systems have been engineered to achieve greater NO_x abatement during warm-up where a large emission is made. Direct Injection types (DISI) adopted both TWC and LNT technology since this combustion system allowed lean mixtures to be run over some of the operating map. It should be noted that, in comparison to Diesel, the Gasoline regulations for NO_x, CO and HC have remained fairly static from Euro 4 onwards reflecting that development of emissions mitigation systems had reached a high level of maturity over 40 years of development. However for Euro 5 a PM regulation was introduced for the first time at the same level as Diesel (DISI only). For Euro 6 Particulate Number (PN) was also introduced in common with the Diesel regulation, again DISI only, both engine systems being required to meet the same target. There was no harmonisation in the required NO_x, CO and HC levels for the two different systems although this has been a target for some time by policy makers.

ON-BOARD DIAGNOSTICS

On-Board-Diagnostics (OBD) regulations have been progressively introduced since Euro 2 (Gasoline), Euro 3 (Light Duty Diesel) and Euro 4 (Heavy Duty Diesel). With each subsequent legislation level the requirements for OBD have become more stringent. Certain system requirements are mandated. There are on-the-road limits set for the system compliance. As an example the following tables show the development of Passenger Car OBD threshold values for comparison with the legislation tables:

Table 8. European Passenger Car Diesel: Development of OBD emissions threshold values

Legislation	Date	CO g/km	HC g/km	NO_x g/km	PM g/km
Euro 3 / Euro 4	Jan-2000 / Jan-2005	3.20	0.40	1.20	0.18
Euro 5	Sep-2009	1.90	0.32	0.54	0.05
Euro 6	Prior Sep-2014	1.90	0.32	0.24	0.05
Euro 6	Up to Sep-2017	1.75	0.29	0.18	0.025
Euro 6	"Final"	1.75	0.29	0.14	0.012

[1] HCs were THC ("Total Hydrocarbon" for Euro 3 / Euro 4 and NMHC ("Non-Methane Hydrocarbon") subsequently.

[2] From Euro 5 emissions were expressed in mg/km in the legislation but g/km is shown to be compatible with previous tables.

Table 9. European Passenger Car Gasoline: Development of OBD emissions threshold values

Legislation	Date	CO g/km	HC g/km	NO_x g/km	PM g/km
Euro 3 / Euro 4	Jan-2000 / Jan-2005	3.20	0.40	0.60	none
Euro 5	Sep-2009	1.90	0.25	0.30	0.05
Euro 6	Sep -2014 to Sep-2017	1.90	0.17	0.15	0.025
Euro 6	"Final"	1.90	0.17	0.09	0.012

[1] HCs were THC for Euro 3 / Euro 4 and NMHC subsequently.

[2] From Euro 5 emissions were expressed in mg/km in the legislation but g/km shown.

[3] PM DISI only.

EXHAUST GAS RE-CIRCULATION (EGR)

On the face of it exhaust gas re-circulation would appear to be a very strange approach. However it is used for NO_x reduction quite extensively in modern engines. In a Diesel engine there will be a fuel consumption penalty which will vary according to operational modes. On a Gasoline engine there is actually an improvement in engine efficiency / fuel consumption as it allows optimisation at a higher compression ratio yet avoiding "knock". This is the primary reason for EGR in modern Gasoline engines as the extremely high NO_x conversion efficiency of the Three-Way-Catalyst has obviated the need for EGR as the primary NO_x mitigation system. However in the Diesel engine, EGR has been extensively used for NO_x mitigation for many years in Light Duty and has now been introduced to most European Heavy Duty engines for Euro 6 (not IVECO) and some from Euro 4 (Scania and MAN).

If EGR is used in the lower speed / light load sector of the operating map than the fuel consumption penalty will be quite low. Indeed, in Berlin where some buses were specially adapted with OEM end-of-line fitted EGR no penalty was recorded for the fleet compared to the non-EGR buses.

Intrinsically one would be concerned with accelerated engine wear with the use of EGR but tests using the "tritium trace" technique have revealed that the wear, although slightly higher initially, settles down to the same rate as a non-EGR engine. In any case there is now a vast accumulated mileage with engines using EGR and the experience has been extremely positive.

Under valve "overlap" conditions there is a degree of "internal" EGR inherent in all engine designs without the fitment of a supplementary system since for a short time there is likely to be a direct connection in-cylinder between the inlet and exhaust manifolds. Engines with variable valve timing are able to exploit this feature to a greater degree and this technology is present in an ever increasing number of engines.

In the Diesel engine the major source of NO_x is from this "Thermal" source (sometimes known as "Zeldovich" NO_x). The other sources are termed "Prompt NO_x" (formed rapidly in rich mixtures) and "Fuel NO_x" (formed from any nitrogen compounds present in the fuel).

Two mechanisms allow EGR to reduce peak cylinder temperature and thus abate NO_x. EGR decreases the oxygen proportion and thus increases the inert gas content of the intake charge whereby the existing N_2 is supplemented by the products of combustion CO_2 and H_2O to give an overall increase in the mix's specific heat which modifies the combustion heat release such that lower peak cylinder temperatures are attained.

There are two design approaches for EGR popularly known as "long route" (or "high pressure") and "short route" (or "low pressure") systems. The short route system relies on the natural pressure differential between the exhaust and the inlet manifolds to drive the re-circulating gas round. This occurs naturally over the lower speed and load areas of the operating map. The EGR gas path takes a direct route from the exhaust manifold to the inlet manifold. As speed and / or load increases the percentage of EGR versus intake air mass must decrease due to the reduced pressure differential and the need to minimise PM emissions as the mixture richens. This system had previously been suitable for Light Duty engines homologated on the NEDC and currently for some heavy duty engines fitted with SCR; the SCR dealing with NO_x at the higher speeds and loads and EGR at the lower speeds and loads. For non-SCR heavy duty engines the long route system was developed whereby a variable geometry turbocharger increases the exhaust manifold pressure to effect EGR at the high speed / high load part of the operating map. Short and long route systems may also be used in combination. Examples of this may be found in both Light Duty and Heavy Duty modern engines.

EGR flow may be increased at light speeds and loads by use of a throttle in the inlet manifold to increase the pressure differential relative to the exhaust manifold. There may be a combustion tolerance of as much as 75% EGR at idle which may only be effected by use of a throttle. An EGR cooler further increases EGR mass flow and the decreased temperature at the inlet manifold gives additional assistance in the treatment of NO_x since the cycle starts from a lower initial temperature. The EGR "rate" may be determined by the following formula:

EGR Rate = (((Non-EGR intake mass flow) – (With-EGR intake mass flow)) / (Non-EGR intake mass flow)) x 100

The rate may be implied by use of a sensor measuring the net inlet air mass flow. Care must be taken to map the engine to take into account the NO_x / PM trade-off characteristics. When the injection timing is advanced to give minimum fuel consumption PM will be minimised but NO_x will be maximised. However, with retard of timing the trend is reversed. Addition of EGR is then likely to increase PM to an unacceptable level. The skill of the calibrator is to optimise these effects for best balance of emissions and efficiency. Note that with many engines there would be a DPF fitted which gives a little extra flexibility in the calibration but care must be taken not to overload the trap with soot emissions.

An interesting aspect of calibrating an EGR system is that the gas throughput of the engine is being modified from the non-EGR mode. It is possible to increase the emissions concentration at tailpipe yet with the reduced mass flow one may end up with a lower gramme / kilometre contribution on a Light Duty homologation cycle.

In early systems the EGR valve was of a simple "on / off" design. This was run open loop and effected by reference to a mechanical signal from the fuel injection pump. Two-stage valves which had an intermediate lift compared to on / off types were also used. This progressed to systems with valves that hovered between two stops and closed loop control using by signals from EGR valve position, air mass flow sensor and speed / load information from the electronic fuel injection system.

The EGR must be evenly distributed cylinder to cylinder otherwise an over-rich cylinder will produce a significant PM emission. In development this may be checked by measurement of CO_2 concentration at each inlet valve. An effective and simple device to achieve good distribution is to introduce the EGR against the incoming flow in the air inlet so that good mixing may occur. A more sophisticated mixing unit may also be fitted.

There are some limitations on operating EGR. It is not normally applied until the engine water jacket temperature has achieved circa 60° C. Some systems have a time-out after long periods of idling or even avoid EGR altogether at idle. A by-pass may be fitted around the EGR cooler to accelerate warm-up. This device is also a tool to avoid "fouling" of the cooler which may be exacerbated by running "cold" EGR through it which leads to condensation of HCs etc. that build up and modify the heat transfer rate. Some examinations of fouling suggest that it stabilises after a certain number of hours but this phenomenon is still not fully understood and remains a relatively weak link in the system.

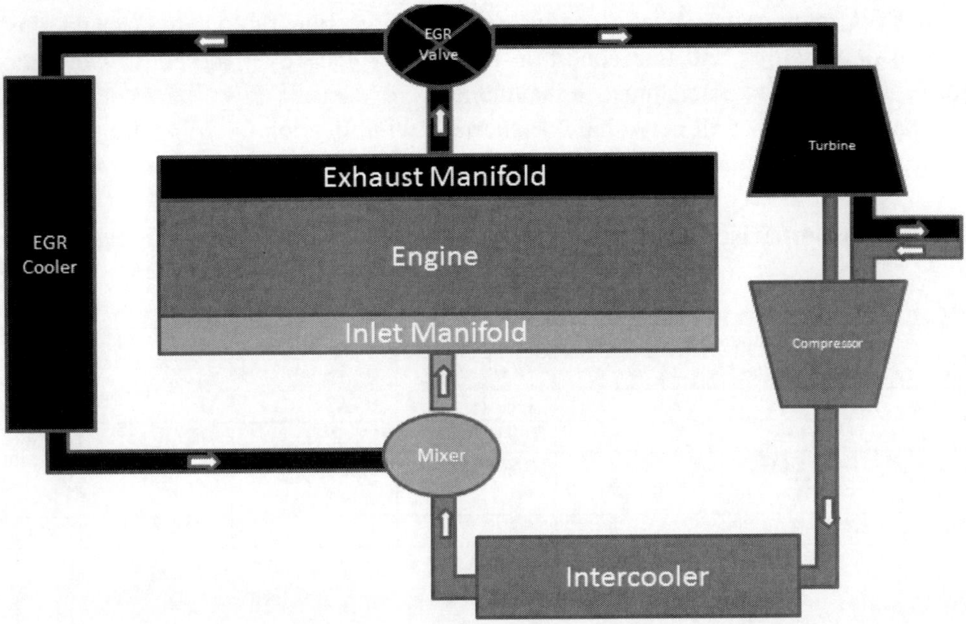

Figure 1. Gas handling on the "Short Route" EGR system.

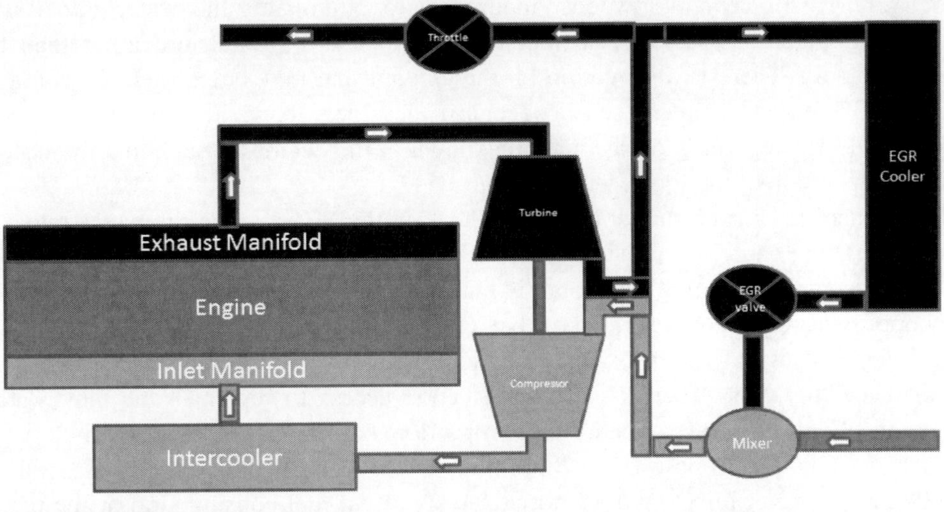

Figure 2. Gas handling on the "Long Route" EGR system.

EGR should be switched out or heavily modified under acceleration to control PM emissions. The transient control is a key element of the calibration and is further complicated by the use of a VGT in the long route system which has a longer time constant. This is another weakness in the use of EGR since measurements taken at roadside at air quality "hotspots" has revealed high NO_x emissions from Diesel cars relative to expectations under congested traffic "stop-start" conditions.

High EGR rates, rather like excessive retard of injection timing, may cause unstable combustion or "misfire" and this condition must be avoided over the entire calibration as good driveability of the vehicle must be maintained.

Oil change intervals will potentially be shortened. Lubrication oil will suffer a significant loss in viscosity if soot contaminates the sump which will damage the engine.

Table 10. Comparison of specific heats and gas constants and their ratios ratios (1 bar / 300⁰K)

Gas	Chemical formula	C_p kJ/Kg^{0}K	C_v kJ/Kg^{0}K	R kJ/Kg^{0}K	C_p / R
Air	76.7% N_2 + 23.3% O_2	1.005	0.718	0.287	3.502
Nitrogen	N_2	1.039	0.743	0.296	3.510
Oxygen	O_2	0.918	0.658	0.260	3.880
Carbon Dioxide	CO_2	0.846	0.657	0.189	4.476
Water (steam)	H_2O	1.872	1.411	0.461	4.061

WATER INJECTION

Many test results have shown that if water is injected into the cylinder then less NO_x is formed. A fairly simple and low cost method is to "fumigate" the water into the inlet manifold. The water may also be included in the fuel as an emulsion rather than being injected separately. There is a limitation on the amount that may be carried. On automotive trials up to 13% water has been used in specialist road fuels for niche applications but there have been concerns regarding occasional breakdown of the chemistry such that the water has affected the operation of the fuel injection pump.

The mechanism is such that some of the heat is used to vaporise the water rather than raise cylinder temperature. As previously stated, Thermal NO_x formation is a function of peak cylinder temperature. A useful mnemonic is that 1% water / fuel ratio (by mass) will reduce NO_x by approximately 1%. Systems have been fitted to large engines in development where there is usually plenty of room for the water tank, which by definition must be as large or larger than the fuel tank. There is, of course, good access to replenish the tank water on marine applications. Up to 90% water / fuel proportion has been demonstrated which suggests this system, in theory, is as powerful a NO_x mitigation tool as any other proposed designl.

Unlike EGR, water injection does not adversely affect fuel consumption or the potential development of extra PM emissions. In fact, many studies have shown significant improvements in PM. This technology has not yet found an automotive application but has been used on vehicles for performance enhancement some years ago. Note that the technology has also been used for many years for temporary performance enhancement of aircraft engines in the form of water methanol injection.

There is a growing likelihood that some OEMs will introduce water injection for marine applications to meet the International Maritime Organisation (IMO) Tier III regulations which affect Emission Control Areas (ECAs) where stringent NO_x control has been deemed necessary. These ECAs include the North Sea, Baltic Sea and North American coastline etc. For information the following table describes the development of the legislation which is based around engine speeds (“n” rev/min) rather than engine power or vehicle weight as in automotive applications.

Table 11. IMO NO_x Legislation (g/kWh)

Level	Date of introduction	n < 130	130 <= n < 2000	n >= 2000
Tier I	2000	17.0	$45 \times n^{-0.2}$	9.8
Tier II	2011	14.4	$44 \times n^{-0.23}$	7.7
Tier III	2016 (ECAs only)	3.4	$9 \times n^{-.02}$	1.96

[1] As Tier III legislation only applies to ECAs Tier II will continue to apply elsewhere.

[2] Not all countries have signed up to Tier III but encouragingly those that have represent circa 82% of shipping tonnage.

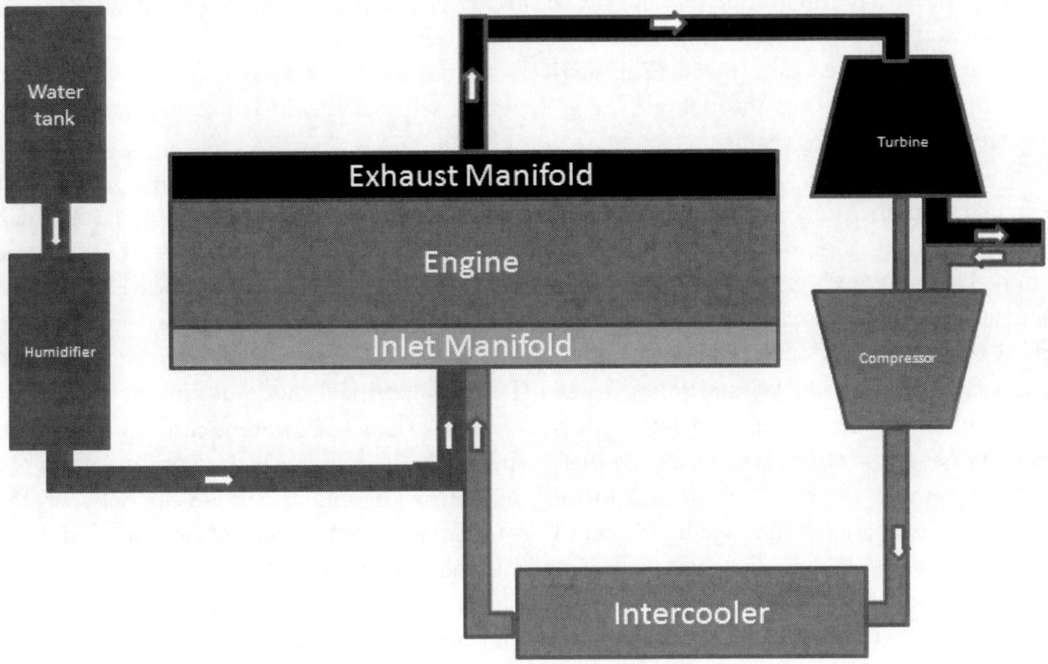

Figure 3. Water injection by fumigation.

Catalysts

A catalyst is a chemical that modifies or accelerates a reaction without itself being modified or consumed. The skill in applying this technology to exhaust abatement systems is to match the catalyst performance to the available exhaust temperature profile without untoward side effects such as increased emissions of unregulated substances.

Location of the device is very important with all catalytic systems. The minimum operating temperature is typically between 150° C and 200° C. Peak conversion efficiency is likely to be well above the minimum operating temperature. To improve the temperature profile one may move device nearer to the engine (“close-coupled”) and / or lag the exhaust down pipe. Whatever the light-off temperature it is possible for catalysts to have a mechanism whereby noxious emissions are stored ("adsorbed") and converted once light-off temperature has been exceeded. Substances known as Zeolites typically provide this function. These are alumina silicate minerals which may also be synthesised. The term Zeolite was coined from the Ancient Greek zeo (ζέω) "I boil" and lithos (λίθος) "stone" and thus gained the meaning "boiling stone".

The gas distribution is important to enable the reactive area to be fully utilised. The chemical formulation may be changed according to the exhaust temperature profile. The precious metal content that provides the catalytic assistance has a cost / performance trade-off. A "washcoat" enhances performance by providing more effective area for the reaction to take place. Aluminium Oxide (Al_2O_3) is used extensively for this purpose. There is, however, a trade-off with exhaust back pressure that must be optimised to remain within the engine manufacturer's recommendations.

The frontal area should be maximised within the packaging restraints. The space velocity (equivalent to the residence time in the catalyst body) should be sufficient to allow the chemical reactions to take place. The device volume and frontal area must be optimised regarding the engine gas throughput. "Vee" engines will have twin banks of cylinders and thus twin exhaust pipes requiring a catalyst in each.

Many automotive catalysts are Sulphur sensitive. Sulphur in the exhaust may either originate from the fuel or from the lubrication oil and may cause emission of sulphates which are measured as additional particulate. Sulphur may also retard or even stop the catalytic action. This may be irreversible. That is the reason that road fuels have progressively reduced their sulphur content to less than 10 ppm under current European guidelines. Sulphur from the lubrication oil may pass by the piston rings especially in stop-start operation such as in buses, inner city delivery vehicles and refuse trucks. There is a critical mode of circa 50% speed / 25% load where the presence of lubrication oil in the exhaust HCs is maximised due to the phenomenon of "ring-shake" where the piston rings provide a less effective seal between the piston and the cylinder. The selected loading of the piston ring is a trade-off between the desire for low friction and the requirement of providing an effective seal. Other lubrication oil derived "poisons" include Calcium (Ca), Zinc (Zn) and Phosphorus (P).

The vehicle duty cycle must be taken into consideration as this will determine the exhaust operating temperature and thus the level of engine-out emissions. Of course, some countries have severe cold winter weather to accommodate which makes the application of catalytic devices even more challenging. A good transient response of the catalyst is also essential.

Catalysts may lose performance in the very early stages of usage. A process of "de-greening" is usually applied in production to settle the emission conversion at its durable efficiency.

Considerable advances have been made in the production of catalyst substrates such that they are offered with increasingly denser cells packing with progressively thinner walls. Ceramic types are now offered up to 1200 cells per square inch (cpsi) with as low as 0.05mm wall thickness. Metallic types are offered up to 1000 cpsi with wall thicknesses as low as 0.025mm. These features allow greater performance and durability within the same package or, of course, downsizing of the package which is useful in an increasingly tight available space on the vehicle. This allows the designer to situate the catalyst nearer the exhaust manifold as a technique for rapid warm-up which is now down to single figure seconds for a typical TWC.

OEMs have made a concerted effort to improve their fleet CO_2 by, amongst other things, improvements to the engine efficiency (higher boost, lower friction etc.) and harvesting of thermal energy. This has led to the risk of lower exhaust temperatures available at the catalyst. Cascading of catalytic devices (e.g. DPF + SCR) also presents problems as the one furthest away from the engine will be at a disadvantage. Development is ongoing to address these issues.

THREE WAY CATALYST (TWC)

The three-way catalyst must rank as one of the most cost-effective and socially desirable inventions of the 20th century. It has held supreme for emissions control in Gasoline engines for several decades. Unfortunately for the Diesel engine the device requires a reducing atmosphere to deal with the NO_x; i.e. where there is no excess oxygen. Diesels run with excess oxygen at all times and well away from in-cylinder stoichiometric air / fuel ratios so other means for control of NO_x are needed. The stoichiometric air / fuel ratio is that whereby theoretical complete combustion of the fuel is achieved with the minimum oxygen required. This varies from fuel to fuel according to its chemical make-up. For Gasoline it is approximately 14.7:1. Actual air / fuel ratio divided by the stoichiometric air / fuel ratio is typically represented by the Greek letter lambda (λ). For $\lambda < 1$ the mixture is termed "rich". For $\lambda > 1$ the mixture is termed "lean" (i.e. excess oxygen). This ratio lends its name to the controlling sensor for the TWC.

The TWC consists of essentially two systems, one for the reduction of NO_x using platinum (Pt) and rhodium (Rh) when the A/F ratio is at stoichiometric or slightly richer (a"reducing" atmosphere) and one for oxidation of the HCs and CO using platinum and palladium (Pd) when the mixture is slightly lean; i.e. with some excess oxygen. Oxygen may also be stored in the catalyst to assist this reaction. The characteristics of the TWC may be altered by the Pt / Rh and Pt / Pd content ratios.

The reducing chemical equations are:

$2NO \Rightarrow N_2 + O_2$
$2NO_2 \Rightarrow N_2 + 2O_2$

The oxidation chemical equations are:

$2CO + O_2 \Rightarrow 2CO_2$
$C_xH_{2x+2} + ((3x+1)/2)O_2 \Rightarrow xCO_2 + (x+1)H_2O$

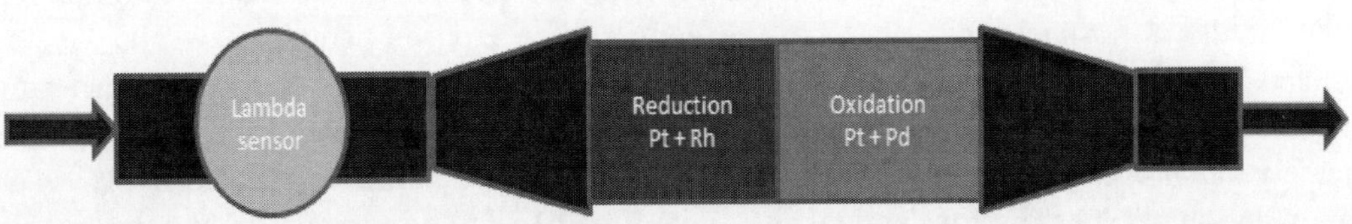

Figure 4. Three Way Catalyst system layout.

Diesel Oxidation Catalyst (DOC)

Oxidation catalysts have been fitted to European diesel light duty passenger car and car-based vans since Euro 2 (1996) and are now fitted to light duty trucks (< 3.5 tonnes). There has been an option for OEM fit on some heavy duty applications but generally has been replaced by DPFs now following the intent of Euro 6 (2014) legislation regarding particulate mass and particulate number. The catalyst is coated on a porous substrate that allows the exhaust gas to pass but maximises the residence time and area for the chemical reaction to take place. A washcoat, as previously described, further increases the effective reactive area.

The catalytic action is realised by platinum and also perhaps latterly, jointly with palladium (around the time of Euro 4). Early examples, especially with underfloor locations had a high loading of platinum. At elevated temperatures emissions of sulphates could be made which impaired attainment of low PM on the homologation chassis dynamometer test. Palladium is more suited to the temperature profile of Gasoline exhaust. Additionally it was also more costly than Platinum at one time. However, latterly, the cost differential reversed and Palladium was adopted for some DOCs. The Palladium assists in reducing light-off temperature and in suppressing the creation of sulphates and thermal ageing that may be a problem in a pure Platinum catalysed DOC.

The DOC as a retrofit is particularly suited to the older heavy duty engine which has a greater lubrication oil contribution to the particulate that precludes the use of a catalytic particulate trap where low sulphur content of the exhaust gases is essential to ensure regeneration.

An oxidation catalyst will remove circa 90% of HC and CO emissions in laboratory conditions. Some particulate may also be removed mainly by stripping out of the "wet" hydrocarbons. The level of reduction will be dependent on the type and age of the engine. Some soot reduction has also been observed - more so in DI than IDI engines and may depend on the degree of NO_2 production and the residence time within the device. Up to 10% has been recorded. Some heavy duty engines have shown 50% overall PM removal levels.

One data set for retrofitted vehicles gave an average of 0.5% fuel consumption increase from emissions tests on the NEDC (New European Drive Cycle for light duty vehicles < 3.5 tonnes). This compared the 3 chassis dynamometer tests without aftertreatment and 3 with the DOC.

Note that there is no maintenance requirement with oxidation catalysts and should the device stop working there are no detrimental effects to the vehicle or engine and the driveability should not be impaired.

Oxidation catalysts have been criticised for not removing ultrafine particles; indeed the removal of HCs from around the particulate soot kernel will potentially reduce the emitted particle size. Primary NO_2 production; i.e. direct from the exhaust rather than from oxidation of NO in the atmosphere has also been a concern for environmentalists as NO_2 is the legislated air quality metric and may be very likely to affect readings at roadside metering systems that are used to track local emissions. Note that the dilution rate in the atmosphere is extremely rapid – approximately 1000:1 in 1 second. Atmospheric NO_2 is normally made from oxidation of NO emissions and takes a finite time measured in minutes and hours. There have been no reported kerbside health effect problems for primary NO_2 and the general medical opinion is that ultrafine particles are a far greater risk to human health. Nevertheless there is impending legislation to limit NO_2 emissions. The ability of the DOC to produce NO_2 has actually been useful in the design of DPF and LNT systems.

The oxidation chemical equations are:

$$2CO + O_2 \Rightarrow 2CO_2$$
$$C_xH_{2x+2} + ((3x+1)/2)O_2 \Rightarrow xCO_2 + (x+1)H_2O$$

A beneficial effect of the DOC has been a significant contribution to the reduction in exhaust odour (together with fuel desulphurisation). Not only did this make the Diesel engine more socially acceptable but the aldehyde (and some aromatic) compounds that are the primary cause of the odour are particularly harmful to human health.

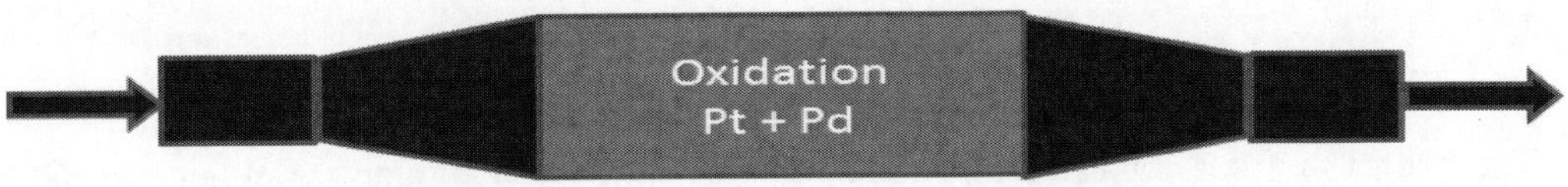

Figure 5. Oxidation catalyst layout.

DIESEL PARTICULATE FILTER (DPF)

There is essentially no real difficulty in filtering out the particulate in the exhaust stream – it is well established technology. The important factor is how to regenerate the trap once it has accumulated a certain amount of matter. Carbon ignites at circa 550° C. This temperature is rarely attained on light duty vehicles, or certain heavy duty applications such as city buses, inner-city delivery vehicles and refuse trucks.

The following methods have been used to assist regeneration of the trap by an increase of the exhaust temperature; fuel burner, electric heater, increased engine back pressure and electronic fuel system "post" injection. These types are often referred to as "active" (regeneration).

Other methods to have been used to assist re-generation of the trap are by reduction of the carbon ignition temperature; catalysis of the exhaust gas components by placement of an oxidation catalyst upstream of filter or actually coating the filter itself and / or by a fuel borne catalyst which may also modify combustion and give a reduction in engine-out emissions and a possible fuel consumption improvement. These include cerium, iron / strontium and platinum. These types are often referred to as "passive" (regeneration).

With active generation types (usually factory fitted) the system relies on a sensor such as exhaust gas back pressure or differential pressure across the filter to indicate the loading and thus when re-generation is required. With passive types (catalytic or fuel borne catalyst) it is expected that regeneration occurs often enough such that the soot loading is never too high. An exhaust gas back pressure sensor is now fitted as a diagnostic aid.

If regeneration is attempted with too low a soot loading it is difficult to get a successful ignition of the carbon and thus fuel or electrical energy is wasted. If regeneration is left too late there may be excessive fuel consumption in the meantime and a possible thermal runaway under regeneration conditions whereby the filter body may be destroyed.

Normally the particulate trap replaces the silencer. Note that in laboratory comparisons a silencer would be likely to have a superior performance in isolation from the vehicle. However, drive-by tests have shown that the particulate filter achieves a similar noise attenuation to a silencer and importantly, subjectively no difference is likely be perceived.

The chemical reaction for soot removal is:

$$C + O_2 \Rightarrow CO_2$$

For a catalytic trap the chemical reactions are:

$$2NO + O_2 \Rightarrow 2NO_2$$
$$2NO_2 + 2C \Rightarrow N_2 + 2CO_2$$
$$2CO + O_2 \Rightarrow 2CO_2$$
$$C_xH_{2x+2} + ((3x+1)/2)O_2 \Rightarrow xCO_2 + (x+1)H_2O$$

The wall flow filter consists of passages that are alternately plugged at either end. Gases may flow through the walls to escape through the tailpipe but particulates are trapped by the plugged ends.

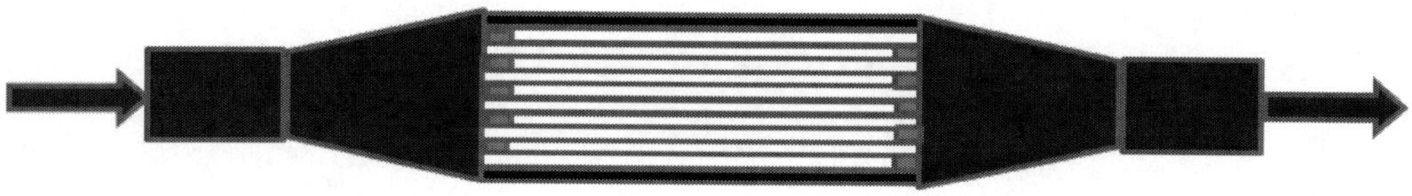

Figure 6. Wall flow filter layout.

There are options for the material of the filter body with cordierite, silicon carbide and sintered metal being the most popular. Cordierite has a lower temperature tolerance compared to silicon carbide and sintered metal. This is not usually a problem unless there is a malfunction in the engine or system failure (such as lack of catalytic action) to lead to severe plugging of the filter, in which case there is risk of a thermal runaway to an excessive temperature during the regeneration process. There are claims that certain arrangements of sintered metal filter bodies are able to require less frequent de-ashing and that the process may be more easily facilitated in-house with standard equipment rather than using a specialist oven technique which has been most successful with cordierite filter bodies.

The following table summarises some of the filter material characteristics that have been or are about to be used for DPFs.

All particulate traps require “de-ashing”; i.e. the removal of the combustion residue, most of which is attributable to lubrication oil components. Low ash and synthetic lubrication oils are available at higher cost. The maintenance interval depends on the type of trap, vehicle usage and oil consumption but should be at least 30,000 km. This is an extra operating cost but field experience suggests an average of once a year de-ashing is sufficient. The filter is removed and ash is blown out under controlled conditions. Regeneration in an oven greatly facilitates the ash removal. Note that vehicles undergoing stop / start operation which risks more oil passing the piston rings into the exhaust do far less mileage and may then still achieve an annual de-ashing. It is possible to utilise “low ash” oils. There are also synthetic oils that contain no sulphur. Attention to these details may enhance the prospect of trouble free operation of catalytic particulate filters.

Table 12. Filter material characteristics

Material	Chemistry	Max. Safe Temperature	Thermal Expansion	Comments
		^oC	$10^{-6}/^oC$	
Cordierite	$2MgO\text{-}2Al_2O_3\text{-}5SiO_2$	1200	1.7 - 3.0	Lowest cost
Silicon Carbide	SiC	2700	4.4 - 5.1	Highest melting point
Sintered Metal	Cr-Ni-Steel	1360	10.0 - 13.0	Simpler de-ash process
Aluminium Titanate	Al_2TiO_5	1500	0.1 - 1.0	Latest technology

There are certain health checks that may be carried out. An inspection of tailpipe deposits and / or exhaust gas colour would reveal if the trap has failed as in normal operation the exhaust gases are colourless. An exhaust back pressure sensor would indicate any excessive build-up of deposits and is now invariably fitted as a standard item. A free acceleration smoke opacity test may be undertaken (this test is only valid for full particulate traps (not partial filter). As a guide, the VERT compliance test sets a certification limit of 0.12 m^{-1} and an in-service limit of 0.24 m^{-1} for this type of device.

There are examples of good practice in the maintenance of a vehicle fitted with a trap; e.g. avoidance of over-filling the sump with lubrication oil. Also it is important to check the system following engine failures; e.g. turbocharger failures may send debris and oil through the exhaust; the catalyst and filter body should be checked following such an event.

Catalysed traps remain the dominant particulate treatment system in Europe both as an OEM fit and for retrofit of suitable vehicles. Typical results: >80% HC – >80% CO – >95% PM (including ultrafines).

Catalysed traps have also been criticised for street level “primary” NO_2 production. The NO_2 created in the trap may be removed by installing the trap upstream of an SCR system. The SCR chemistry would then need to be able to withstand trap re-generation temperatures however.

There are emerging catalyst formulations which include Ceria (Cerium Dioxide - CeO_2) and / or Zirconia (Zirconium Dioxide - ZiO_2) that accelerates oxidation of the soot by 70% at a temperature as much as 75^{0}C lower.

Catalysed Partial Filters

Partial filters have filtration efficiencies in the intermediate range between oxidation catalysts (circa 25%) and particulate trap (> 90%). There has been an increasing use on passenger car and light vans and some heavy duty applications; e.g. MAN Euro 4 with EGR. The devices reduce ultrafine particle numbers and do not require de-ashing. Their performance has been criticised for being cycle specific as unlike a full DPF the particulate abatement varies according to application and operating characteristics. When fully loaded and if not re-generated, the performance reverts to that of a standard oxidation catalyst.

Fuel Borne Catalyst

A fuel borne catalyst may be used to lower the regeneration temperature rather than a catalyst coated filter or installing a DOC pre-catalyst. There will be possible greater fluctuations in exhaust back pressure compared to catalysed traps causing possible objection from engine suppliers. FBCs may also slightly improve fuel consumption and give reductions in engine-out emissions. Examples of FBCs include cerium (Ce a "rare earth"), iron / strontium (Fe / Sr) and platinum (Pt). Copper (Cu) has been shown to have unwanted unregulated emissions and is no longer likely to be considered. In conjunction with post-injection assistance from the fuel system dosing rates may be a little as 3 ppm for Fe, 10 ppm Pt and 30 ppm Ce. This allows for the installation of enough stored FBC for the vehicle life.

Peugeot introduced a DPF with an FBC (cerium) injected into the fuel for their largest Diesel car for Euro 3. The regeneration was assisted by a post-injection from the common rail fuel pump that temporarily increased the exhaust temperature.

Regeneration Systems Schematics

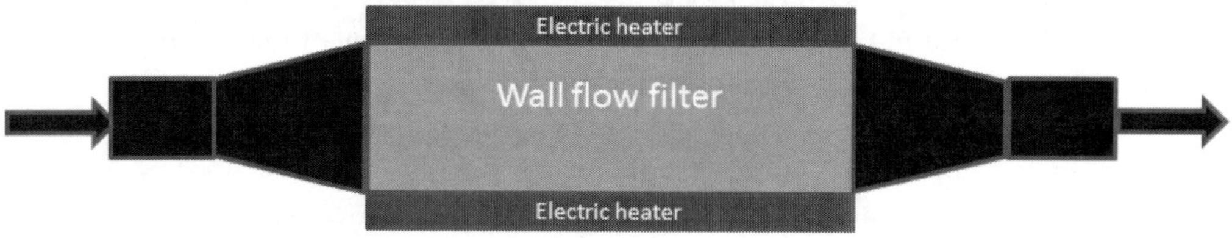

Figure 7. Active trap regeneration - Electric heater.

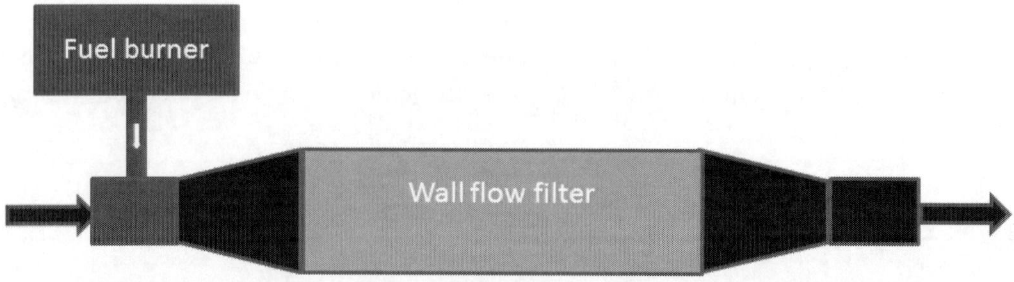

Figure 8. Active trap regeneration – Fuel burner.

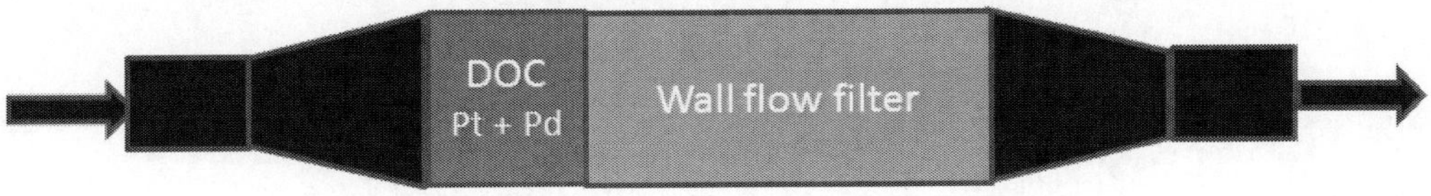

Figure 9. Passive trap regeneration – Upstream DOC.

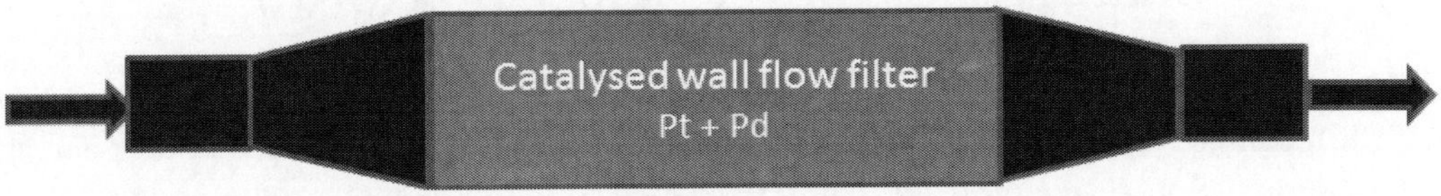

Figure 10. Passive trap regeneration – Catalysed filter.

Figure 11. Passive trap regeneration – Fuel borne catalyst.

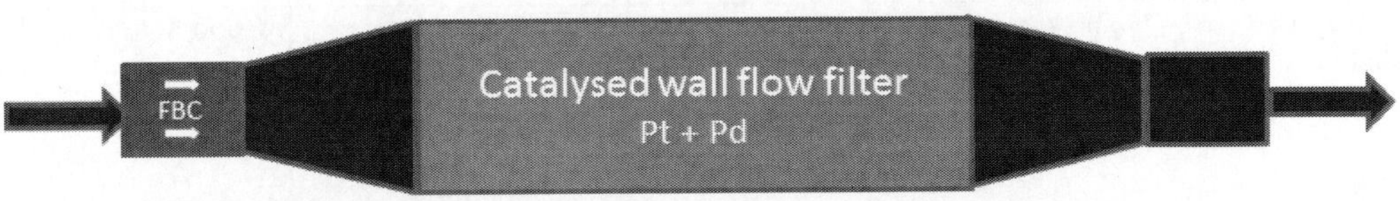

Figure 12. Passive trap regeneration – Fuel borne catalyst + Catalysed filter.

Other Systems

Another approach has the trap in a by-pass which extracts particulate matter from the gas stream using an electrostatic precipitator methodology with separation and then agglomoration of matter. Ash migrates through the filter body which therefore does not require a de-ashing maintenance process. There are no fuel composition or temperature sensitivities. The re-generation is by electrical heater in the only production example which has found a niche as a retrofit item.

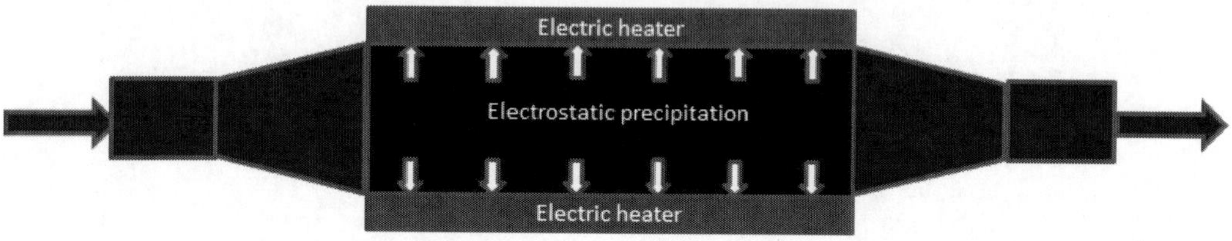

Figure 13. Electrostatic precipitation.

SELECTIVE CATALYTIC REDUCTION (SCR)

Selective Catalytic Reduction systems only abate one emission species - in this case NO_x. The SCR technology uses ammonia as the reductant and the NO_x is converted to nitrogen and water in the reaction. The European industry decided to avoid using stored ammonia on the vehicle and instead opted to create it from a 32.5% solution of urea in water (marketed as "AdBlue"). This decision was made due to health and safety concerns regarding ammonia handling and that urea was freely available on the market due to its use as a fertiliser. Urea has no significant handling issues as it is stable and non-flammable. Nevertheless, the direct use of ammonia represented the lowest cost and best engineering solution. According to the catalyst chemistry employed the reaction may commence at between 150° C and 300° C. The catalyst used may vary according to the desired properties. The following table summarises the trade-off for each composition.

Table 13. SCR chemistry overview

Chemistry	Low temperature performance	High temperature performance	Thermal durability	De-activants
Vanadium + Titania	High	High	Up to 600°C only Thus not suitable for use with DPF in combination due to regeneration temperatures	Good Sulphur tolerance
Copper + Zeolite	High	High	High (up to 900°C) Suitable for use with DPF in combination	Sulphur (reversible)
Iron + Zeolite	Low Unless NO_2 levels high (as with upstream DOC)	Very high	High	Soot + HC (reversible)

It should be noted that it is possible to run SCR with hydrocarbons (fuel) as the reductant. However with current technology the required exhaust temperature profile is above that typically found in Diesel exhaust and the degree of NO_x abatement is lower than the production types using urea as the reductant. From the legislators point of view urea is unattractive since there are difficulties in determining whether urea is actually present in the tank rather than, say, pure water, thus defeating the NO_x abatement system whereas HC as a reductant could be taken from the engine fuel tank.

Care must be taken in calibration not to develop excessive ammonia "slip" at the tailpipe. The World Health Organisation limit is 25 ppm and European legislation has recently added a 10 ppm limit for Euro 6. The typical catalyst has a final DOC section to oxidise any ammonia. Care must also be taken not to produce excessive N_2O which as a greenhouse gas has 315 times CO_2 equivalence by mass and therefore excessive emissions of N_2O may seriously affect the target GHG emissions for an engine. In OEM applications SCR is capable of over 90% NOx abatement. In retrofit systems perhaps 75% is more likely.

SCR technology requires several system components. These comprise a catalyst can, urea tank, urea injection system, NO_x sensor and a controller which may be built into the engine ECU. The urea tank needs replenishment at less frequent intervals than the fuel tank and is typically around $^1/_3$ the size of the fuel tank. The urea tank filter may be changed at normal service intervals. Reported dosing rates proportional to fuel consumption have been up to 4% for Euro 4 and up to 8% for Euro 5 but the rate will vary considerably according to the type of vehicle useage. Due to the high efficiency of SCR catalysts for NO_x control it has been possible to "advance" the diesel injection timing which gives certain performance and emissions improvements. There is potential for improved fuel consumption to offset the cost of the reductant. In turn there should be improvements in engine-out PM, HC & CO. However there is potential for higher levels of combustion noise to be generated. This may be ameliorated by use of special functions in the fuel injection system such as "pilot" injection or injection rate "shaping". There is likely to be no reduction in ultrafine particles number but some additional soot reduction may be possible in the presence of NO_2 within the SCR catalyst (the principle of operation of the Johnson Matthey "CRT" catalysed trap).

The chemical decomposition to extract ammonia from the urea solution is as follows:

$$CO(NH_2)_2 + H_2O \Rightarrow 2NH_3 + CO_2$$

Many chemical reactions may occur in an SCR catalyst. The following are for the NO_x emissions:

$$6NO + 4NH_3 \Rightarrow 5N_2 + 6H_2O$$
$$4NO + 4NH_3 + O_2 \Rightarrow 4N_2 + 6H_2O$$
$$6NO_2 + 8NH_3 \Rightarrow 7N_2 + 12H_2O$$
$$2NO_2 + 4NH_3 + O_2 \Rightarrow 3N_2 + 6H_2O$$
$$NO + NO_2 + 2NH_3 \Rightarrow 2N_2 + 3H_2O$$

The DOC to clean up the ammonia may also oxidise HC and CO emissions and these reactions are:

$$4NH_3 + 3O_2 \Rightarrow 2N_2 + 6H_2O4$$
$$NH_3 + 5O_2 \Rightarrow 4NO + 6H_2O$$
$$2CO + O_2 \Rightarrow 2CO_2$$
$$C_xH_{2x+2} + ((3x+1)/2)O_2 \Rightarrow xCO_2 + (x+1)H_2O$$

There are certain undesirable reactions that occur that may produce N_2O as follows:

$2NH_3 + 2O_2 \Rightarrow N_2O + 3H_2O$
$8NO_2 + 6\,NH_3 \Rightarrow 7\,N_2O + 9\,H_2O$
$4\,NO_2 + 4\,NH_3 + O_2 \Rightarrow 4\,N_2O + 6\,H_2O$

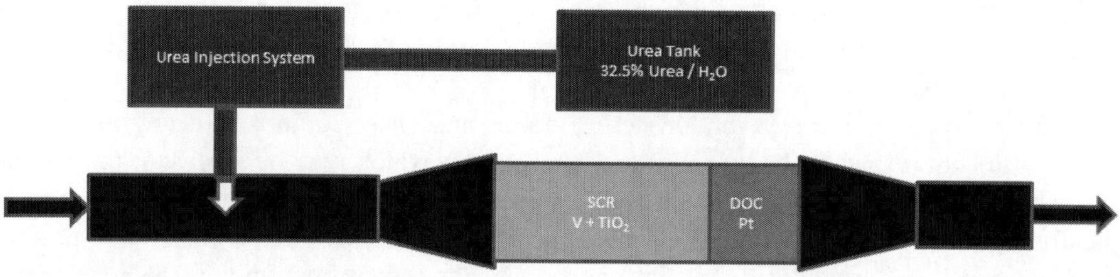

Figure 14. SCR system layout.

Lean NOx Trap (LNT)/ NOx Storage Catalyst (NSC) / NOx Adsorber Catalyst (NAC)

The Lean NOx Trap is also now known as a NO_x Storage Catalyst or NO_x Adsorber Catalyst. It collects NO_x using compounds that form nitrates under stable conditions in lean operation. These materials may be of the alkaline, alkaline earth or rare earth families. Barium carbonate is a typical example. In the adsorption mode nitrites and nitrates are formed. It then regenerates very regularly for some seconds by the use of a reducing atmosphere effected by using fuel as the reductant. This is a great advantage compared to SCR which requires an extra tank to carry its urea solution. The LNT was originally used on Gasoline Direct Injection (DISI) engines which could switch between normal Gasoline operation (at or around stoichiometric air / fuel ratios) and lean mixtures. Any Sulphur build-up is exhausted by running at an elevated temperature of between 600°C and 700°C. This is rather more easily achieved on Gasoline engines which are able to run up to 900°C compared to the Diesel engine's 700°C. In a Diesel engine use may be made of the very flexible "Common Rail" fuel system to create a "post" injection to effect the required temperature rise. Use of fuel as the reductant inherently increases the fuel consumption of the vehicle. This may be of the order of 2%. The peak efficiency of the device is circa 90% NO_x abatement.

An unwanted emission from the LNT is that of ammonia. This requires an oxidation catalyst to keep within the European limit of 10 ppm at tailpipe. This feature may, however, be turned to advantage by use of a passive SCR system downstream which may use the ammonia as a reductant without the need for the extra tank and injection system.

The absorption equations are as follows with Barium as an example:

$Ba + 2NO + O_2 \Rightarrow Ba(NO_2)_2$ (barium nitrite)
$Ba + 2NO_2 + O_2 \Rightarrow Ba(NO_3)_2$ (barium nitrate)

Similar to a TWC, the LNT uses both platinum and rhodium in reducing conditions. The reducing chemical equations are:

$Ba(NO_2)_2 \Rightarrow Ba + N_2 + 2O_2$
$Ba(NO_3)_2 \Rightarrow Ba + N_2 + 3O_2$

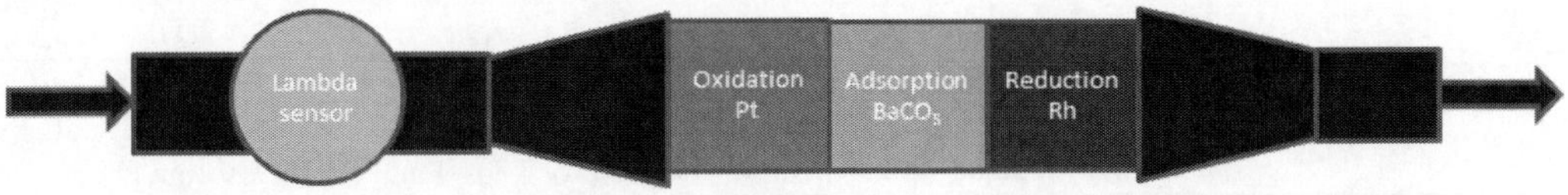

Figure 15. Lean NO_x Trap.

SYSTEM COMBINATIONS

Individual technologies may be combined. Note that it is perfectly possible in some cases to combine catalysts within one can and thus save expense and space. When used in combination however, there may well be issues to resolve and the following table summarises these.

Table 14. Combination systems

Combination system	Notes
DPF + SCR	If the SCR unit is placed downstream of the DPF it will be very effective in abating any primary NO_2 from any upstream oxidation catalysis. However, when the DPF is regenerating, temporary very high temperatures would exceed that which allowed the use of a Vanadium based SCR technology. If the DPF is fitted downstream of the SCR then primary NO_2 will be emitted. Future legislation is likely to set a limit on the NO_2 / NO_x ratio which will probably force the pathway to be DPF then SCR in the exhaust stream.
DPF + EGR	Care must be taken to avoid excessive soot emissions that may block the DPF before it can be regenerated. This could lead to failures of the filter at a later stage when excessive temperatures may be met under regeneration. Catalysed traps work by ensuring the engine-out NO_x / PM ratio is greater than 25:1. Usually a much greater ratio is achieved, circa 50:1 typically. The use of EGR will push engine-out NO_x significantly lower and is likely to increase PM thus adversely affecting the NO_x / PM ratio and the margin required for consistent re-generation.
EGR + SCR	This may utilise EGR where low exhaust temperatures preclude the use of SCR and then switch to the more effective SCR where possible i.e. at higher speeds and loads (and thus temperatures) - this avoids an excessive fuel consumption penalty compared to the use of EGR at all appropriate points in the map. Additionally, there is likely to be a saving in urea reductant by sharing the NO_x abatement task.
EGR + SCR + DPF	A DPF may be added to the EGR + SCR system for treatment of particulates. This is the system adopted by European Heavy Duty Diesel OEMs for Euro 6. Note that IVECO have avoided use of EGR in their system for this level of legislation presumable in the interests of fuel economy.

Table 14. (Continued)

Combination system	Notes
LNT + SCR	The SCR may utilise any NH_3 emanating from the Lean NOx system to eliminate further NO_x from the tailpipe. This has been used, for instance, on a Mercedes-Benz "Bluetec" vehicle system and may become a much more general approach as the Diesel engine OEMs are faced with ever more stringent NO_x legislation.
LNT+SCR +DPF	Additionally a DPF may be added to the LNT + SCR system for treatment of particulates. DPFs will become necessary for Euro 6 and beyond as partical number legislation has been introduced for Diesel and DISI Gasoline types.
FBC + Catalysed DPF	An FBC may be used in combination with a catalysed DPF to assist in reducing the carbon ignition temperature. This should achieve a greater regularity in re-generation and avoid blocked filter bodies or excessively high peak temperatures that may damage the filter body.
Double DOC + DPF	Two upstream DOCs have been used to assist lowering of the carbon ignition temperature beyond that of a single DOC. The Johnson Matthey example is designated "CCRT" and reduces the carbon ignition temperature by a further 15° C compared to the standard "CRT" with single upstream DOC.

Retrofit Examples

Although the emissions abatement business is driven by the very large numbers of production systems required there is a steady market for retrofit options which remain highly practical. Sometimes this has been from the wish of an operator to "green" their fleet but there are also mandated low emission zones (LEZs) that have proliferated in Europe which have created a demand. Some examples of retrofit case studies follow.

TWC

At first sight it would appear unusual for the TWC to have penetrated at retrofit level. However in Germany at the turn of the millennium such an exercise was carried out on "pre-Euro" Gasoline cars with standard carburettors. The control was effected by use of air slides to enable the air / fuel ratio to be kept within the limits for good emissions conversion ratios with feedback from the usual "lambda" sensor. Driveability also appeared to improve with the greater control of mixture strength. The system was extremely low cost. The scheme was voluntary and over 500,000 members of the public had their cars modified! The vehicles were brought up to an official Euro 1 level of emissions attainment but were only just short of Euro 2 which was extant at the time for new vehicles.

DPF

A DPF retrofit exercise was initialised in the UK sponsored by the Department for Transport (DfT). This was driven initially by a voluntary scheme that offered a reduction in Vehicle Excise Duty for compliance with Euro 4 PM emissions on the ETC cycle (0.03

g/kWh) and the operator was awarded a "Reduced Pollution Certificate" (RPC) for the vehicle. The revised scheme was launched on 5 January 2001 and ran to 30 September 2006 whereafter new vehicles were Euro 4 homologated. This allowed a maximum of £500 VED rebate on a sliding scale according to GVW of the vehicle. The PM RPC award is for life of the vehicle but subject to an annual vehicle check. Grants to assist purchase and fitment were available at that time from the DfT sponsored "CleanUp" programme.

After the commencement of Euro 4 the RPC scheme has been used to incentivise early introduction of the next legislation tier (both NO_x and PM). The certificate lapses from the date when the relevant tier is mandated for all new vehicles.

There then followed a requirement by the Mayor of London for a Low Emission Zone (LEZ) which mandated for PM emissions compliance in the whole of the greater London area for heavy duty vehicles. The following table summarises the timetable.

Table 15. London LEZ timetable

Phase	Date	Vehicle categories	PM target	NO_x target
1	4 February 2008	Lorries > 12t GVW	Euro 3	None
2	7 July 2008	Lorries > 3.5t < 12t GVW Buses & Coaches > 5t GVW	Euro 3	None
3	3 January 2012 (Delayed from 4 October 2010)	Larger vans and minibuses	Euro 3	None
4	3 January 2012	Lorries > 3.5t Buses & Coaches > 5t GVW	Euro 4	None
5	2015 (tbd)	TfL Buses only	Euro 4	Euro 4

Similar exercises have been carried out elsewhere and one source claims that 250,000 vehicles worldwide have been retrofitted with a DPF.

SCR

As may be seen from the previous table the Phase 5 of the London LEZ is introducing a NO_x compliance target for the first time. This will be at Euro 4 level and will only include buses operating under the umbrella of Transport for London (TfL). It is anticipated that circa 900 Euro 3 buses will be retrofitted with SCR to achieve the regulation. These buses already have DPF fitted for the Euro 4 PM compliance. Strict NH_3 limits apply (10 ppm) and GHG emissions may not increase by more than 1% (N_2O emissions).

SUMMARY OF TECHNOLOGIES

Table 16. Summary of emission mitigation systems and initial useage dates

System	HC	CO	NO_x	PM	PN	Undesirable	European Production Application
EGR			✓			Fuel consumption	Euro 1 Diesel PC

Table 16. (Continued)

System	HC	CO	NO_x	PM	PN	Undesirable	European Production Application
TWC	✓	✓	✓			NH_3	Euro 1 Gasoline PC
DOC	✓	✓		✓		NO_2	Euro 2 Diesel PC
Catalysed DPF	✓	✓		✓	✓	NO_2	Euro 6 HD
DPF				✓	✓		Euro 4 PC
Partial DPF				✓	✓		Euro 4 HD
FBC + DPF				✓	✓		Euro 3 PC
FBC + Catalysed DPF						NO_2	Retrofit
SCR			✓			NH_3; N_2O	Euro 4 HD
LNT			✓			NH_3	Euro 3 Gasoline PC (DISI)
EGR + DOC	✓	✓	✓	✓		NO_2	Euro 2 Diesel PC
SCR + Catalysed DPF	✓	✓	✓	✓	✓	NH_3; N_2O; NO_2	Euro 6 HD
Partial DPF + EGR			✓	✓	✓	Fuel consumption	Euro 4 HD
EGR + SCR			✓			Fuel consumption	Euro 6 HD
EGR + SCR + Catalysed DPF	✓	✓	✓	✓	✓	NH_3; N_2O; NO_2	Euro 6 HD
LNT + Passive SCR			✓				
LNT + Passive SCR + DPF			✓	✓	✓		Euro 6 Diesel PC
Water Injection			✓				IMO Tier III?

In: Automotive Exhaust Emissions and Energy Recovery
Editor: Apostolos Pesiridis
ISBN: 978-1-63321-493-4

Chapter 2

EXPERIMENTAL INVESTIGATION OF IN-CYLINDER NOX AND SOOT FORMATION BY MEANS OF OPTICAL TECHNIQUES IN A CR DIESEL ENGINE FUELLED WITH OXYGENATED FUEL

Silvana Di Iorio, Ezio Mancaruso and Bianca Maria Vaglieco*
Istituto Motori – CNR Naples, Italy

ABSTRACT

Diesel engines are the main sources of particulate matter (PM) and nitrogen oxides (NOx) emissions in urban areas. The alternative fuels allow the reduction of the pollutant emissions. In particular, rapeseed methyl ester (RME) is a promising biodiesel fuel for compression ignition (CI) engines. Generally, the combustion of biodiesel fuel in CI engines results in lower PM. On the other hand, the effect on NOx emissions is not yet fully understood.

In this chapter, the effect of RME on combustion process and pollutant emissions was analyzed. In particular, the in-cylinder soot formation/oxidation process was associated to the particle emissions at the exhaust. Moreover, the flame temperature was correlated to the NOx emissions.

The investigation was carried out on a single cylinder research optical engine equipped with the head of a Euro 5 multi-cylinder engine and a last-generation common rail (CR) injection system. The investigation was carried out at 1500 rpm and 2 bar break mean effective pressure (BMEP) and 2000 rpm and 5 bar BMEP. These engine points were chosen as representative of the typical urban driving conditions. Moreover, they are included in the new European driving cycle (NEDC). The natural flame emission chemiluminescence was measured by means of the 2D digital imaging and the UV–visible flame emission spectroscopy techniques. The 2D digital images allow the analysis of the combustion process evolution. Moreover, the flame temperature and the soot concentration were evaluated applying the theory of the two-color pyrometry. The in-cylinder broadband UV–visible flame emission spectroscopy measurements were carried out to characterize the soot formation and oxidation process. Furthermore, a new

* Instituto Motori – CNR, Naples, Italy; Email: b.m.vaglieco@im.cnr.it.

approach based on the elaboration of flame combustion in the UV–visible wavelength range was applied. In particular, the extinction spectrum was computed and the soot particle size function distribution was determined by means of an inversion procedure of the optical data. An opacimeter was used to measure the smoke opacity and evaluate the particulate matter concentration. The sizing and the counting of the particles were performed by means of an Electrical Low Pressure Impactor™ (ELPI).

Keywords: Soot particles; Optical techniques; In-cylinder/exhaust particle size distribution function; Alternative fuel

INTRODUCTION

In the last years, diesel engines have become widespread because of the better efficiency, resulting in low fuel consumption and CO_2 emissions, and good drivability. On the other hand, they are the main source of particulate matter (PM) and NOx emissions in urban areas.

In diesel exhaust, NOx is mainly composed of NO. NOx formation mechanisms have been deeply studied [1]. It can be mainly ascribed at three processes: Thermal NOx, Prompt NOx and Fuel NOx. The first process is driven by the temperature. N_2 and O_2 can, in fact, react through a series of chemical reactions known as the Zeldovich mechanism at high temperatures. In particular, NOx formation occurs at temperatures above 1500°C, and the rate of formation increases with the increasing of the temperature. The prompt NOx, also known as Fenimore NOx, are formed because of the intermediate hydrocarbons, such as CH and CH_2, which react with N_2 in the combustion chamber. The resulting C\N species is then involved in reaction with O_2 producing NOx. This mechanism is prevalent under fuel rich conditions. In the last mechanism, the NOx formation is due to the oxidation of nitrogen contained in the fuel. This formation process is generally negligible as the natural nitrogen level in diesel fuel is low.

PM is formed from the locally fuel rich mixture as a result of incomplete combustion. In-cylinder particle formation is driven by the following processes: pyrolysis, nucleation, surface growth, coalescence and agglomeration [2]. The pyrolysis prevails in the first phase, in presence of high temperature without significant oxidation. During this phase soot precursors, such as polycyclic aromatic hydrocarbons (PAH), are produced. The nucleation or soot inception regards the formation of particles from the gas phase reactants. During this phase small particles called nuclei are formed. The nucleation and surface growth are concurrent processes. During this latter process, the hydrocarbons are adsorbed on to the surface of the soot particles. This leads to an increase in soot mass, while the number of particles remains constant. During the coalescence, called also coagulation, the particles collide and coalesce. The agglomeration occurs when primary particles stick together forming large groups of primary particles, typically chain-like structures. In these cases the particle number decreases, the diameter increases, while the mass does not change. Simultaneously at each point of soot formation, soot oxidation occurs. The rate of the formation/oxidation processes depends on several parameters such as the fuel composition, the oxygen content, the in-cylinder temperature, the pressure, etc. Most of particles are oxidized during the combustion process [3-5], the residue is exhausted in the form of solid agglomerates [6-8]. These particles when diluted in the atmosphere are subjected to complex transformation processes [9]. A typical, particle size distribution function measured at the exhaust shows two modes: the nuclei mode, mainly due to volatile particle condensation, and the accumulation mode consisting of carbonaceous particle with adsorbed organic material [10, 11].

The great concerns on environmental and health issues due to the NOx and particle emissions leads to even more stringent emissions regulation. Several studies found out that the number of particles to which the individual is exposed is more important than their mass in terms of adverse effect [12-15]. The toxicity of the particles is strictly linked to their diameter. For this matter, from the current emission legislations a limit also on particle number emissions [16] was introduced.

Great efforts were devoted to reduce pollutant emissions. Some of the most interesting solutions are: the high pressure fuel injection system, common rail (CR), the exhaust gas recirculation (EGR) and the after-treatment devices. The CR injection system allows a better mixing and then a more efficient combustion process. Nevertheless, the application of higher injection pressure caused a reduction in particle size [17]. Good results in terms of NOx emissions reduction was obtained with the EGR technology. On the other hand, soot formation increases because of the lower temperature and oxygen content due to the exhaust recirculation. Diesel particulate filters (DPFs) are very effective in reducing particle emissions both in terms of mass and number. Nevertheless, the regeneration of filter causes larger fuel consumption and particle emissions [18-20].

Currently, great attention was paid to the advanced combustion as well as the alternative fuels. The adoption of non-conventional combustion, such as Low Temperature Combustion (LTC), Premixed Charge Compression Ignition (PCCI) and Homogeneous Charge Compression Ignition (HCCI), allows a simultaneous reduction of NOx and soot emissions [21-24]. On the other hand, the combustion efficiency is lower with respect to the standard diesel combustion mode and a stable combustion can be achieved only at low engine regime. The oxygenated biofuels, instead, allow an effective reduction of the particle emissions [25-27]. On the other hand, the effect on NOx emission is not yet fully understood [28] as the results depend upon numerous factors, such as the engine type and configuration, fuel injection strategy, the presence of EGR. Several studies were performed to understand the effect of biodiesel on NOx emissions [29-31]. These studies highlight that the complexity of the combustion processes make the description of the effect of biodiesel on NOx emissions [28] very difficult.

Despite the good results obtained in terms of NOx and particle emissions reduction, big efforts are necessary to comply with the more stringent emission regulations. To further reduce their emissions it is necessary to better understand the processes responsible for their formation within the combustion chamber [10, 32-34].

In-cylinder soot measurements can be performed through a fast sampling valve [35-37]. In this case, in-cylinder gases were extracted and accumulated over a number of cycles in steady-state engine operating conditions. The obtained aggregates were typically analyzed through a particle sizer, for particle size and number measurements, and a High-Resolution Transmission Electron Microscope (HR-TEM) for the primary particles and the fractal agglomerates sizing. This kind of measurement allows the evaluation of the soot particle size distribution and the particle mass as a function of crank angle. On the other hand, the sampling can influence the formation and oxidation processes. Moreover, it is very important to find a proper dilution factor which allows to avoid the post reactions and at same time does not influence the sampling.

The optical techniques allow the characterization of the in-cylinder NOx and particle emissions with a good temporal and spatial resolution. Furthermore, they are in situ measurements, non-intrusive and provide a comprehensive analysis of the chemical and physical structure of the particles without perturbation of the combustion process. Several techniques can be applied according to the aim of the study. Generally a light source, typically a laser beam, excites the species under study, and then the light emitted from these excited species is detected by a photomultiplier tube (PMT) or a high speed camera. A brief description of the main optical techniques used for NOx and soot particle detection in the combustion chamber is given in the follow.

In-cylinder concentrations of nitric oxide can be obtained by laser-induced fluorescence measurements [38-45]. A laser was used to excite the species in a short UV wavelength range. From the analysis of two-dimensional imaging it is possible to have information about the location of NO formation.

Laser induced incandescence, LII, allows to quantitative measure the soot volume fraction and the primary soot particle diameter. Soot is irradiated by a high energy laser beam with duration of few nanoseconds. It absorbs the energy and its temperature increases from the surrounding temperature to approximately 3500 to 4000 K. The incandescence signals emitted from the heated soot particles are detected by means of optical devices. LII technique was successfully applied in different types of flames [46-54].

Ultraviolet-visible scattering and extinction measurements [3, 25, 51, 55-57] can be used to measure particle size and to have information about soot volume fraction. In this case, a broadband pulsed light source was used to excite the particles.

Natural flame chemiluminescence allows the characterization of the combustion process as well as the pollutant formation. In this case no light sources are used, as suggested by the name. The natural emission of flame can be collected on a spectrograph coupled to an intensified high speed camera. The presence of radical intermediate species of combustion and soot particles is observable by the analysis of the emission spectra. Ad hoc pass band filters can be also used to detect the most interesting radical emissions [58-61] such as OH*, marker of combustion ignition and soot oxidation [58, 62], and CH*.

The temporal and spatial distribution of the soot concentration and the temperature in the combustion chamber can be measured by means of the color pyrometry [48, 63-68]. In particular, the light radiated from hot particles during combustion is recorded at two or more wavelengths and used to determine the particle temperature and the KL factor, which is proportional to soot concentration [68].

The investigations of in-cylinder soot and NOx have rarely been performed because of the high-pressure and high-temperature typically of diesel combustion. Several studies, instead, on the structural properties of flame-generated soot [4, 5] have been carried out. Nevertheless, the diesel in-cylinder soot can differ significantly from laboratory flame generated soot [34] as the engine environment is much more complex due to the turbulent nature of combustion, temperature and pressure fluctuations. These kinds of studies can only give information on the structural properties of soot emitted from the engines but does not characterize the actual soot formation process. In-cylinder diagnostics, instead, can contribute significantly in the understanding of soot processes.

In this chapter a comprehensive analysis of the effect of RME on in-cylinder soot formation and oxidation and the correlation between the flame temperature and the NOx formation was provided. The investigation was carried out on a single cylinder research optical engine equipped with the head of a Euro 5 multi-cylinder engine and a last-generation common rail (CR) injection system. The investigation was carried out at 1500 rpm and 2 bar break mean effective pressure (BMEP) and 2000 rpm and 5 bar BMEP. These engine points were chosen as representative of the typical urban driving conditions. Moreover, they are included in the new European driving cycle (NEDC). The natural flame emission chemiluminescence was measured by means of the 2D digital imaging and the UV–visible flame emission spectroscopy techniques. 2D images of combustion evolution were detected. Moreover, they were processed by two-color pyrometry technique to assess both the flame temperature and the soot concentration. In-cylinder broadband UV–visible flame measurements were carried out to characterize soot formation and oxidation process. A new approach based on the elaboration of flame combustion in the UV–visible wavelength range was proposed. The extinction spectrum was computed and the soot size function distribution was determined by means of an inversion procedure of the optical data. An opacimeter was used to measure the particulate mass concentration. The sizing and the counting of the particles were performed by means of the electrical low pressure impactor™ (ELPI).

EXPERIMENTAL APPARATUS

Engine

The analysis was performed on an optical single-cylinder research engine equipped with the combustion architecture and the injection system of a Euro5 four-cylinder engine. The engine and injection system specifications are listed in Table 1.

Table 1. Engine and injection system specifications

Engine type	4-stroke single cylinder
Bore	8.5 cm
Stroke	9.2 cm
Swept volume	522 cm^3
Combustion bowl	19.7 cm^3
Vol. compression ratio	16.5:1
Injection system	Common Rail
Injector type	Solenoid driven
Number of holes	7
Cone angle of fuel jet axis	148°
Hole diameter	0.141 mm
Rated flow @ 100bar	440 cm3/30s

The engine is equipped with an open electronic control unit (ECU) that allows the setting of the main calibration parameters, such as the injection timing and duration and the injection pressure. The exhaust gas recirculation (EGR) and swirl are managed by external devices. In particular, for EGR regulation a back pressure valve is fitted in the pipe line in order to increase the pressure at the exhaust and bring the exhaust gas into the pressurized intake manifold. The variable swirl actuator (VSA) system consists of a set of blades mounted in the intake manifold in front of the helical shape intake duct of the engine head. The high value of the VSA position induces high swirl motion to the intake air. The engine is operated in continuous mode. More details and specifications are reported in [69].

Fuels

All the measurements were performed using a commercial European low sulfur diesel fuel (REF) and a rapeseed methyl-ester (RME), representative of the most widespread FAME fuel in Europe. The main properties of the fuels are shown in Table 2.

Table 2. Fuels specifications

Feature	Method	REF	RME
Density @ 15°C [kg/m^3]	EN ISO12185	840.1	883
Viscosity @ 40°C [mm^2/s]	EN ISO3104	3.141	4.254
Oxidation Thermal Stability @110°C [h]	EN 14112	-	8.6
Cetane Number	EN ISO5165	51.8	52.3
Low Heating Value [MJ/kg]	ASTM D3338	43.1	37.3
Distillation [°]	IBP		322
	10% vol.		333
	50% vol.	280	337
	90% vol.	338	343
	95% vol.		347
	FBP	362	360
Carbon [%, m/m]	ASTM D5291	86.5	78.5
Hydrogen [%, m/m]	ASTM D5291	13.5	10.8
Nitrogen [%, m/m]	ASTM D5291	-	0.2
Oxygen [%, m/m]	ASTM D5291	-	10.5

Gaseous Emissions Measurement Systems

Gaseous emissions were measured at the exhaust by a commercial analyser. CO, CO_2 and HC were measured by non-dispersive infrared detectors. NOx were measured by means of electrochemical sensors. An opacimeter was used to measure the opacity. It is a partial flow system that measures the attenuation of the visible light (λ=550 nm) in the measuring chamber. The opacity value was correlated to the FSN by means of an empirical relationship [70]. An electrical low pressure impactor (ELPI) was used to measure particle size distribution in real-time. It combines a cascade low-pressure impactor with a diffusion charger and an electrical detection [71, 72]. It operates in the size range from 30 nm up to 10 μm, and on the size range from 7 nm to 10 μm when equipped with an additional Faraday's cage-type filter stage. ELPI can give underestimated apparent size of particles due to fractal-like structure, hence overestimating the number concentration. Before entering in the ELPI, the exhaust gas was diluted in two steps by a Fine Particle Sampler (FPS), which allows sampling and controlling the dilution ratio and the temperature. During all the experiments, the exhaust for the particle analysis was sampled one meter after the exhaust manifold and the sampling frequency was set at 1 Hz. The first dilution temperature was set around 250°C and the secondary was around the ambient temperature. The dilution ratio was set at 30:1.

Optical Experimental Apparatus and Theory

The optical experimental layout is depicted in Figure 1. The natural flame emission was detected through a 45° mirror.

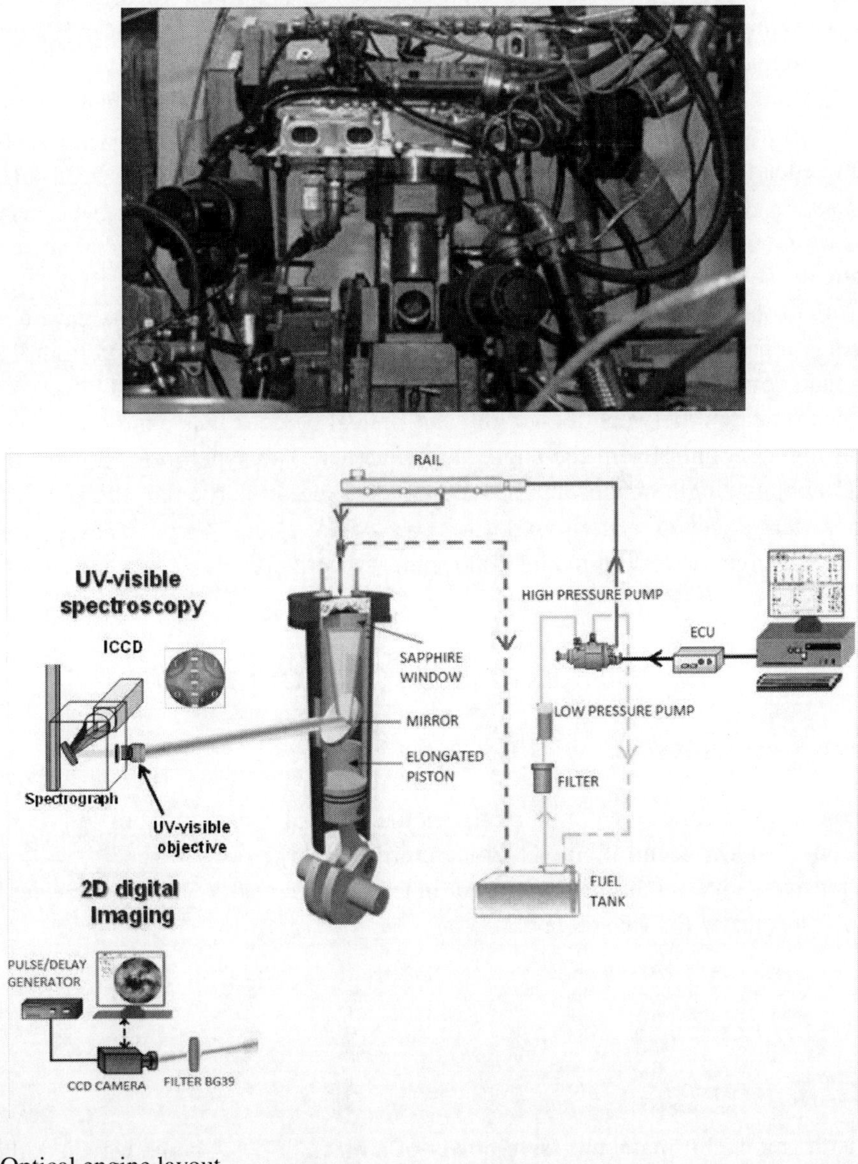

Figure 1. Optical engine layout.

The UV-visible flame emissions were collected and focused on the entrance slit of a spectrograph through an UV lens (Nikon 78 mm f/3.8). Spectrograph has 15 cm focal length, f/4 luminous, and is equipped with a grating of 300 grooves/mm, blazed at 300 nm, with a dispersion of 3.1 nm/mm. An entrance slit width of 100 μm was used. The spectral image formed on the spectrograph exit plane was matched with a gated intensified CCD (ICCD) camera (512 x 512 pixels) with 24x24 μm^2 pixels. The ICCD has high sensitivity in the UV-Visible range. The spectrometer's field of view is due to the magnification of the optics, the slit height (1 mm) and variable width of spectrograph, so as the ICCD pixel size. Data were detected with the spectrograph placed at a central working wavelength of 350 nm and with the intensifier-gate duration of 55 s in order to have a good accuracy in the timing of the combustion onset. The combustion chamber was divided into nine regions of interest; each one was made of 14x200 pixels and had an area of 3.1x42 mm^2. In order to reduce the statistical uncertainty, the flame emission measurements were carried out over 50 consecutive cycles using a frequency repetition of 20 Hz.

Digital imaging analysis was performed by a CCD camera. The CCD camera with 640 x 480 pixels (pixel dimensions of 9.9 x 9.9 μm^2) and high sensitivity over a wide visible range was used in order to acquire the visible combustion. Visible lens, Nikon 55 mmf/3.5, was used. Due to speed limitation of CCD camera, only one image was detected at a given cycle. Ten images from ten separate cycles at fixed crank angle were captured; we analyzed images and subjectively select the one that better represented the whole set. A BG-39 filter was placed in front of the CCD in order to shield it from the IR stimulation. This gave a detection window from approximately 300-600 nm. Thus it was possible to determine the soot temperature and concentration by means of the two-color pyrometry method [73].

The synchronization between the engine and optical devices was controlled by the delay unit with the signal coming from the angle shaft encoder. The synchronization system could be adjusted to obtain single images at a desired crank angle. In particular, the UV and Visible flame images were detected with an exposure time of 55 μs and 42 μs. They correspond to 0.5 crank angle degrees at 1500 rpm and 2000 rpm, respectively.

Theoretical Background

Two-Colour Pyrometry

The two-colour pyrometry technique utilizes the thermal radiation from soot particles for the calculation of soot concentration and temperature [74, 75].

The intensity radiation emitted from a blackbody depends on the wavelength and temperature according to the Planck law:

$$I_{b,\lambda}(T) = \frac{C_1}{\lambda^5 \left[\exp\left(\frac{C_2}{\lambda T} \right) - 1 \right]} \tag{1}$$

Where $I_{b,\lambda}$ is the monochromatic emissive power of a black body; λ is the wavelength; C_1 and C_2 are the first and second Planck constants; T is the flame temperature.

The monochromatic emissivity of a non-black body is defined as the fraction of the black body radiation emitted by a surface at wavelength λ:

$$\varepsilon_\lambda = \frac{I_\lambda(T)}{I_{b,\lambda}(T)} \tag{2}$$

where $I_\lambda(T)$ and $I_{b,\lambda}(T)$ are the monochromatic emissive power of a non-black body and a black body respectively, at the same temperature, T, and wavelength, λ.

The equation (2) can be rewritten as

$$\varepsilon_\lambda = \frac{I_{b,\lambda}(T_a)}{I_{b,\lambda}(T)} \tag{3}$$

where T_a is the apparent temperature, defined as the temperature of a black body which will emit the same radiation intensity as a non-black body at temperature. It depends on the wavelength:

Combining the equations (1) and (3):

$$\varepsilon_\lambda = \frac{\exp\left(\frac{C_2}{\lambda T}\right) - 1}{\exp\left(\frac{C_2}{\lambda T_a}\right) - 1} \tag{4}$$

The ε_λ is evaluated from the empirical correlation of Hottel and Broughton [69]:

$$KL = -\lambda^\alpha \ln\left[1 - \frac{\exp\left(\frac{C_2}{\lambda T}\right) - 1}{\exp\left(\frac{C_2}{\lambda T_a}\right) - 1}\right] \tag{5}$$

where K is the absorption coefficient, it is proportional to the soot concentration; L is the optical path length or flame thickness; α is the absorption index, it depends on the physical and optical properties of soot in the flame and depends on the wavelength. Matsui et al. [76] carried out a validation study of the above correlation by performing measurements of soot emissivity in a diesel engine. They concluded that in the visible range this is the correct functional relationship between emissivity and wavelength.

Combining the equations (4) and (5):

$$KL = -\lambda^\alpha \ln\left[1 - \frac{\exp\left(\frac{C_2}{\lambda T}\right) - 1}{\exp\left(\frac{C_2}{\lambda T_a}\right) - 1}\right] \tag{6}$$

In the two-colour method, the thermal radiation is detected at two different wavelengths then it is possible to write:

$$\left[1-\frac{\exp\left(\frac{C_2}{\lambda_1 T}\right)-1}{\exp\left(\frac{C_2}{\lambda_1 T_a}\right)-1}\right]=\left[1-\frac{\exp\left(\frac{C_2}{\lambda_2 T}\right)-1}{\exp\left(\frac{C_2}{\lambda_2 T_a}\right)-1}\right] \quad (7)$$

Known the apparent temperature at the two measured wavelengths this equation can be solved for the flame temperature using a calibrated two-colour optical pyrometer system.

Once known the flame temperature it is possible to determine through the equation (6) the KL which is proportional to soot concentration. The volumetric density of soot and the soot gravimetric density can be also evaluated.

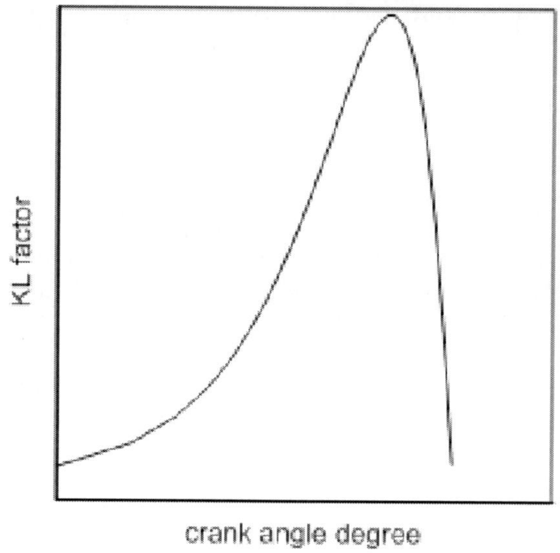

Figure 2. Typical KL distribution for Diesel combustion.

A typical temporal evolution of KL factor is depicted in Figure 2. The rising slope is mainly due to soot formation whilst during the soot oxidation the KL factor decreases.

Procedure for In-Cylinder Particle Size Distribution Function from the Natural Flame Luminosity

The natural flame luminosity measurements give information about soot formation and oxidation processes. Moreover, the in-cylinder size distribution of soot particles can be evaluated.

The flame emission was calibrated with a Tungsten lamp and DeVos data [74, 76-77]. By the knowledge of the absolute flame emission intensity and the in-cylinder flame temperature, the soot emissivity was determined applying the Planck law (1). The extinction is considered as the attenuation of electromagnetic wave by the scattering and the absorption as it transverses a cloud of particles in a gas or a liquid. The extinction cross-section of the particle produced in a burning flame is a complex function of the chemical and physical properties. It can be interpreted in the framework of electromagnetic theory of the light by:

$$K_{ext}(\lambda) = N \cdot C_{ext}(\lambda, D_p, n, k) \tag{8}$$

Assuming that the particles are spherical and constitute poly-dispersed system, the extinction coefficient is function of the particle diameter and the Mie theory can be applied. However the Mie theory converges to the Rayleigh approximation in the case of small particle ($D_p \ll \lambda$) [78]. The Rayleigh extinction coefficient is proportional to wavelength as $K_{ext} \propto \lambda^{-1}$. This assumption permits the employment of the Rayleigh approximation and to neglect the scattering contribution to extinction.

Finally, a numerical procedure for retrieving the optical data was used. The inversion procedure was based on the minimization of the difference between experimental and theoretical spectrum. It was widely applied in a previous paper [79].

The procedure for the analysis of the spectrum of the natural flame has a lower accuracy as it depends on several parameters such as the refractive index, the in-cylinder conditions and so on. Moreover, the procedure assumes that the particles have a spherical morphology. It was seen that the approximation of the spherical particles for agglomerates yields errors around 17%.

RESULTS AND DISCUSSION

The measurements were carried out at 1500rpm and 2 bar BMEP and at 2000rpm and 5bar BMEP under steady-state condition. These engine operating points were chosen as representative of the typical urban driving conditions. Moreover, they are representative of the engine behavior on the new European driving cycle (NEDC) when installed on a D-class vehicle. The injection strategies consist of two injections per cycle, pilot and main. For both the investigated fuels, the start of pilot and main injections were kept constant. The pilot energizing time also was fixed; while the main energizing times of RME was properly adjusted in order to obtain the same BMEP. In particular, injection duration was increased because of the lower LHV of RME. The amount of EGR was set at 57% and 35% for 1500 rpm and 2000 rpm, respectively. The VSA system was set at 66% and 27% for 1500 rpm and 2000 rpm, respectively. The main engine parameters are listed in Table 3.

Table 3. Main engine parameters

Fuel	Speed [rpm]	SOI Pilot [cad]	ET Pilot [μs]	SOI Main [cad]	ET Main [μs]	Prail [bar]	EGR [%]	VSA [%]
REF	1500	-16	350	-5.5	600	500	57	66
RME	1500	-16	350	-5.5	640	500	57	66
REF	2000	-18.5	257	-2	556	730	35	27
RME	2000	-18.5	257	-2	594	730	35	27

Thermodynamic Analysis

The in-cylinder pressure, the rate of heat release (ROHR) and the injector energizing current versus the crank angle degree (cad), over the end of compression and the early expansion stroke are depicted in Figure 3.

The effect of RME on injection duration is more evident at 2000 rpm, as possbiler to see looking at the injector energizing curve. For both fuels, the in-cylinder pressure shows overlapped peaks at 1500 rpm, due to the main combustion. At 2000 rpm, the longer fuel injection results in to a longer combustion during the early expansion phase and peak pressure retarded. From the analysis of the ROHR the pilot combustion is well distinguishable only at 1500 and for REF fuelling. For both the investigated conditions the main combustion is delayed for RME fuelling, despite its CN is slightly higher. Moreover, it is possible to distinguish three stages for the main combustion: the premixed combustion phase, the mixing-controlled combustion phase and the late combustion phase. Particles are mainly formed during the mixing-controlled combustion phase [80, 81]. For both the investigated conditions the REF fuel shows a narrower peak with respect to RME fuelling, likely because of the longer RME main injection duration. The start of combustion (SOC) is determined by the analysis of the heat release rate (ROHR) trace. From a general point of view the start of combustion was detected when the energy release begins to exceed the energy losses due to the fuel evaporating process [80]. The SOC of the pilot injection occurred at 8° before top dead center (BTDC) while the SOC of main injection occurred at 1° after top dead center (ATDC) at 1500 rpm and for both fuels. At higher engine speed, the SOC of main injection occurred at 7° ATDC for both fuels.

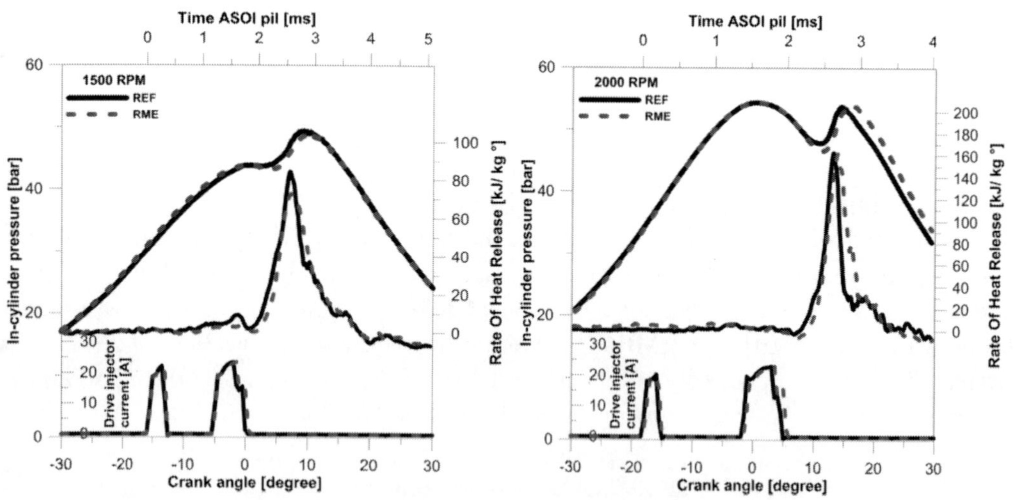

Figure 3. In-cylinder pressure, ROHR curve and energizing current signal measured for REF and RME fuels at 1500 rpm (left) and 2000 rpm (right).

2D Digital Imaging

The thermodynamic analysis gives real-time cycle-resolved information over all the combustion processes but they do not allow a local analysis. The optical techniques, instead, are a powerful tool for detailing the thermal and fluid dynamic phenomena occurring in the combustion chamber with high spatial and temporal resolution. In particular, 2D digital imaging allows the analysis both of the spray phase and combustion processes.

Only the data at 2000 rpm will be presented as fully representative of the effect of biofuel on the combustion process. Figure 4 shows the visible flames detected in the combustion chamber. In particular, the visible images for REF and RME fuels are reported. The main evidence in the analysis of the images is the low luminosity of the flames for the engine running with RME. For both fuels, the combustion evolves in the same way: it starts from the periphery of the jet, where more favorable conditions are available, and spreads along the jets. The evaporation of the main fuel jets is favoured by the pilot combustion; moreover, the vapour fuel is transported by the swirl motion in anti-clockwise direction and both the mixing and ignition of main injection occur.

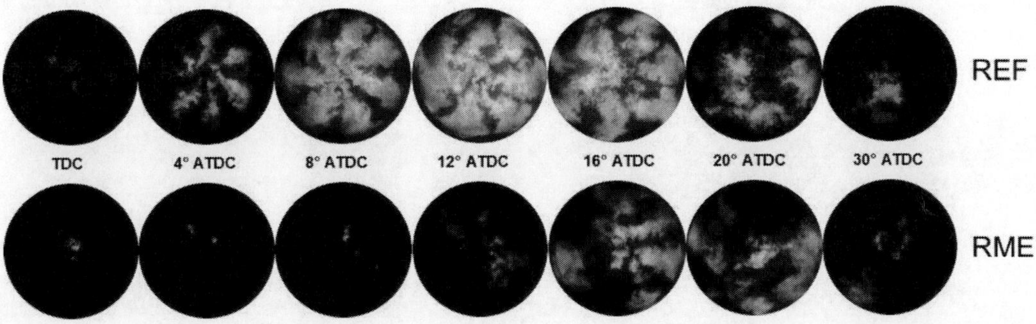

Figure 4. Selection of the visible images detected during the REF (up) and RME (bottom) combustion at 2000 rpm.

Figure 5 shows the false-colour maps of flame temperature and KL factor that is proportional to the soot concentration, related to the REF images of Figure 4. The spatial temperature and KL distribution in the combustion chamber is non-uniform during the combustion evolution. In particular, homogeneous distribution of areas at high temperature is detected from 8° up to 16° ATDC. While the highest soot concentration is noted along the jet axis where the fuel burns in rich condition.

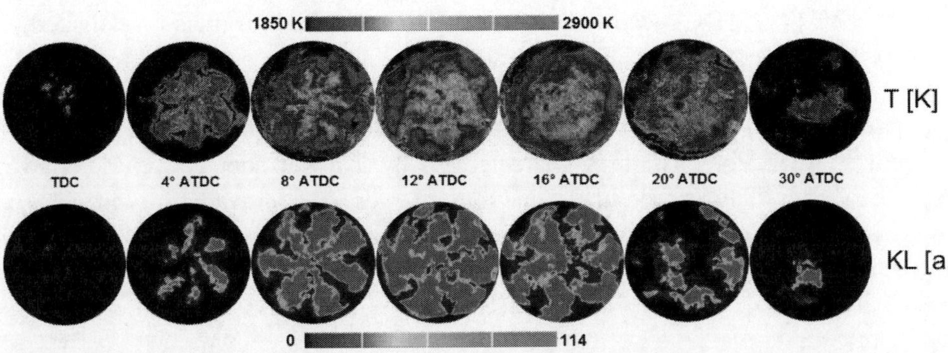

Figure 5. Temperature and KL spatial evolution at 2000 rpm for REF.

At each crank angle the number of pixel corresponding to an assigned temperature was counted in order to better analyze the effect of fuel on the temperature. This procedure allows the evaluation of the combustion area at typical temperatures. A pixel can be, in fact, interpreted as a combustion chamber area of about 0.1 x 0.1 mm^2. The analysis of the in-cylinder flame temperature distribution gives interesting information about the thermal NO formation [74]. The reactions that are the basis of thermal NO formation occur at temperatures over 2200 K [80]. The flame temperature distribution for REF and RME fuels and the NOx exhaust emissions are shown in Figure 6.

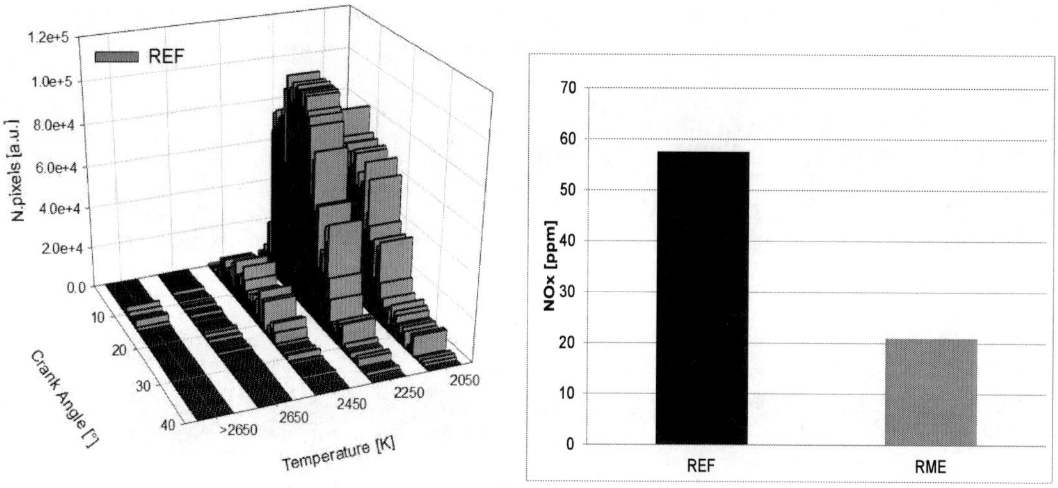

Figure 6. Flame temperature distribution for REF (left) and RME (right) during the combustion and NOx emission at the exhaust (bottom) at 2000 rpm.

The in-cylinder flame distribution shows for RME a smaller area at the typical temperatures of the NOx reaction activation during the premixed combustion. The temperature values are quite similar during the last phase of the combustion. At the same time a lower NOx emission is measured at the exhaust.

The temporal evolution of integral soot concentration for the tested fuels at 2000 rpm is depicted in Figure 7.

Even if the SOC (7° ATDC) occurred at the same time for the two fuels investigated, the start of soot formation happened in different time depending on the chemical characteristics of the adopted fuel. In particular, REF fuel start to form soot at 11° ATDC, 4° after the start of combustion of main injection. The start of soot formation for RME is noted later, at 13° ATDC. Moreover, RME combustion is characterized by a lower soot concentration, with respect to the REF ones. This can be likely due to the larger oxygen availability in rich-zone and the lower aromatics content that as known are the main soot precursors [82-84]. At the same time, the soot oxidation is enhanced for RME fuel because of the larger oxygen content and soot reactivity [85, 86].

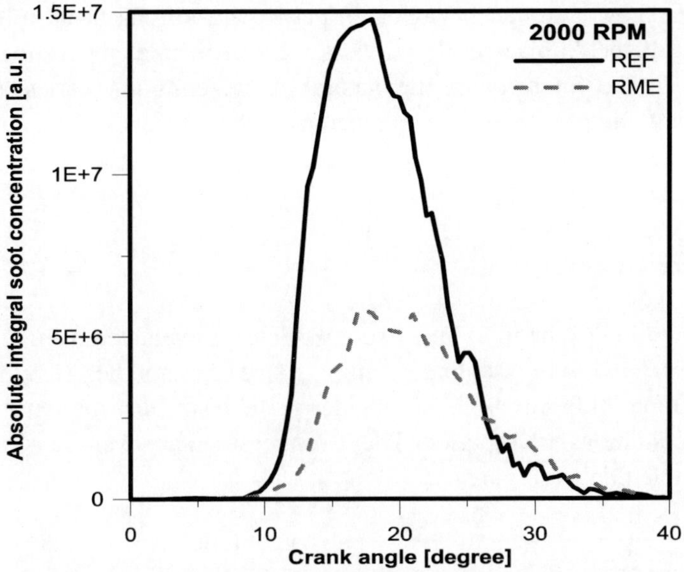

Figure 7. Integral soot concentration for REF and RME fuels at 2000 rpm.

Natural Flame Emissions

The natural flame emission spectra detected at several crank angles after the top dead center (ATDC) during the mixing-controlled and late combustion, are depicted in Figure 8 for REF fuelling.

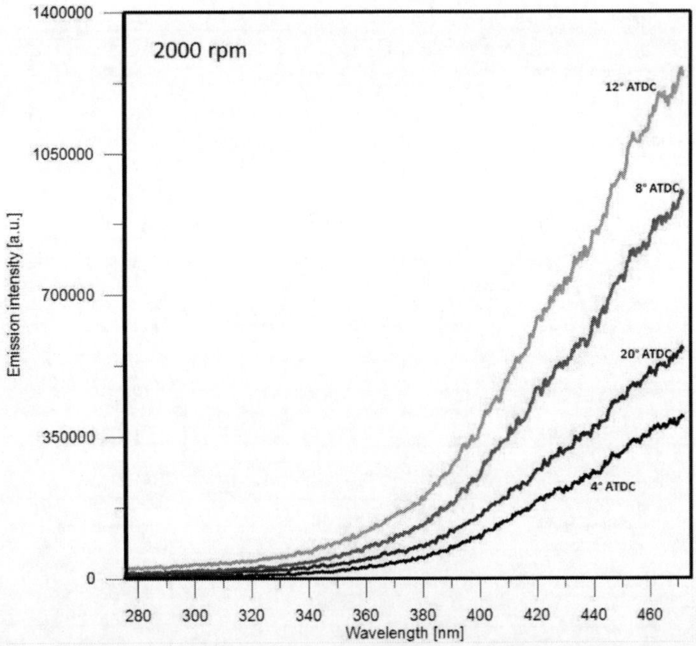

Figure 8. Natural flame emission spectra detected during the REF combustion at different crank angles.

All the spectra show a similar behavior. In particular, a broad band in the visible region very similar to the Planck's black body curve is well resolvable, suggesting the presence of soot particles [75]. The intensity of the natural flame emission varies according to the combustion phasing. In particular, it first increases then reaches a maximum value and, finally, decreases.

In-Cylinder Size Distribution

The size distribution of the in-cylinder soot particles is evaluated using the new approach based on the elaboration of the in-cylinder natural flame emission described in the theoretical background paragraph. In Figure 9 the in-cylinder particle size distribution for both fuels and engine operating conditions are reported. They are measured at sever crank angles during the combustion of visible flames (Figure 4).

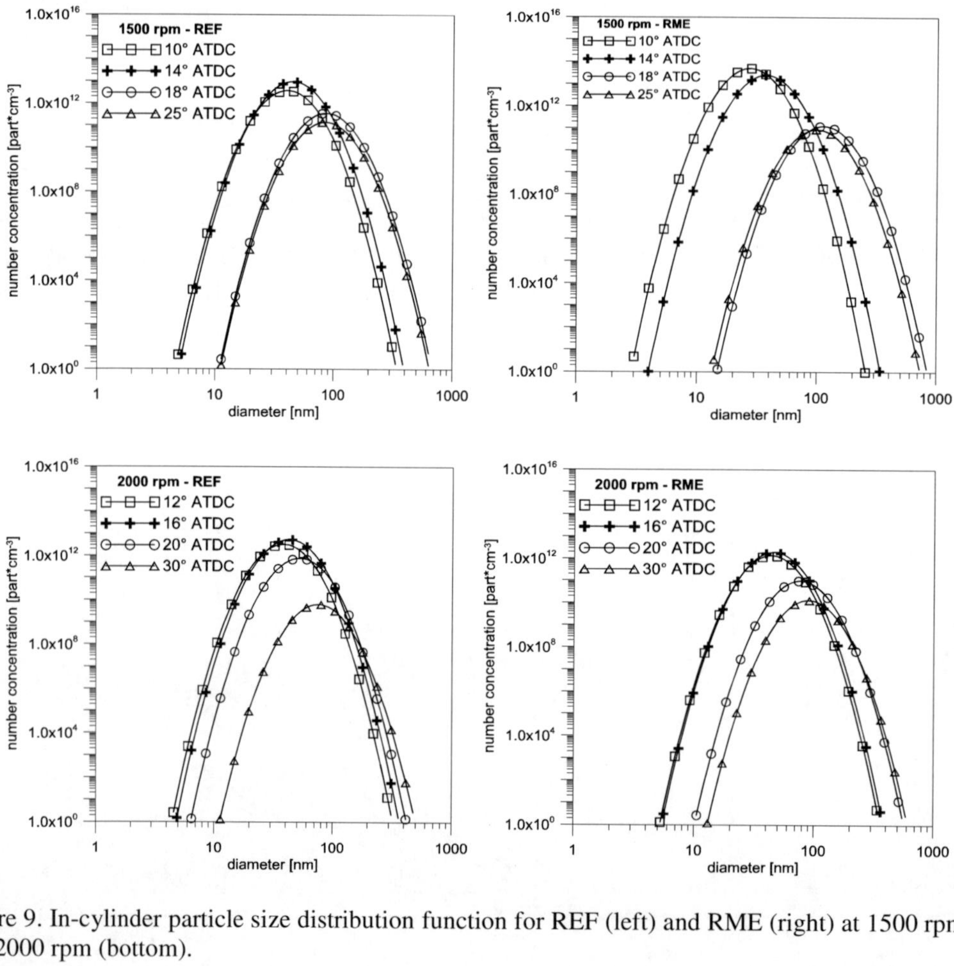

Figure 9. In-cylinder particle size distribution function for REF (left) and RME (right) at 1500 rpm (up) and 2000 rpm (bottom).

The size distribution ranges from few nanometer till to 600 nm. Here are presented only the data related to the spectra of natural flame emission intensity detected for REF and RME at the most representative phase of the soot formation/oxidation process at 1500 and 2000 rpm. In particular, it was chosen the crank angles in correspondence of the formation process, the absolute peak intensities for the integral soot concentration and of oxidation process. At these crank angles the soot is at the end of its formation. For both the fuels, at the increase of the crank angles, the particle number concentration decreases and particle diameter increases. This is due to the oxidation and agglomeration processes. It is worth of noting the sharper decrease of soot particle number due to the enhanced oxidation highlighted in the KL factor results. Moreover, it is important to take in to account the greater biofuel soot reactivity [85, 86].

A comparison of the in-cylinder soot particle distribution is depicted in Figure 10.

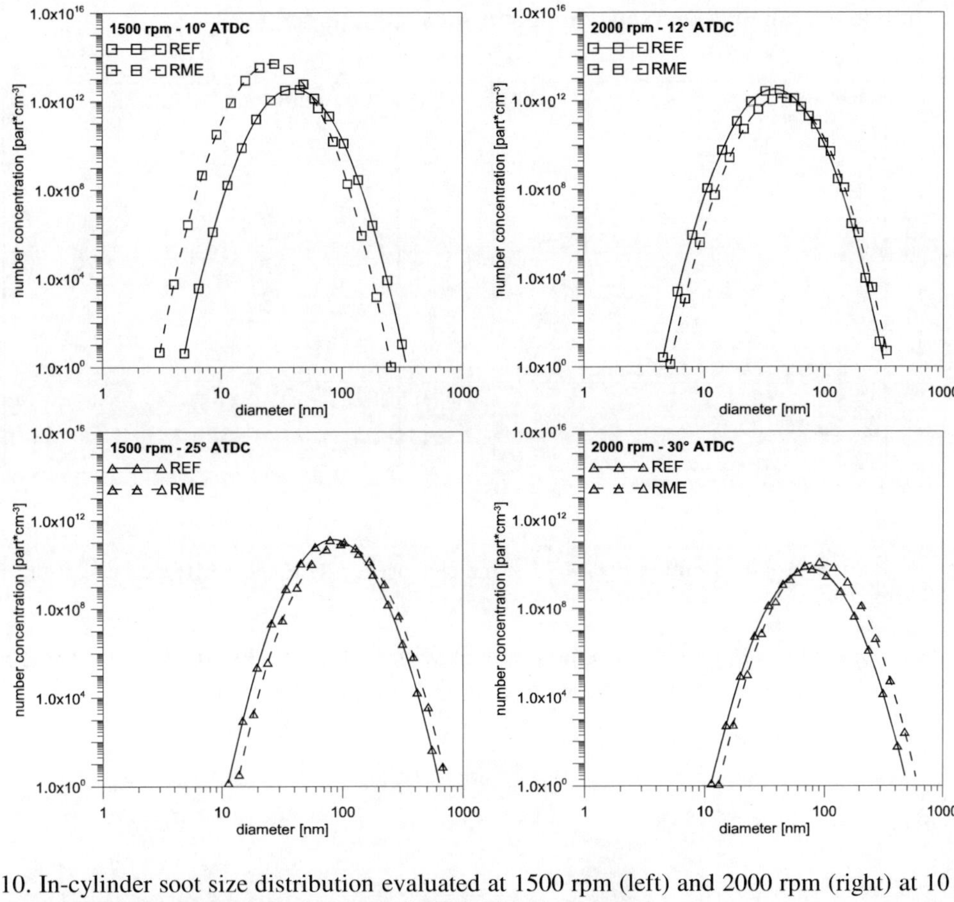

Figure 10. In-cylinder soot size distribution evaluated at 1500 rpm (left) and 2000 rpm (right) at 10 cad and 12 cad ATDC (up) and 25 cad and 30 cad ATDC (down).

At 1500 rpm, RME is characterized by a larger soot particle concentration at the first crank angle. This result can be due to the different fuel chemical and physical properties. In particular, because of the higher viscosity and density of RME, larger droplets with respect to diesel fuel are produced. This effect is more evident at 1500 rpm because of the lower injection pressure. For both the tested engine conditions, the soot particles have the same diameter during the formation phase, around 30 nm. On the other hand, the RME soot particle diameter is slightly larger at the late crank angle, during the last phase of oxidation. In particular, at 1500 rpm and 25 cad ATD soot particle size distribution is centered around 80 nm for REF and around 100 nm for RME. At 2000 rpm and 30cad ATDC the soot particle size distribution in centered around 80 nm for REF and around 90 nm for RME. It is important to highlight that the RME combustion is longer with respect to the REF and that the oxidation process lasts longer.

The particle size distributions measured at the exhaust for the tested fuels and engine operative conditions are shown in Figure 11. For the two fuels it is interesting to observe that particle number concentration is 4 orders of magnitude lower than the in-cylinder soot concentration. As said before the oxidation process occurs until the in-cylinder temperature is sufficient high for the oxidation. At the same time, the particle size distribution is shifted towards larger diameter because of the agglomeration process. Nevertheless, as observed for in-cylinder data the REF combustion is characterized by a larger particle number concentration overall the size range. Similar results are reported in [84].

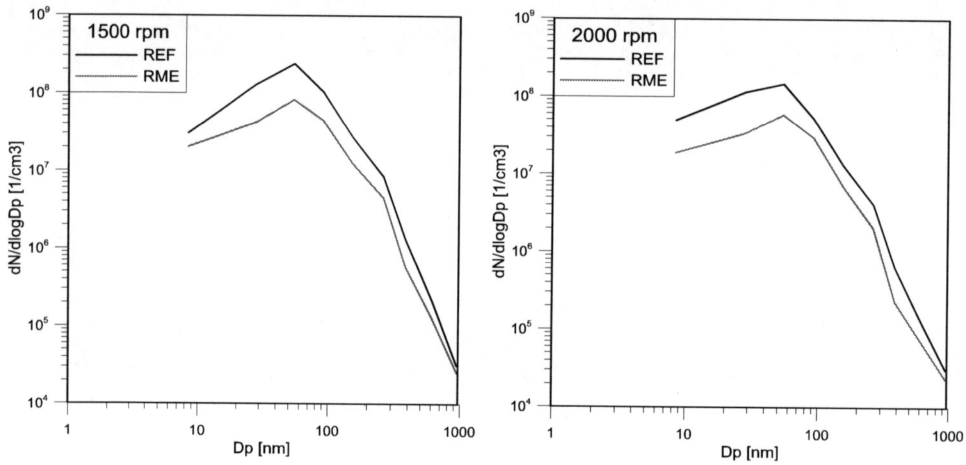

Figure 11. Particle size distribution measured at the exhaust for REF and RME fuels at 1500rpm (left) and 2000 rpm (right).

The opacimeter data are in good agreement with PSDF results. In Figure 12 is depicted the FSN evaluated for both the fuels and operative conditions from the opacity.

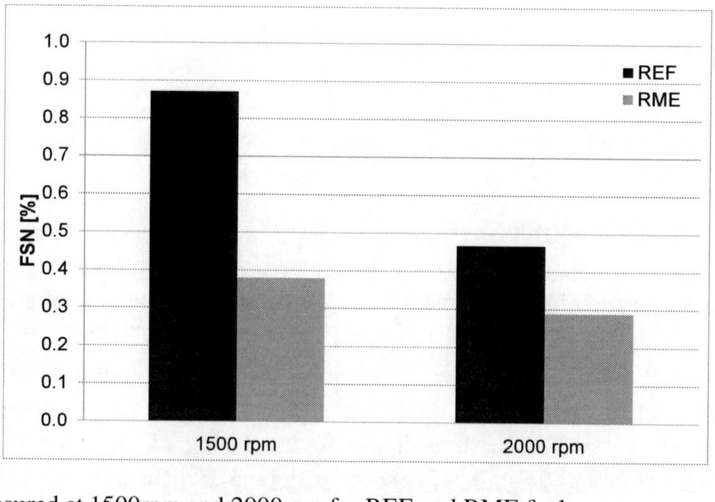

Figure 12. FSN measured at 1500rpm and 2000rpm for REF and RME fuels.

In particular, RME shows a smoke opacity reduction of around 60% and 40% with respect to REF at 1500 rpm and at 2000 rpm, respectively.

REFERENCES

[1] Miller, J.A., Bowman, C.T. 1989. Mechanism and modeling of nitrogen chemistry in combustion. *Progress in Energy and Combustion Science* 15 287–338.

[2] Tree, D.R., Svensson, K.I. 2007. Soot processes in compression ignition engines. *Prog. Energy Combust. Sci.* 33, 272–309.

[3] Li, Z., Song, C., Song, J., Lv, G., Dong, S., Zhao, Z. 2011. Evolution of the nanostructure, fractal dimension and size of in-cylinder soot during diesel combustion process, *Combustion and Flame* 158, 1624–1630.

[4] Vander Wal, R.L., Tomasek A.J. 2004. Soot nanostructure: Dependence upon synthesis conditions. *Combust. Flame* 136 (1–2), 129–140.

[5] Vander Wal, R.L. 2005. Soot Nanostructure: Definition, Quantification and Implications. *SAE Technical Paper* 2005–01–0964

[6] Myung, C., Lee, H., Choi, K., Lee, Y., Park, S.2009. Effects of gasoline, diesel, LPG and low-carbon fuels and various certification modes on nanoparticle emission characteristics in light-duty vehicles. *International Journal of Automotive Technology,* 10(5), 537–544.

[7] Bérubé, A., Jones, T.P., Williamson, B.J., Winters, C., Morgan, A.J., Richards, R.J. 1999. Physicochemical characterization of diesel exhaust particles: factors for assessing biological activity. *Atmospheric Environment,* 33, 1599–1614.

[8] Kittelson, D.B. 1998. Engines and nanoparticles: a review. *Journal of Aerosol Science,* 29, 575–588.

[9] Kittelson D.B. 2001. Recent Measurements of Nanoparticle Emissions from Engines, Proceedings of the meeting on Current Research on Diesel Exhaust Particles. Japan Association of Aerosol Science and Technology.

[10] Lepperhoff, G., Houben M. 1990. Particulate Emission and Soot Formation Processes by Diesel Engines. *IMechE,* Chester.

[11] Pischinger, F., Lepperhoff, G., Houben, M. 1991. Soot Formation and Oxidation in Diesel Engines. International Workshop 'Mechanisms and Models of Soot Formation, University Heidelberg.

[12] Donaldson K., Li X.Y., MacNee W. 1998. Ultrafine (nanometer) Particle Mediated Lung Injury, *Journal of Aerosol Science* 29, 553–560.

[13] Oberdörster, G., Utell, M. J. 2002. Ultrafine Particles in the Urban Air: to the Respiratory Track and Beyond, Environmental Health Perspectives, 110, A440–A441.

[14] Pope C.A., Dockery D.W. 2006. Health Effects of Fine Particulate Air Pollution: Lines That Connect, *Journal of the Air & Waste Management Association,* 56,709–742.

[15] Kennedy I.M. 2007. The Health Effects of Combustion-Generated Aerosols. *Proceedings of the Combustion Institute*, 31, 2757–2770.

[16] Regulation (EC) No 715/2007 of the European parliament and of the council, *Official Journal of the European Union* (2007).

[17] Lapuerta, M., Martos, F., Herreros, J. 2007. Effect of engine operating conditions on the size of primary particles composing diesel soot agglomerates. *Aerosol Sci.* 38, 455–466.

[18] Millo, F., Vezza, D., Vlachos, T., Fino, D. et al., 2010. Particle Number and Size Distribution from a small displacement automotive diesel engine during DPF regeneration. *SAE paper* 2010-01-1552.

[19] Beatrice, C., Di Iorio, S., Guido, C., Napolitano, PP., 2012. Detailed characterization of particulate emissions of an automotive catalyzed DPF using actual regeneration strategies. *Experimental Thermal and Fluid Science* 39, 45–53.

[20] Maricq, M. 2007. Chemical Characterization of Particulate Emissions from Diesel Engines: A Review. *Aerosol Sci.* 38, 1079–1118.

[21] Imarisio, R., Ivaldi, D., Lisbona, M. G., and Tonetti, M. 2006. Technologies Towards EURO 6 Passenger Car Diesel Emissions Standards. ATA Conference, Siracusa, Haly, 2006

[22] Kimura, S., Ogawa, H., Matsui, Y., and Enomoto, Y. 2002. An Experimental Analysis of Low-Temperature and Premixed Combustion for Simultaneous Reduction of NOx and Particulate Emissions in Direct Injection Diesel Engines. *Int. J. Engine Res.* 3(4):249–259.

[23] Denbratt, I., and Helmantel, A. 2004. HCCI Operation of a Passenger Car Common Rail DI Diesel Engine with Early Injection of Conventional Diesel Fuel. *SAE Paper* 2004-01-0935. SAE, Warrendale, PA.

[24] Choi, D., Miles, P. C., Yun, H., and Reitz, R. D. 2005. A Parametric Study of Low-Temperature Late-Injection Combustion in a HSDI Diesel Engine. *JSME Int. J.,* Serie B 48(4):656–664.

[25] Johnson, T.V. 2009. Diesel Emission Control in Review. *SAE Int. J. Fuels Lubr.* 1(1):68-81.

[26] Mueller, C. J., Martin. G. C. 2002. Effects of Oxygenated Compounds on Combustion and Soot Evolution in a DI Diesel Engine: Broadband Natural Luminosity Imaging. SAE 2002-01-1631.

[27] Klein-Douwel, R.J.H., Donkerbroek, A.J., van Vliet A.P., Boot, M.D., Somers L.M.T., Baert, R.S.G., Dama, N.J., Meulen, J.J. ter. 2009. Soot and chemiluminescence in diesel combustion of bio-derived, oxygenated and reference fuels. *Proc. Combust. Inst.* 32 2817– 2825.

[28] Hoekman S. K., Robbins C. 2012. Review of the effects of biodiesel on NOx emissions. *Fuel Processing Technology* 96, 237–249.

[29] Mueller, C.J., Boehman, A.L., Martin, G.C. 2009. An experimental investigation of the origin of increased NOx emissions when fueling a heavy-duty compression ignition engine with soy biodiesel. *SAE International,* 2009-01-1792 pp. 1–28.

[30] Cheng, A.S., Upatnieks, A., Mueller, C.J. 2006. Investigation of the impact of biodiesel fuelling on NOx emissions using an optical direct injection diesel engine. *International Journal of Engine Research* 7 297–318.

[31] Ban-Weiss, G.A., Chen, J. Y., Buchholz, B.A., Dibble, R.W. 2007. A numerical investigation into the anomalous slight NOx increase when burning biodiesel; a new (old) theory. *Fuel Processing Technology* 88, 659–667.

[32] Lepperhoff, G., Houben M. 1990. Soot Formation and Oxidation in IDI DI Passenger Car Diesel Engines. SIA Congress, Lyon.

[33] Wölfle, M., Strömberg, S., Houben, Lepperhoff, M., G. 1993. Influence of Engine Operating Parameters on Pollutant Formation during Diesel Engine Combustion. IMechE, Birmingham.

[34] Clague, A.D.H., Donnet, J.B., Wang, T.K., Peng, J.C.M. 1999. Comparison of diesel engine soot with carbon black, *Carbon* 37 (10), 1553–1565.

[35] Mosbach S., Celnik M.S., Raj A., Krafta M., Zhang H. R., Kubo S., Kim K-O. 2009. Towards a detailed soot model for internal combustion engines, Combustion and Flame 156, 1156–1165.

[36] Pungs, A., Pischinger, S., Bäcker, H., Lepperhoff, G. 1999. Analysis of the Particle Size Distribution in the Cylinder of a Common Rail DI Diesel Engine During Combustion and Expansion. 2000-01-1999.

[37] Krüger, V., Wahl, C., Hadef, R., Geigle, K.P., Stricker, W., Aigner, M. 2005. Comparison of laser-induced incandescence method with scanning mobility particle sizer technique: the influence of probe sampling and laser heating to soot particle size distribution. *Meas. Sci. Technol.* 16, 1477–1486.

[38] Verbiezen, K., Donkerbroek, A.J., Klein-Douwel, R.J.H., van Vliet, A.P., Frijters, P.J.M., Seykens, X.L.J., Baert, R.S.G., Meerts, W.L., Dam, N.J., ter Meulen, J.J. 2007. Diesel combustion: In-cylinder NO concentrations in relation to injection timing, *Combustion and Flame* 151, 333–346.

[39] Brugman, Th.M., Klein-Douwel, R., Huigen, G., van Walwijk, E., Meulen, J.J. ter. 1993. Laser induced fluorescence imaging of NO in an n-heptane and diesel fuel driven diesel engine. *Appl. Phys. B* 57, 405–410.

[40] Brugman, Th.M., Stoffels, G.G.M., Dam, N.J., Meerts, W.L., Meulen J.J. ter. 1997. Imaging and post-processing of laser-induced fluorescence from NO in a diesel engine. *Appl. Phys.* B 64, 717–724.

[41] Dec, J.E., Canaan, R.E. 1998. PLIF Imaging of NO Formation in a DI Diesel Engine. *SAE Tech.* Paper no.980147.

[42] Stoffels, G.G.M., van den Boom, E.J., Spaanjaars, C.M.I., Dam, N., Meerts, W.L., Meulen, J.J. ter, Duff, J.C.L., Rickeard, D.J. 1999. In-Cylinder Measurements of NO Formation in a Diesel Engine. *SAE Tech.* Pap. Series no. 1999-01-1487.

[43] Hildenbrand, F., C. Schulz, J. Wolfrum, F. Keller, E. Wagner, Laser diagnostic analysis of NO formation in a direct injection Diesel engine with pump-line nozzle and common-rail injection *systems. Proc. Combust. Inst.* 28 (2000) 1137–1143.

[44] van den Boom, E.J., Monkhouse, P.B., Spaanjaars, C.M.I., Meerts, W.L., Dam, N.J., ter Meulen, J.J. 2001. V.I. Vlad (Ed.), ROMOPTO 2000, Sixth Conference on Optics, *Proc. SPIE,* vol. 4430, pp. 593–606.

[45] Hildenbrand, F., Schulz, C., Keller, F., König, G., Wagner, E. 2001. Quantitative Laser Diagnostic Studies of the NO Distribution in a DI Diesel Engine with PLN and CR Injection Systems. *SAE Tech. Pap. Series,* no. 2001-01-3500.

[46] Hofmann, M., Bessler, W.G., Schulz, C., Jander, H. 2003. Laser-Induced Incandescence for Soot Diagnostics at High Pressures. *Appl. Opt.* 42, 2052–2062.

[47] Liu, F., Daun, K.J., Snelling, D.R., Smallwood, G.J. 2006. Heat conduction from a spherical nano-particle: status of modeling heat conduction in laser-induced incandescence. *Appl. Phys.* B 83, 355–382.

[48] Boiarciuc, A., Foucher, F., Mounaim-Rousselle, C. 2006. Soot volume fractions and primary particle size estimate by means of the simultaneous two-color-time-resolved and 2D laser-induced incandescence. *Appl. Phys.* B 83, 413–421.

[49] Dreier, T., Bougie, B., Dam, N., Gerber, T. 2006. Modeling of time-resolved laser-induced incandescence transients for particle sizing in high-pressure spray combustion environments: a comparative study. *Appl. Phys.* B 83, 403–411.

[50] Ni, T., Gupta, S.B., Santoro, R.J. 1994. Suppression of soot formation in ethene laminar diffusion flames by chemical additives. *Symp. (Int.) Combust.* 2, 585–592.

[51] Vander Val, R.L., Jensen, K.A. 1998. Laser-Induced Incandescence: *Excitation Intensity. Appl. Opt.* 37 (9), 1607–1616.

[52] Braun-Unkhoff, M., Chrysostomou, A., Gutheil, E., Lückerath, R., Sticker, W. 1998. Experimental and Numerical Study on Soot Formation in Laminar High Pressure Flames. *Symp. (Int.) Combust.* 27, 1565–1572.

[53] Angrill, O., Geitlingea, H., Streibel, T., Suntz, R., Bockhorn, H. 2000. Influence of exhaust gas recirculation on soot formation in diffusion flames. *Proc. Combust. Inst.* 28, 2643–2649.

[54] Cignoli, F., De Iuliis, S., Manta, V., Zizak, G. 2001. Two-Dimensional Two-Wavelength Emission Technique for Soot Diagnostics. *Appl. Opt.* 40, 5370–5378.

[55] Bockhorn, H., Geitlinger, H., Jungfleisch, B., Lehre, T., Schön, A., Streibel, T., Suntz, R. 2002. Bockhorn. *Phys Chem. Chem. Phys.* 3780–3793.

[56] Park, J., Ryoo, H.R., Chun, H.S., Song, S., Hahn, J.W., Chun, K.M. 2006. A Study on Time-Resolved Laser Induced Incandescence Analysis Method for the Measurement of Primary Particle Size in Diesel Exhaust. *JSME Int. J. Series B* 49, 1–7.

[57] Filippov, A.V., Markus, M.W., Roth, P. 1999. In-Situ Characterization of Ultrafine Particles by Laser-Induced-Incandescence (LII): Sizing and Particle Structure. *J. Aerosol Sci.* 30, 71–87.

[58] Dec, J.E., Espey, C. 1998. Chemiluminescence imaging of autoignition in a DI diesel engine. *SAE Technical Paper* 982685.

[59] Edwards, C.F., Siebers, D.L., Hoskin, D.H.1992. A study of the autoignition process of a diesel spray via high speed visualization. *SAE Technical Paper* 920108.

[60] Espey, C., Dec, J.E. 1993. Diesel engine combustion studies in a newly designed optical-access engine using high-speed visualization and 2d laser imaging, *SAE Technical* 930971.

[61] Higgins, B., Siebers, D.L. 2001. Measurement of the flame lift-off location on DI diesel sprays using OH chemiluminescence. *SAE Technical Paper* 2001-01-0918.

[62] Kirchen, P., Boulouchos, K., Obrecht, P., Bertola, A. 2009. Exhaust-stream and in-cylinder measurements and analysis of the soot emissions from a common rail diesel engine using two fuels. *ICEF* 2009-14085.

[63] Menkiel, B., Donkerbroek, A., Uitz, R., Cracknell, R., Ganippa, L. 2012. Measurement of in-cylinder soot particles and their distribution in an optical HSDI diesel engine using time resolved laser induced incandescence (TR-LII). *Combustion and Flame* 159, 2985–2998.

[64] Bougie, B., Ganippa, L.C., van Vliet, A.P., Meerts, W.L., Dam, N.J., ter Meulen, J.J. 2007. Soot particulate size characterization in a heavy-duty diesel engine for different engine loads by laser-induced incandescence. *Proc. Combust. Inst.* 31, 685–691.

[65] Kock, B. F., Eckhardt, T., Roth P. 2002. In-cylinder sizing of Diesel particles by time-resolved laser-induced incandescence (TR-LII). *Proc. Combust. Inst.* 29, 2775–2781.
[66] Kock, B.F., Tribalet, B., Schulz, C., Roth, P. 2006. Two-color time-resolved LII applied to soot particle sizing in the cylinder of a Diesel engine. *Combust. Flame* 147, 79–92.
[67] Ryser, R., Gerber, T., Dreier, T. 2009. Soot particle sizing during high-pressure Diesel spray combustion via time-resolved laser-induced incandescencel. *Combust. Flame* 156, 120–129.
[68] Hottel, H. C., Broughton, F. P. 1932. Determination of True Temperature and Total Radiation from Luminous Gas Flames. *Ind. Eng. Chem. Anal. Ed.*, 4(2), pp. 166–175.
[69] Mancaruso, E., Vaglieco, B.M. 2012. *Appl. Energy,* 385:394.
[70] Mörsch, O., Sorsche P. 2001. Investigation of Alternative Methods to Determine Particulate Mass Emissions. UNECE/WP. 29/GRPE report 2001.
[71] Marjamäki, M., Keskinen, J., Chen, D.R., Pui, D.Y.H., 2000. *J. Aeros. Sci.,* 249:261.
[72] Maricq, M.M., Xu, N., Chase, E. 2006. *Aeros. Sci. Technol.,* 68:79.
[73] Lee, J., Oh., H., Bae, C., 2012. Combustion process of JP-8 and fossil Diesel fuel in a heavy duty diesel engine using two-color thermometry. *Fuel.* 102: 264-273.
[74] Zhao, H. Ladommatos, N. 2001. Engine Combustion Instrumentation and Diagnostics. SAE International, Warrendale.
[75] Zhao *Energy Comb.* Vol. 24 pp 221-255
[76] Matsui, Y., Kamimoto, T., and Matsuoka, S. 1979. A Study on the Time and Space Resolved Measurement of Flame Temperature and Soot Concentration in a D. I. Diesel Engine by the Two-Color Method. *SAE Technical Paper* 790491, doi:10.4271/790491.
[77] Vaglieco, B.M., Beretta, F., D'Alessio, A. 1990. *Comb. Flame* 259:271.
[78] Bohren, C. F., Huffman, D. R. 1983. Absorption and Scattering of Light by Small Particles. *Wiley-Interscience,* New York.
[79] Di Iorio, S., Mancaruso, E., Vaglieco, B.M. 2012. *Aeros. Sci. Technol.,* 272:286.
[80] Heywood, J.B. 1990. Internal Combustion Engine Fundamentals, *McGraw-Hill,* New York, USA.
[81] Genzale, C.L., Reitz, R., Musculus, M.P.B. 2009. *Proc. Combust. Inst.* 32, 2767–2774.
[82] Johansson, M., Yang, J., Ochoterena, R., Gjirja, S., Denbratt, I. 2013. NOx and soot emissions trends for RME, SME and PME fuels using engine and spray experiments in combination with simulation. *Fuel.* 106: 293-302.
[83] Bagley, S.T., Gratz, L.D., Johnson, J.H., McDonald, J.F. 1998. *Environ. Sci. Technol.* 1183:1191.
[84] Lapuerta, M., Armas, O., Rodriguez-Fernandez, J. 2008. *Progress in Energy Comb. Sci.,* 198:223.
[85] Jung H., Kittelson D.B, and Zachariah M.R., 2006. *Environ. Sci. Technol.* 4949:4955.
[86] Song, J., Alam, M., Boehman, A.L., Kim U. 2006. *Combust. Flame* 589:604.

In: Automotive Exhaust Emissions and Energy Recovery
Editor: Apostolos Pesiridis
ISBN: 978-1-63321-493-4

Chapter 3

EMISSIONS IN DIESEL ENGINE WITH DIFFERENT RATES OF EGR

Lucas Lázaro Ferreira Squaiella[1], Cristiane Aparecida Martins[2*] and Pedro T. Lacava[3]

[1] Departamento de Engenharia Mecânica,
Universidade Estadual de Campinas, Campinas, SP, Brasil
[2,3]Departamento de Propulsão, Instituto Tecnológico de Aeronáutica,
São José dos Campos, SP, Brasil

ABSTRACT

Diesel engines are not only among the most applicable internal combustion engines today, but they are also one of the biggest polluters. There is great concern, in particular, with the emissions of NOx and particulates; EGR (*Exhaust Gas Recirculation)* is among the techniques used to reduce NOx emissions. This technique involves, besides a detailed study of integrated devices, accurate calibration regarding the achievement of the ideal EGR rate. This is because in addition to the NOx emissions, particulate matter emissions should also be evaluated without losing sight of their performance parameters. This work will present a detailed experimental study carried out with ACTEON, a four-cylinder engine that meets Euro III emission standards. This engine has an urban application, i.e., it works most of the time at low rotational speeds. In this important, operating range, a high rate of EGR is required for emission levels to be met. Different EGR configurations were studied by varying the EGR rate from 2.5 to 28 %. The values of emissions and performance will also be presented. The definition of the study conditions was carried out after the application of the Design of Experiments (DoE) technique. Findings are detailed for the most critical operating conditions.

Keywords: Diesel engine emissions, EGR

* Departamento de Propulsão, Instituto Tecnológico de Aeronáutica, ITA, 12228-210, São José dos Campos-SP, Brasil; Email: cmartins@ita.br.

Introduction

The Diesel engine, or compression ignition engine - CI, has some important operational and combustion differences in comparison to the spark ignition engine - SI. First of all, during the charge admission and part of the compression phase, only air is present; it is a great advantage, because it allows higher compression ratio than SI, for example 17. In the case of fuel/air mixture admission (SI), it is possible for auto-ignition or detonation to occur during the engine compression phase; this requires the use of a high octane fuel and a limited compression ratio (up to approximately12). Thus, in practical systems, CI is more efficient than SI. Additionally, another important CI advantage is the high torque at low speeds moreover it is possible keep this high torque constant in the range between 1200 and 1800 rpm, for example.

As a consequence of the high air temperature, when the fuel is injected and vaporized into the chamber at the end of the compression, the ignition of a primary mixture is initiated and there is a sequence of events that characterizes the combustion process in this kind of engine, which will be discussed later. However, the higher pressure and temperature in the combustion phase of a CI are critical for both NO_x emissions and particulate matter - PM. Basically, the most important pollutant emissions, in terms of control and restrictions of agencies, as stipulated by, for example, the Environmental Protection Agency, – EPA, in the United States - are carbon monoxide, unburned hydrocarbons, PM and NO_x. The first two pollutants are in general a consequence of incomplete combustion inside the chamber and their control is less critical than the other two. On the other hand, NO_x and PM formation is much more complex in terms of physical and chemical processes and their strategies of control have modified the design and the operational arrangements for CI engines, since the introduction of NO_x and PM emissions standards for diesel cars with EURO I in 1992.

The present chapter regards to the control of pollutants emissions, especially NO_x, based on exhaust gas recirculation – EGR. This strategy has great potential to control NO_x; however, some especial attention shall be given to PM emissions, more specifically to the soot formation during combustion with the presence of burned gases. Then, to better explain the effects of EGR on emissions, in the next topic some comments are made about NO_x and PM formation, CI combustion and EGR strategy.

Emissions, Combustion and EGR Aspects

In general, there are two applicable combustion methods in CI engines; direct or indirect fuel injection. In the first one, liquid fuel is injected directly inside the combustion chamber, normally sculpted on the top of the piston head (Figure 1a). In this case, the combustion process depends intensively on the atomization process. On the other hand, in the indirect injection, the liquid fuel is injected into a heated pre-chamber to first vaporize and immediately afterwards the fuel vapor feeds the main chamber (Figure 1b). In general, the modern diesel engine designs have opted for fuel direct injection because it is more appropriate to pollutant control, especially in order to comply with the pollutant emissions standards; so, the next comments will be focused in the combustion dynamics for this configuration.

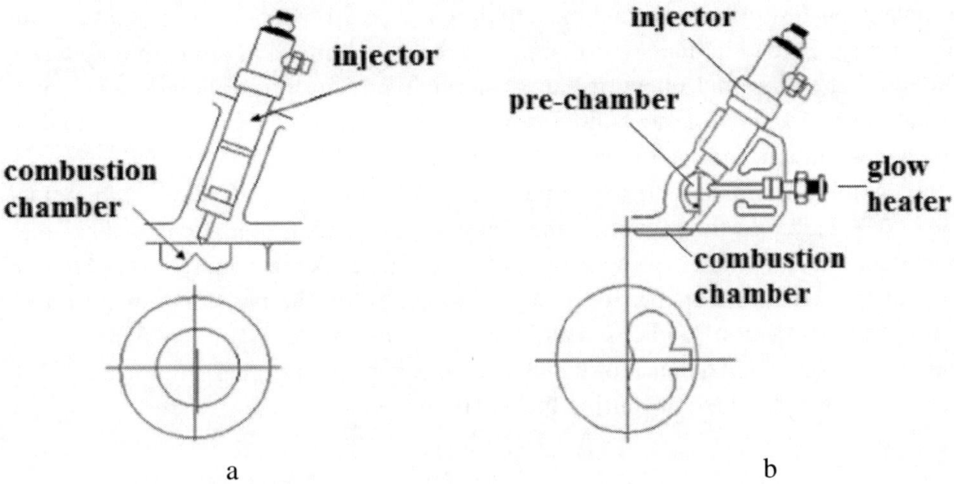

Figure 1. Different configuration for fuel injection and combustion chamber: (a) direct fuel injector, (b) indirect fuel injector.

The direct fuel injection is based on fine spray characteristics, which means droplets of around 20 μm in size, and high droplet penetration (Argachoy, 2001). In the first case, small droplets vaporize rapidly and, consequently, the fuel vapor also mixes rapidly with air; then part of the combustion is in the premixed mode. On the other hand, the droplet penetration is important to spread the fuel throughout the combustion chamber. The first situation is important for the initial combustion, and is called the premixed combustion phase; the second one is important for the later combustion phase, when normally the fuel injection stops and is called the mixing-controlled combustion phase. Figure 2 summarizes these comments.

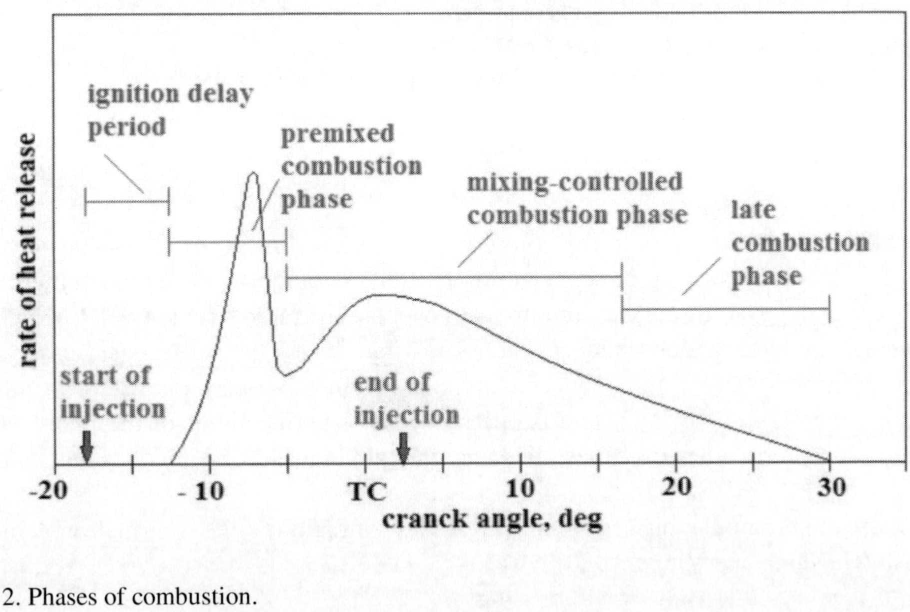

Figure 2. Phases of combustion.

To obtain small droplets and high penetration, the liquid fuel is atomized passing through the injector orifices, with a diameter of approximately 0.15 mm and a pressure between 1200 to 2000 atm. When the fuel injection starts, the high velocity of the liquid layer causes low pressure at the boundary of the jet fuel; as a consequence, air entrainment takes place in the regions of atomization and vaporization. Therefore, there is a mixture of fuel vapor and air when the ignition starts and the first part of combustion is controlled by the fuel rich premixed combustion. In the sequence, the unburned fuel and the fuel rich products burn in a diffusion flame. In the inner region of the diffusion flame, the concentration of soot is high, like a candle flame; however, part of soot is oxidized when the particles move to the flame front. The NO_x formation happens mainly on the outer sections of the flame, but at the extremities, due to the dependence of the thermal mechanism in higher temperatures. Figure 3 illustrates the sequence of events during the CI combustion.

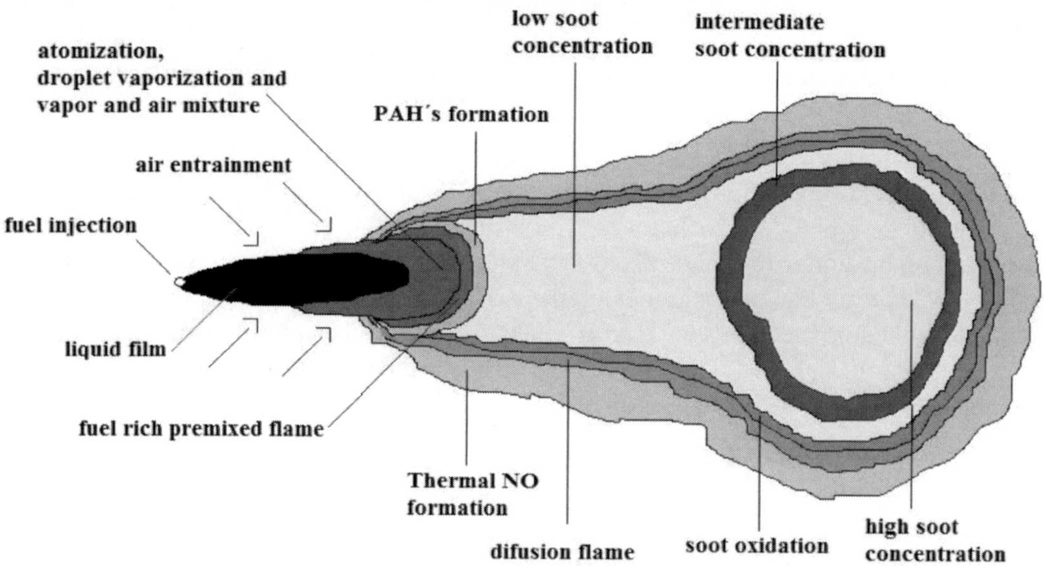

Figure 3. Sequence of events during the combustion of CI.

In nomenclature NO_x refers to the combined emissions of nitrogen monoxide – NO and nitrogen dioxide – NO_2. In fact, the concentration of NO in burned gases is much higher than NO_2; however, the NO dispersed into the atmosphere is rapidly converted to NO_2. As environmental and health consequences, it is possible to notice: smog, acid rain, ozone formation on troposphere, pulmonary edema and Methemoglobin disease or Cyanosis. Additionally, the presence of NO_x and ozone reduces the permeability of the plants leaves' membranes. As a consequence of these impacts, this pollutant has been controlled since the eighties.

The most important chemical mechanism of NO formation is the Thermal or Zeldovich mechanism (Hayhurst and Vince, 1980):

$$N_2 + O \leftrightarrow NO + N \quad (1)$$

$$N + O_2 \leftrightarrow NO + O \quad (2)$$

$$N + OH \leftrightarrow NO + H \quad (3)$$

The first reaction is the initialization reaction and its reaction rate is quite slow compared to the combustion chemical mechanism; consequently the mechanisms are not linked and the NO formed by the thermal mechanism normally takes place in a region where the combustion products are in chemical equilibrium. Additionally, the activation energy required by the initialization equation is relatively high, 75 kcal/mol, and the NO formation has an exponential dependence on temperature. The rate of NO formation can be calculated by equation (4) assuming that the combustion gases are in chemical equilibrium.

$$\frac{d[NO]}{dt} = 1{,}3.10^5.\exp\left(-\frac{67650}{T}\right)[O_2]^{1/2}.[N_2] \quad (4)$$

where t is the time of reaction in s, T is the temperature in K and the concentrations [ ] are in mol.cm^{-1}. In general, at least 70% of the total NO formed comes from the thermal mechanism (Kitamura et al., 2005). As a result, basically three parameters should be taken into account to prevent or to minimize the NO formation: combustion product temperature, time of exposure to higher temperatures and oxygen availability. So, the strategies to control NO emissions have to act preferentially on these three aspects, and the recirculation of combustion gases needs to address these aspects.

There are other mechanisms of NO formation. Among these is the "Prompt" mechanism, or Fenimore mechanism (Blauwens et al., 1976; Miyauchi et al., 1976; Hayhurst and Vince, 1980), which involves several complex reactions with the presence of some radicals from the hydrocarbon oxidation mechanism; so, in this case, the NO formation takes place close to the narrow flame region. In addition, it is possible to obtain NO formation through the presence of nitrogen in the fuel composition. However, as mentioned before, most of the NO is formed by the thermal mechanism and the action to minimize its formation should control its parameters of dependence.

The principle of the EGR, Exhaust Gas Recirculation, is to mix air and combustion products during the admission phase and three types of effects are observed in the combustion process and NO formation: thermal, chemical and dilution effects. The thermal effect is the absorption of energy due to the presence of burned gases, controlling the combustion temperature; the molecules of CO_2 and H_2O have more capacity of heat absorption than O_2 and N_2. The main chemical effect is the endothermic dissociation of CO_2 and H_2O molecules, also reducing the temperature. Additionally, the oxygen and nitrogen dilution has two consequences: the first one is the reduction of O_2 and N_2 concentration, as can be seen in the equation of NO formation rate (equation (4)). Finally, as a consequence of lower oxygen diffusion, the flame is more distributed, avoiding temperature peaks.

In fact, several studies have been conducted and have been researched to understand the mechanism of EGR action aiming to map the degree of importance of each of them. In the 1990s, for example, the group led by Ladommatos (1996a, 1996b, 1997a and 1997b) from Brunel University in London, England, conducted different experiments in order to identify the phenomena responsible for reducing the formation of NOx with the use of the EGR technique. Among the effects mapped and studied as responsible for the reduction of NOx, are the thermal and chemical dilution and the increasing air mass admitted. Of these, the greatest impact on NOx reduction was the dilution effect at inlet of the exhaust gas, which provides a delayed onset of ignition, reducing the flame temperature. Hawley et al. (1999) and Mellow (1999) also showed similar results. Fukuda et al. (1998), show the results of experiments with the selective EGR. It evaluated individual effects on the whole of the dilution gas inlet.

The recirculation of combustion gases can be done in three ways: internal EGR, High Pressure EGR and Low Pressure EGR. In the first case, internal EGR, the exhaust valve is reopened during the intake (air admission) phase, when the air is coming into the cylinder. At this moment the pressure inside the cylinder is lower than in the exhaust duct; then, part of the volume of burned gases reverts its direction back into the cylinder. This is the simplest way to obtain EGR; however the rate of gas recirculation (percentage of burned gases mass inside the cylinder after admission) is limited.

The High Pressure EGR uses the pressure differential between the exhaust and inlet systems to inject and to mix the combustion gases with air. Figure 4 represents a typical High Pressure EGR architecture. In this setup the turbo-compressor plays an important role to keep the pressure in the exhaust system, before the turbine, higher than the intake, independent of the engine operation condition. The EGR valve controls the recirculation rate.

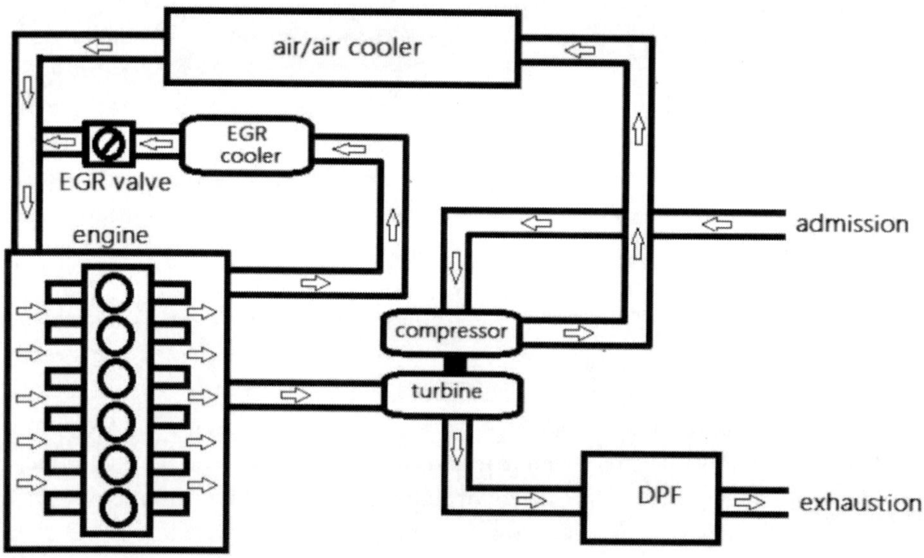

Figure 4. High Pressure EGR.

The final option, Low Pressure EGR, uses the pressure differential between the exhaust system after the turbine and the intake system before the compressor. Figure 5 shows a typical architecture. In this case, the control of differential pressure is less complex than the High Pressure, because the pressure after the turbine is always higher than the pressures before the compressor. Another advantage in terms of control is that all exhaust gases pass through the turbine.

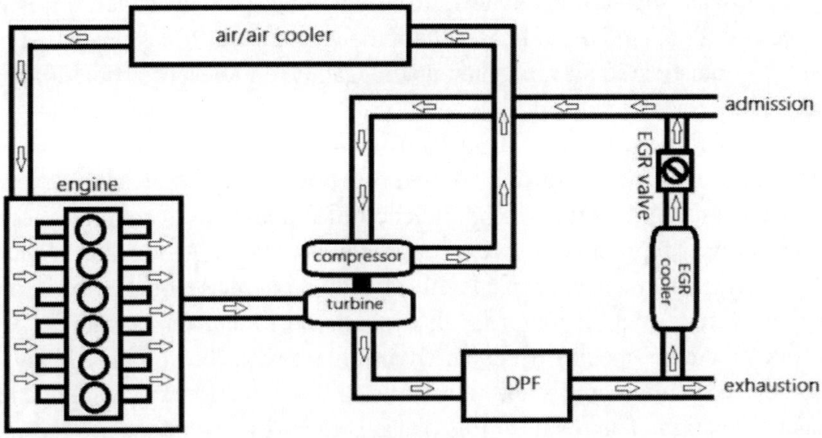

Figure 5. Low Pressure EGR.

Despite the capacity to control NO formation, it is important to observe the consequences of EGR on the emissions of carbon monoxide (CO), unburned hydrocarbons (UHC) and particulate matter (PM). In the case of CO and UHC, the reduction of the combustion temperature and oxidant dilution decreases the oxidation rates of these compounds. Therefore, there is a limit of recirculation rate not to compromise the CO and UHC emissions. As shown by Arcoumanis et al. (1996), increasing the EGR rate leads to reduced O_2 concentrations. The reduction of the mass of O_2 decreases the ability of oxidation of HC and CO which increases the concentration of particulate matter contained in the engine exhaust. In Wagner (2000) can be seen that high EGR rates can cause the chemical alteration of the combustion process, abruptly increasing HC emissions, which generate an increase in the formation of particulate matter. In the same study was showed that EGR rate changes affect even the size and concentration of particles forming the particulate matter. Particulate matter is divided into three components:

- Fraction Solid Inorganic (FSI): it is the part of "dry" insoluble particulate matter. Composed of elemental carbon and ash coming from the metal lubricants and metal friction.
- Solid Organic Fraction (SOF): composes the "wet" part of the particulate matter. It is comprised of hydrocarbons derived from fuel and lubricating oil.
- Sulfate Particles (SOx): are derived from the sulfur in the fuel. More than 95% of the sulfur forms SO2, while a small portion of about 2% to 5% forms SO3 in the presence of water, which it reacts with to form sulfuric acid.

The limiting values for emissions of particulates include the total sum that includes FSI, SOF and SOx. In analyzing particulate matter with the use of the EGR system Khair (1997) shows that SOF remained constant; it indicates that the FSI, mainly composed of carbon is the main component responsible for the increase of PM. Stratakis and Stamatelos (2003) described the "soot" as exclusively composed of dry particulate carbon. This leads to the conclusion that in considering PM emissions, soot formation is the most critical.

In order to mitigate the soot originating from PM emissions, in general, it is necessary a for a filter to be installed, called the Diesel Particulate Filter – DPF, as presented in Figures 1 and 2, or a more sophisticated system, such as the Catalyzed Diesel Particle Filter, where the carbon present in the PM is oxidized.

Soot is the name of particles with a diameter less than 100 nm and have movements similar to molecules. It is formed during the hydrocarbon's cracking reaction, especially in an environment with high temperature and deficiency of oxidant. Its formation mechanism has four steps: formation of precursor species, particle nucleation, coagulation and oxidation (Calcote, 1981). The first step is considered an induction phase or pyrolysis phase and it involves high activation energy; therefore, it depends on higher temperatures. Additionally, the formation of precursor species needs an environment with oxygen deficiency; otherwise, the fuel oxidation mechanism will stand in relation to the induction mechanism and there will not be any soot formation. The result of this phase is the presence of species thermally stable, but highly reactive. The Polycyclic Aromatic Hydrocarbons – PAHs (Chomiak, 1990), and the main species formed are the acetylene (C_2H_2), the radical propargyl (C_3H_3), benzene (C_6H_6) and the radical phenyl (C_6H_5) (Glassman and Sidebotham, 1992). This step is considered the most complex as it determines the velocity and the total concentration of soot formation (Durigon et al., 2002). The second step is the particle nucleation or cracking phase, where the PAHs in gaseous phase are converted into tiny particles, the first nuclei. Carbon layers are composed in the form of shelves, which stacked assumes a crystal structure. A group of crystal structures, approximately 103 of them, form the soot particle. The third step happens when the collision between the first particles (nuclei) starts; then, the particles now become agglomerates of tiny nuclei. Finally, if the particles are exposed to high temperature and oxygen, the oxidation step takes place, and part of the soot is converted to CO or CO_2 (Richter and Howard, 2000). Essentially, the total of soot emitted is a balance between what happens in the first three phases and how much soot is converted in the oxidation phase.

In some cases of EGR operation soot formation can be reduced, for example, when fuel vapor is exposed to lower temperature. However, at the same time, the fuel is exposed to an oxygen dilution, favoring PAH formation. Thus, the EGR limit is the compromise between the reduction of NO_x and soot formation and it depends on the combustion dynamics inside the cylinder. Normally, increasing the EGR rate also increases the soot formation and in many practical systems it is necessary to install a particulate filter as mentioned before.

The EGR changes the combustion process, as previously mentioned, by thermal, chemical and dilution effects; this strategy has been considered and evaluated technically as a consequence of more restrictive emissions standards being continually imposed on manufacturers of engines. Figure 6 illustrates the evolution on restrictions of NO_x and PM emissions according the chronological implementation of EURO standards; additionally, some brief comments are presented regarding the technical improvements required to meet these standards.

The minimum required architecture to achieve EURO III, in general, is based on a fuel injection pressure of approximately 1300 bar and a waste-gated turbocharger, without the EGR system. However, it was not enough to comply with a significant reduction of NO_x and PM imposed by EURO IV in 2010; then, the EGR strategy with recirculation rate between 5% and 12% with higher pressure of injection (approximately 1800 bar) and particulate filters were adopted. EURO V was more restrictive with respect to NO_x emissions and additional modifications were necessary in the architecture: incremental increase of EGR exhaust flow rates at between 12% and 18% and the use of a VNT (Variable Nozzle Turbine) turbocharger to control the EGR rates allowing this higher rate of EGR as well (Squaiella et al., 2012).

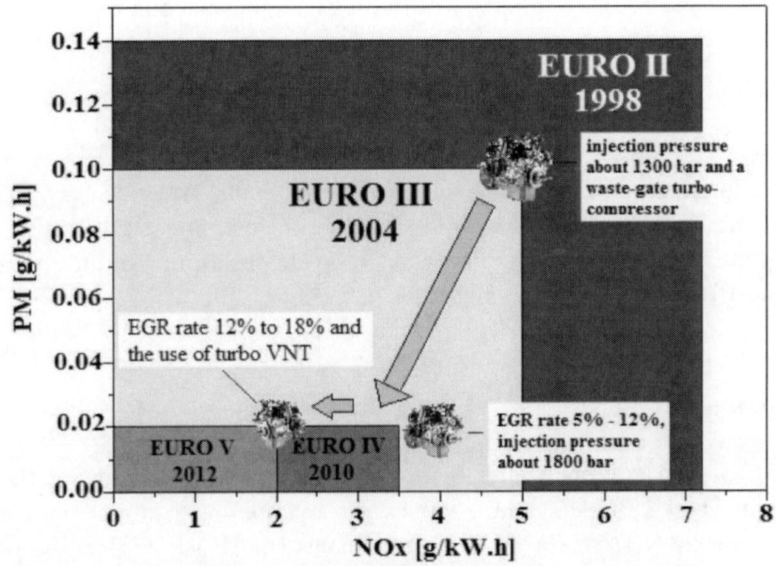

Figure 6. Restrictions of NO_x and PM emissions according the EURO standards and the technical evolution to meet these standards.

Another strategy to control pollutants which is widely-used is the SCR – Selective Catalytic Reduction, based on the gases post treatment. This strategy is as prominent as EGR, but it is not within the scope of the present chapter and some recent comments about it can be found in Lourenço et al. (2013) and in other chapters of the present book.

EGR is among the strategies specifically applied to change the combustion process. Another very attractive technique which is also typically applied, is the delay of ignition. The expansion in the application of EGR, however, appears to have occurred primarily for the benefit of reducing the brake specific fuel consumption (BSFC) compared to the benefit obtained by the effect of delayed ignition which allows only the reduction of emissions of NOx. In the work of Khair (1997) it was shown that when NOx emissions reduced by 30% (from 4 to 2.8 g / BHP) the increase in specific fuel consumption was only 2.3% for the EGR-only system. Equivalently, with the technique of ignition delay the increase of specific consumption was of 13.4%. Both techniques, however, cause an increase in emissions of particulate matter (PM). With respect to particulate matter, EGR causes an increase of 31% when compared to the base value while ignition delay caused a 15% increase, by comparison. Thus it is clear that the ideal solution would by necessity be some sort of compromise.

The focus of the present text is to analyze experimentally the potential of EGR to control NO_x and PM emissions without catalytic root, just control the EGR rate using a VNT turbocharger and a two stage turbo to control and increase this rate. This chapter assesses the effect of EGR rate on the behavior of the engine with the goal of reducing NOx and particulate matter without compromising performance.

In next section, the experimental test bed installation is presented in detail. A description of the procedure followed during the investigation as well as a tabulation of the tests conducted is subsequently given. The next section, also, presents the results of the study separately. In following sections, one test complete is described, which was done with all 13 modes of the ESC cycle before a summary of findings are presented in the conclusion.

THE ENGINE UNDER INVESTIGATION

The MWM International Engines Industria Ltd. in South America was responsible for supplying the engine and the entire infrastructure of instrumentation and measurement equipment for the experimental development. A complete description of the entire workbench is in Squaiella (2010) and Squaiella et al. (2013).

Engine Specification

The test bench was equipped with the engine model ACTEON of 1.2 liters of swept volume per cylinder, 4 cylinders and 4 valves per cylinder with direct injection using an electronic management system. In all the experiments, the basic engine components were kept, namely:

- Engine cyclinder heads, maintaining the same flow coefficients and swirl;
- Engine block;
- Shirts without change in the burnishing processes;
- Pistons;
- Intake and Exhaust Collectors;
- Injection system; the injection system is a Bosch Common Rail composed of pressure and temperature sensors of the intake air, lube oil pressure, fuel pressure, water temperature of the engine intake air flow, ECU (Electronic Control Unit), high pressure pump, rail; these were always kept the same, changing only the hydraulic flow, geometry and number of jets of the injectors.

The characteristics of the engine are shown in Table 1 and the performance curve is in Appendix A. This engine has a high level of torque that from 1200 to 1600 rpm. It is strategically used for vehicles that operate at low speeds; which is advantageous at lower speeds, thereby consuming less fuel.

The dynamometer used was a Schenck W230, an eddy current dynamoter, cooled by water, with a capacity of 230 kW of power and 700 Nm of torque, with an accuracy of +/- 1 kgf. Other instrumentation also used included:

- Instrumentation for measurements of temperature and pressure of the intake system and exhaust, combustion pressure measurement, water temperature and oil etc, as discussed above.

- Bench exhaust gas analysis - The measurements of the concentrations of HC, CO, NOx and other gases found in the exhaust flow were measured with a Horiba MEXA 7100. It is capable of measuring the following components: THC, CO, CO2, O2, NO / NOx, HC, N2O, SO2, CH4. Its method of analysis includes a non-dispersive infrared analyzer, a magneto-pneumatic analyzer, a chemiluminescence analyser and a flame ionization analyser.
- Particulate matter instrumentation - the measurement of the particulate matter was performed using equipment manufactured by AVL, SMART sampler PC. The instrument for particulate matter measurement is a measuring system for gravity of diluted particles, which collects the sample in a partial flow tunnel. From the total flow of gases, only a small portion is collected in a small tunnel and diluted with air which was conditioned internally by the system. After the dilution in the tunnel, the gas passes through the filter that collects the sample. At the end of the test the filter is weighed and the mass of the particulate matter is obtained.

Table 1. Specifications of test engine

	ENGINE PARAMETERS	
1	Manufacturer	MWM International
2	Engine model	ACTEON 412TCE
2	Engine Type Cycle	Common rail 4 cylinder
3	Construction Type	In-line, 4 cylinders
4	Valves per cylinder	4
5	Bore x stroke	105 x 137 mm
6	Unit displacement	1.2 liters
7	Firing order	1-4-3-2
8	Total displacement	4.748 liters
9	Compression ratio	16.8 : 1
10	Idling speed	800 ± 100 min-1
12	Rated power	149 kW @ 2200 min-1
13	Maximum torque	720 Nm @ 1200 to 1600 min-1
14	Fuel	Diesel**
15	Cooling System	60% water + 40% ethylene glycol
16	Application	Heavy-duty truck and bus

* Maximum diesel sulfur content is equal 50 ppm.

METHODOLOGY

Diesel engines can be studied under steady-state or transient operations. This is their first definition. Type-approval testing requires that the engine is submitted to the ESC (European Stationary Cycle) test and the ETC (European Transient Cycle) test as well as the ELR (European Load Response) test, the last two were introduced in early 2000 – Directive 1999/96 /EC. In some countries, like Brazil, the transient tests are required only for engine applications with swept volume of less than 0.75 liters and engine speeds above 3000 rpm or for those engines that use some kind of treatment of the exhaust gases to reduce gaseous emissions. The ACTEON engine used in this work has a maximum speed of 2550 rpm, thus it will be assessed only on the stationary operation using the ESC cycle. Table 2 shows the variations of speed, load factor, weight and ESC cycle duration.

Engine speeds A, B and C can be obtained according to the equations below. The value of high speed (n_{hi}) and low speed (n_{lo}) are obtained from the power curve. Considering that the maximum power (P_{max}) occurs at the reference speed (n_{ref}). So, the high speed (n_{hi}) corresponds to the point where the power value is equal 70% P_{max}, considering values of engine speed above n_{ref}. The low speed (n_{lo}) value is obtained considering the point where the power obtained corresponds to 50% P_{max} and the engine speed is lower than n_{ref}. With respect to the ACTEON 4.12TCE the values are A = 1380, B = 1690 and C = 2000 rpm, respectively:

$$A = n_{lo} + 0.25\ (n_{hi} - n_{lo}) \quad (5)$$

$$B = n_{lo} + 0.50\ (n_{hi} - n_{lo}) \quad (6)$$

$$C = n_{lo} + 0.75\ (n_{hi} - n_{lo}) \quad (7)$$

Table 2. ESC Test Modes

Mode	Engine speed	Load %	Weight %	Duration min
1	Low idle	0	15	4
2	A	100	8	2
3	B	50	10	2
4	B	75	10	2
5	A	50	5	2
6	A	75	5	2
7	A	25	5	2
8	B	100	9	2
9	B	25	10	2
10	C	100	8	2
11	C	25	5	2
12	C	75	5	2
13	C	50	5	2

Tests were performed only at point 2 of the ESC cycle at 1380 rpm with 100% load. ACTEON 4.12 is a diesel engine for urban application, i.e., it works most of the time at low speeds. This is a critical and decisive factor in defining the EGR system components due to the need for large amounts of gases recirculated to meet the desired emissions levels. The component most required under this conditions and which plays a fundamental role, is the turbocharger. If it was sized for operation at this condition only, it would become very restrictive at higher engine speeds, leading to losses due to the high energy used to pump the gases of combustion through the exhaust manifold. Conversely, if the turbocharger is sized to for high engine speed operation, it would not be possible to achieve the required differential pressure for recirculation of exhaust gases at the lower engine speeds,. Another point that also influenced the decision to keep the focus of this study on Point 2 was its high weight factor, equal to 8%, as shown in Table 2.

In the following section the definition of the operating conditions tested are given.

Definition of Test Conditions

The technique used for the elaboration of the tests conditions is based on statistical studies with the analysis of the response surface, a process known as DOE (Design of Experiments). Minitab 15 Statistical Software was, also, used to generate the matrices of the tests, and Calgen for the creation of surface graphics.

The three different setups utilized were named CONFIG I, CONFIG II and CONFIG III, always keeping the main idea which is to evaluate the effect of EGR rate on the behavior of the engine with the goal of reducing NOx and PM without compromising its performance. Table 3 provided information about the three setups. With these setups it was possible to test a range of EGR operation, from 2.5% to 28.8%, as shown in Table 4.

It can be observed that CONFIG I is the same configuration used in applications which meet the Euro IV. So, it will be possible to evaluate the potential of this configuration to the reduction of NOx and Particulate Matter using EGR rate variation.

Before starting the tests to assess performance, it is necessary for the engine to achieve thermal equilibrium. The engine ACTEON used for this chapter was stabilized at the maximum power speed of 2200 rpm, while controlling the parameters described in Table 5.

Table 3. Tests Configurations

Description			Observation
CONFIG I	EGR Cooling System	Modine with 8 passage tubes with 485 mm	Same configuration as in Euro IV engines
	Turbocharger	BW S200 VNT	
CONFIG II	EGR Cooling System	Water used in the EGR cooler from an external FONTE at a very low temperature	Turbo compound provides better air flow, improving the engine performance.
	Turbocharger	Two stage Turbo compound	
CONFIG III	EGR Cooling System	Bigger and more efficient	
	Turbocharger	Two stage Turbo compound	

Table 4. EGR rates in tested configurations

CONFIG I		CONFIG II		CONFIG III	
EGR rate	EGR flow	EGR rate	EGR flow	EGR rate	EGR flow
%	kg/h	%	kg/h	%	kg/h
2.5	12.5	14.9	77.9	27.2	164.5
12.1	54.5	18.2	107.9	28.8	185.2

Table 5. Limits for engine conditioning

Reading Channel	Unit	Definition
Engine Speed	rpm	1380
Dynamometer load	kgf	8

Table 5. (Continued)

Reading Channel	Unit	Definition
PRESSURES		
Oil pressure	kgf/cm^2	4 ± 0.5
P1	mm H_2O	400 ± 50
P4	mm H_2O	450 ± 50
TEMPERATURES		
Lubricant oil Temperature	°C	120 ± 10
Water temperature	°C	95 ± 5
T1	°C	15 ± 2
T22	°C	35 ± 2
T3	°C	limit 760
Fuel temperature	°C	40 ± 2
Observed smoke	BOSCH	limit 6

The tests involved the evaluation of the performance, gaseous emissions and particulate matter by conducting bench tests of the conditions defined previously with the DoE method. Details of the application of the technique for determining the conditions of work of this chapter considering CONFIG I, CONFIG II and CONFIG III are detailed in the Appendix B.

Specifically, the sequence of tests involved the following steps:

1) Test condition definition, using DoE;
2) Engine preparation; achievement of thermal equilibrium;
3) Test runs;
4) Analysis of the results.

Results and Discussion

Considerations of CONFIG I, CONFIG II and CONFIG III

In the first step, CONFIG I, the rated engine follows the configuration used in the normal production Euro III engines. The main idea was to change only the control parameters of the turbocharger and of the EGR valve in order to verify the potential of this set to achieve the lowest NOx. The components used in CONFIG I were:

- EGR Cooler Modine with 8 tubes of gas passages and a length of 485 mm.
- Turbocharger BW S200 VNT, which brings great flexibility of application due to their variable turbine geometry which makes it very efficient at low and high speeds.

Note that this is a configuration engine already in use. Thus, the values for the point of injection and to the injection pressure were kept the same, varying only the position of the EGR valve and also the turbo VNT Vanes. In evaluating CONFIG I, it was observed that the most influential variable affecting the rate of recirculated gases and consequently pollutant emissions, was the opening of the EGR valve because the differential pressure between the intake and exhaust manifolds generated by this set was low in the speed range tested.

Considering the results obtained in CONFIG I, the focus was directed to one assembly that allowed increased EGR rates and further reduced the temperature of the combustion flame. From these considerations in the CONFIG II three changes were made to the engine:

1) Changing the VNT turbocharger settings with a set of two stages. This proposal yielded in an improvement in performance in the engine allowing an increase in torque of 11.31%, from 646.8 to 720 Nm
2) Reduction of the temperature of the water used in the EGR cooler. In CONFIG I the water used was the engine cooling water at 80° C which was replaced for an external source to 35° C leading to the reduction of the gas temperature and an increase in density and the mass of the recirculated air.
3) Installation of a throttle valve in the intake circuit with the goal of increasing EGR rates by increasing the pressure differential between the intake and exhaust manifolds

In CONFIG II, it was noted that the throttle valve was the major component responsible for influencing the reduction in NOx emissions. This valve was used to increase the differential pressure between the inlet and exhaust manifolds, therefore increasing EGR rates. The biggest loss arising from of its use was the increased specific fuel consumption due to the large mass reduction of the oxidant.

In summary, CONFIG I and CONFIG II showed that increased EGR rates were very favorable for further reduction of NOx, but in return, caused an increase in particulate matter. It turns out that the ignition delay caused by the large quantities of recirculated gases is responsible for the alteration in combustion characteristics with the consequence of reducing the formation of NOx and the increase in the formation of PM. Both pollutants are critical to diesel engine operation, therefore it is not possible to consider only one of them as them as the strategic parameter to be targeted for reduction.

Thus, for CONFIG III it was necessary to adopt one additional strategy considering the need to reduce both NOx and PM. This strategy included the application of multiple injections. Benajes et al. (2001) confirms that with the use of post injection, up to 60% of the PM can be reduced. The injection system used in the object of this study allows the use of up to five injections of fuel in a single cycle; in this case it can be used for up to two pre-, one main and two post-injections. Thus, in order to reduce PM emissions and to improve the levels of specific consumption, it was necessary to study the applicability of post injection. Post injection occurs after combustion of the main injected fuel amount and has as a characteristic the use of a very small amount of fuel used to burn the soluble materials present in the PM. During preliminary tests in CONFIG III, the boundary conditions for the main injection and EGR valve position were kept the same, varying only the injection time point and the fuel volume of the post-injection. After all preliminary testing over the ideal conditions to the post-injection was adopted:

- Post SOI = 30 [ATDC].
- Injected fuel quantity = 5% of the fuel mass of the main injection.

After the definition of the ideal point of post-injection, the effect of the change of the original variables was evaluated. This task included varying the pressure and the point of the main injection and EGR valve position. Results showed that the most influential variables for the reduction of NOx had been the association of changes in position of the EGR valve and the main injection point.

Emissions

The main parameters evaluating the performance of diesel engine emissions are compared by analyzing NOx and PM.

The analysis of PM carried out by the traditional means of measurement using the weighing of samples has become slow and expensive. To circumvent this obstacle, the English laboratory "Motor Industry Research Association" created the MIRA (MIRA, 1965), a parameter that is capable of estimating the PM quantity from the degree of blackening from the smoke during combustion. The equivalency of opacity in measurements and the concentration of soot were experimentally determined and tabulated by MIRA and has been adopted, internationally, as a reference. This method of calculating the PM is used only during the development process of the engine but for the certification process of the engine, it is mandatory to carry out the test of weighing.

Thus the results are always referenced to the values of NOx and MIRA. It is worth mentioning that the EGR percentage rates in CONFIG I ranged from 2.5 to 12.1, CONFIG II from 14.9 to 18.2 and CONFIG III from 27.2 to 28.8.

EGR and the Pressure in the Admission Collector (P22)

The advantages of increasing the EGR rate to obtain a reduction of NOx can occur when it is possible to associate it with an increase in turbocharger boost pressure, because with the increase of boost, an increase in the mass of the clean air used during combustion can be obtained. It was noticed that the air pressure in the intake manifold (P22) directly influenced the process of PM formation.

When replacing the VNT turbocharger, used in CONFIG I, with another with two-stages, as used in CONFIG II and CONFIG III, the pressures obtained were much higher, as shown in Figures 7 to 9.

For the results obtained from MAF - Figures 10 and 11 - it is possible to observe the benefits of increased pressure provided by the change of the turbocharger system, where with a 10-fold reduction in NOx emission it was possible to maintain the same level of clean air mass admitted by the engine.

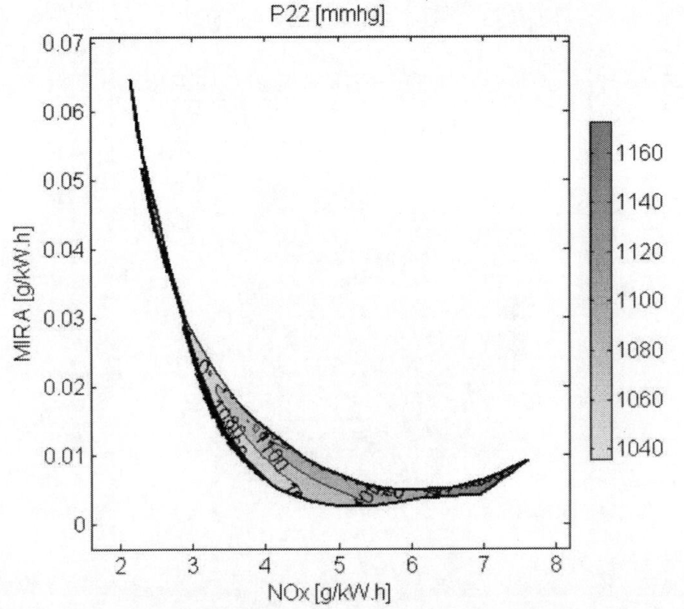

Figure 7. After cooler air pressure in CONFIG I.

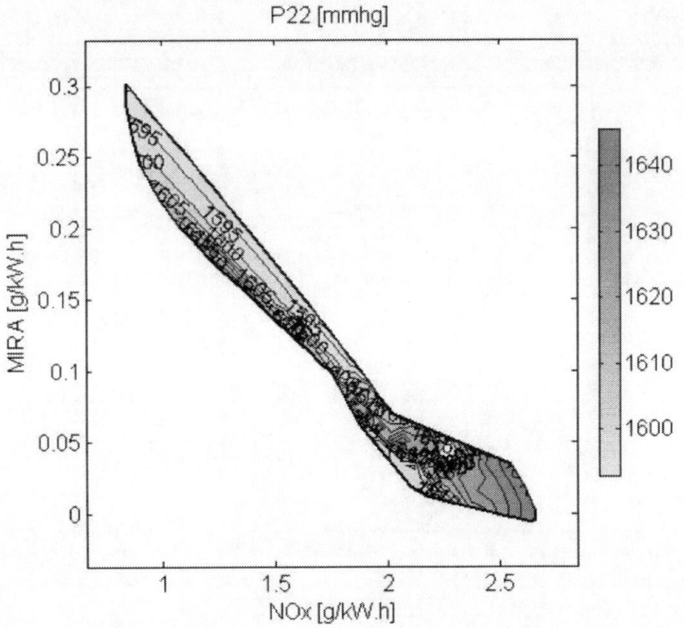

Figure 8. After cooler air pressure in CONFIG II.

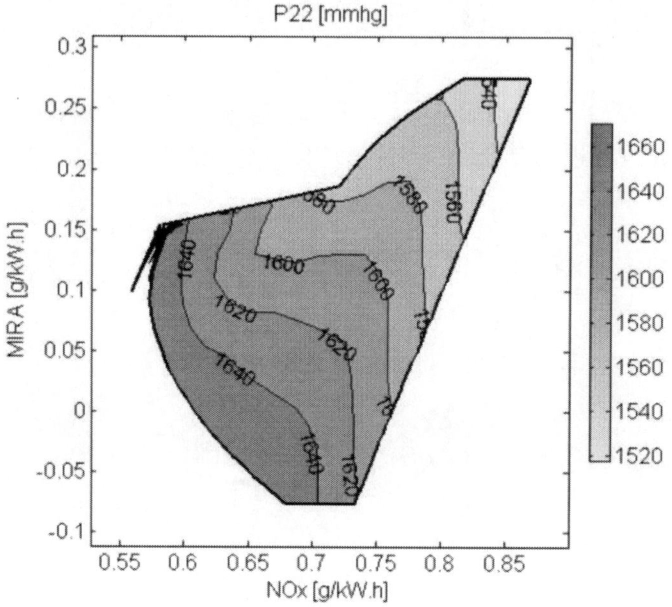

Figure 9. After cooler air pressure in CONFIG III.

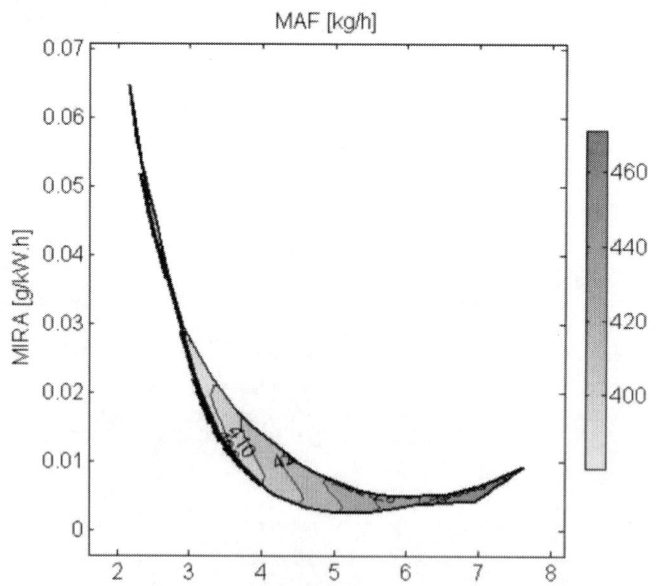

Figure 10. Mass air flow in CONFIG I.

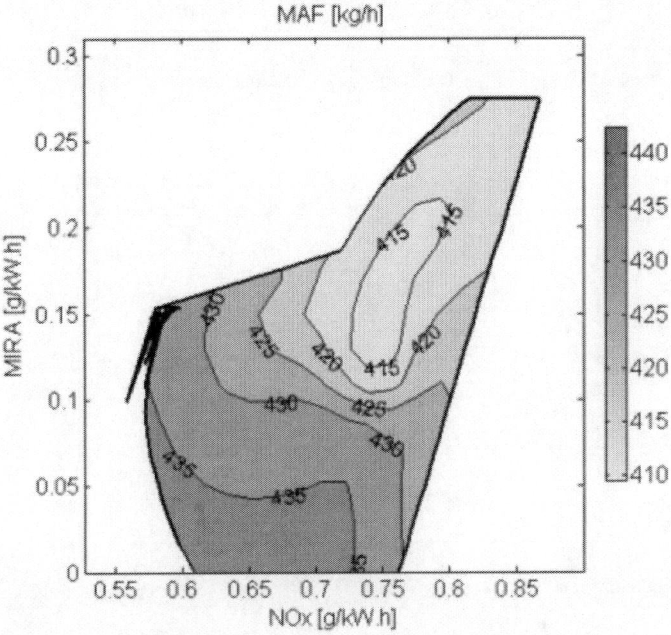

Figure 11. Mass air flow in the CONFIG III.

Another positive effect of the increased pressure can be noticed in CO emissions. It can be observed, from comparison between Figures 12 to 14, that CO levels were maintained even with an increase of the EGR rate by approximately 350%.

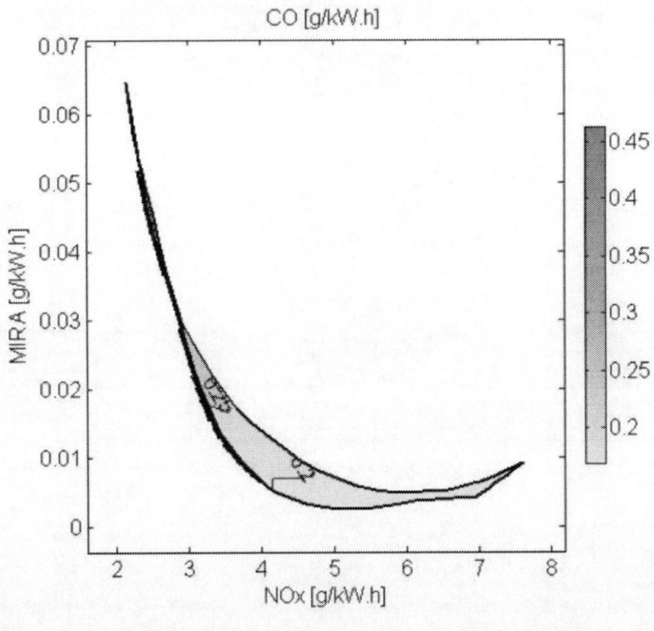

Figure 12. CONFIG I – CO emissions.

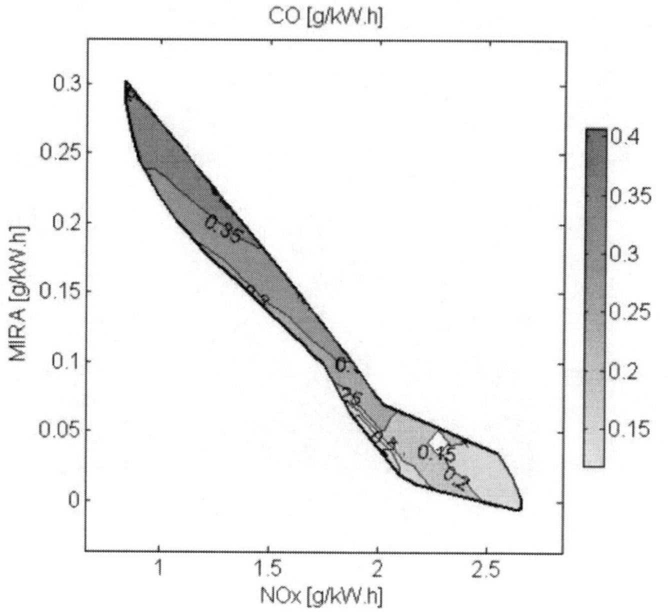

Figure 13. CONFIG II– CO emissions.

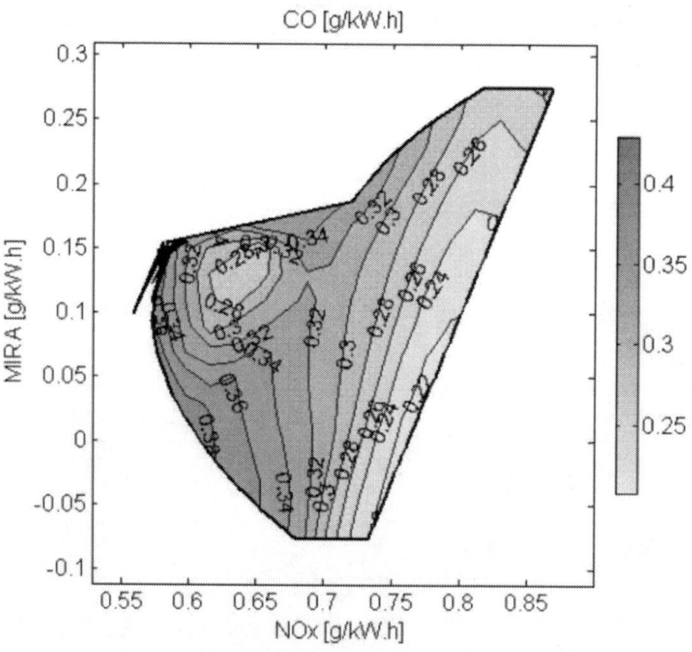

Figure 14. CONFIG III – CO emissions.

EGR versus Combustion Pressure

The combustion pressure is directly linked to the increased mass of air supplied for combustion (Woods and Kamo, 2000). Comparing the combustion pressure that occurred in CONFIG II with CONFIG I, it was observed that an increase of the order of the 13% was obtained, which is believed to be mainly due to the effect of increasing the intake pressure, in this case 47% higher in CONFIG II than CONFIG I. This occurrence can be easily explained considering the change in the compression relative ratio that takes into account two basic factors. The first is the cylinder volume that is formed when the piston reaches the bottom dead center position (BDC). The second explanation is related to the volume of the piston chamber. If the pressure of the intake air is increased, this is equivalent to the cylinder volume expanding. This yields more capacity for compressing a higher mass of air in the same chamber volume of the piston when it is in its highest position (or TDC) in relation to the cylinder head. This directly increases the pressure inside the cylinder which is further enlarged at the time of combustion.

The results showed that regardless of the configuration (I, II or II), emissions of NOx occurred in conditions of lower pressure values. This can be explained due to the reduction of the rate of combustion caused by flame diffusivity.

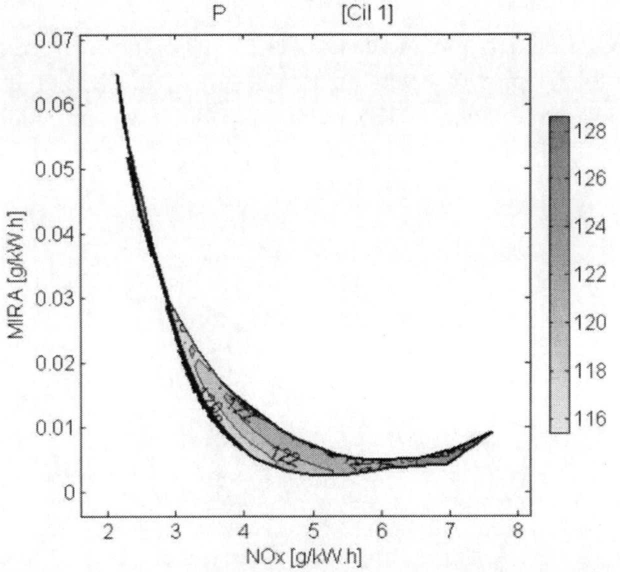

Figure 15. CONFIG I – Combustion pressure.

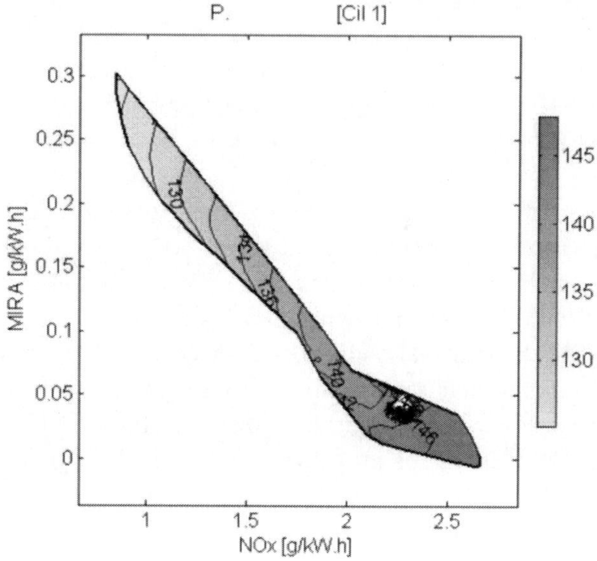

Figure 16. CONFIG II – Combustion pressure.

In Figure 17, CONFIG III, the increase of 12.9% on the average in the combustion pressures in relation to the previous version is a result of the earlier main injection. This increase is only allowed due to a large NOx reduction, which was provided by changes in the rate of EGR in this configuration.

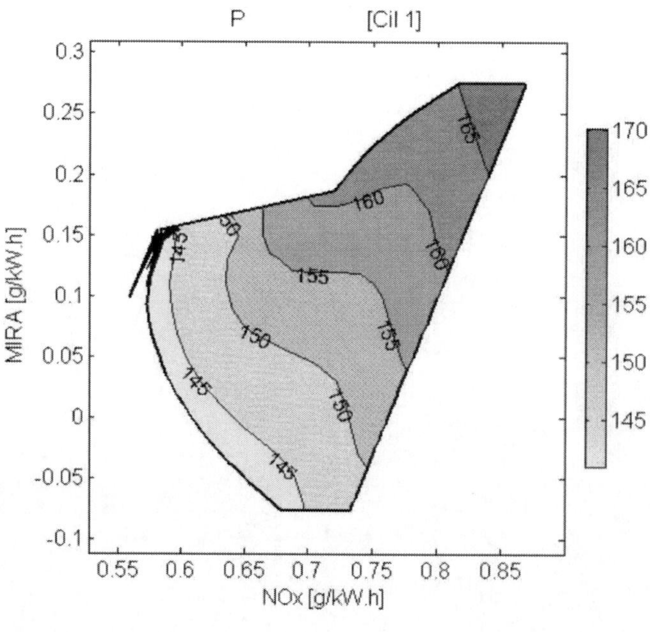

Figure 17. CONFIG III – Combustion Pressure.

EGR versus NOx and PM (MIRA)

This section presents the results of NOx and PM (MIRA) obtained in the different configurations. Note that NOx can vary between extremes of 7.55 g/kW.h to 0.62 g / kW.h while MIRA varied from 0.004 to 0.298 g / kW.h. For EGR rates obtained in CONFIG I, although it has been possible to reduce the amount of NOx to 2.18 g / kW.h, as shown in Figure 13, it is still a high value considering, for example, Euro VI which requires a reduction of NOx emissions to less than 0.5 g / kW.h.

In CONFIG II, Figure 13, there was an increase of about 6.1% of the maximum rate of EGR over CONFIG I, which was effected mainly through changes in the turbocharger and the use of the throttle valve. Another principal factor was the use of an external system for the water used in the EGR cooler which provided a reduction in the temperature of the gases that were recirculated; this recirculated air was diluted with clean air admission favoring the thermal exchange of gases during the combustion process, thereby reducing the flame temperature and NOx production process. Increased EGR rates actually induced the reduction of NOx emissions from 2.59 to 0.88 g / kWh. However, MIRA was multiplied, reaching in CONFIG II the value of the 0.29 g/kWh.

In CONFIG III saw a new gas cooler for EGR gas and also the re-calibration of the injection system. In this configuration was possible to reach a reduction of approximately 10 times the concentrations of NOx emitted by the original engine, see Figure 15. Additionally, in CONFIG III allows more flexibility for work production during the power stroke since NOx is the same for the various points of MIRA. For example, to obtain the same NOx emission of 0.7 g / kWh, MIRA can vary from 0 to 0.17 g / kWh; in this way, another parameter can be used which is defined as a secondary factor, like fuel consumption, combustion pressure or the same exhaust temperature.

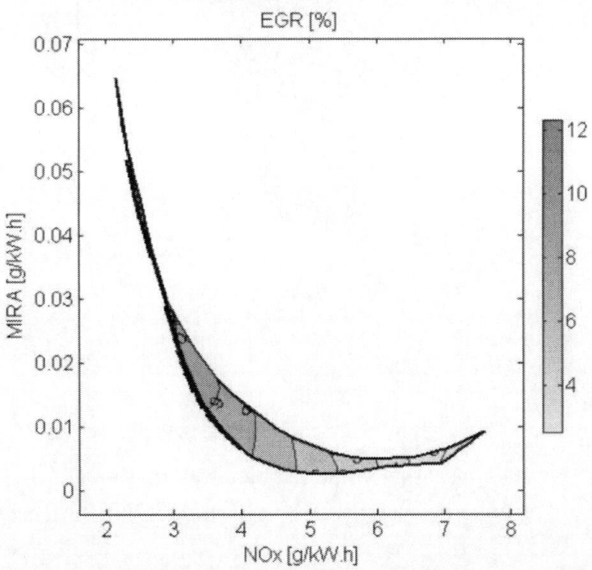

Figure 18. CONFIG I – EGR versus NOx and MIRA.

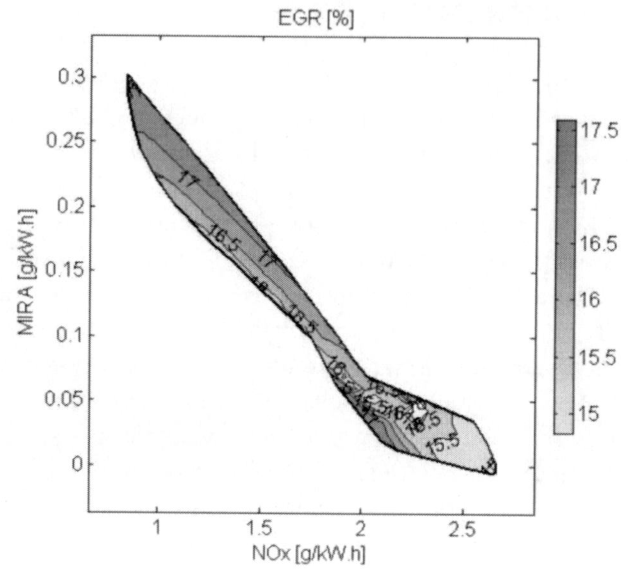

Figure 19. CONFIG II– EGR versus NOx and MIRA.

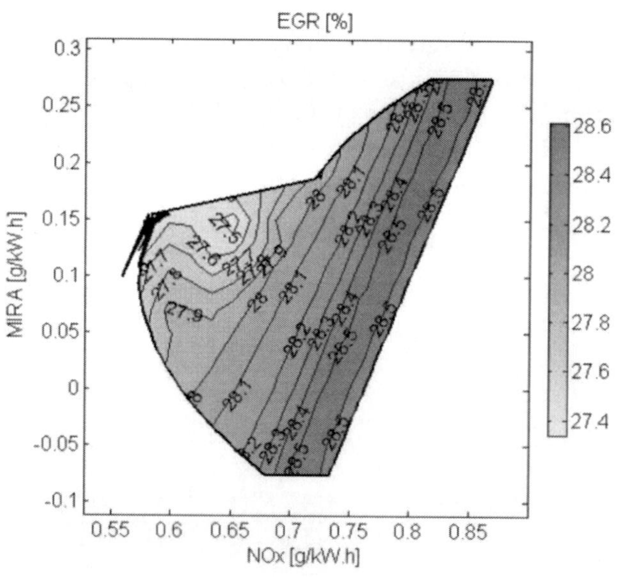

Figure 20. CONFIG III– EGR versus NOx and MIRA.

The NOx reduction caused by the reduction in the flame temperature can be illustrated with the values of T3, which are the values measured in the gas before the turbocharger turbine inlet. Note the difference between CONFIG I and CONFIG III, in Figures 21 and 22.

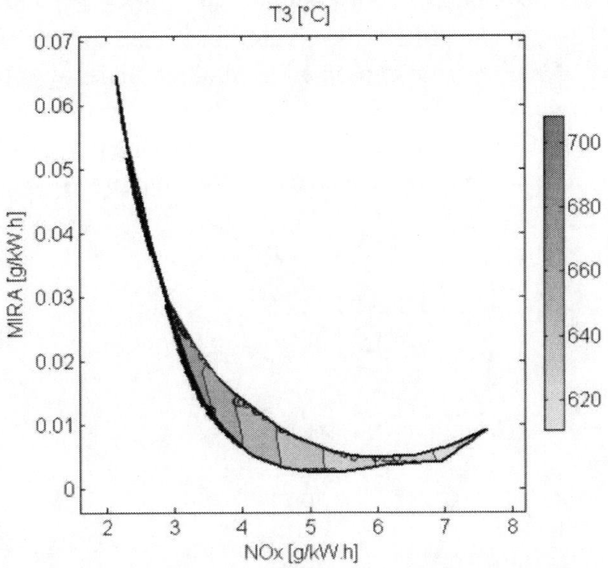

Figure 21. CONFIG I – T3.

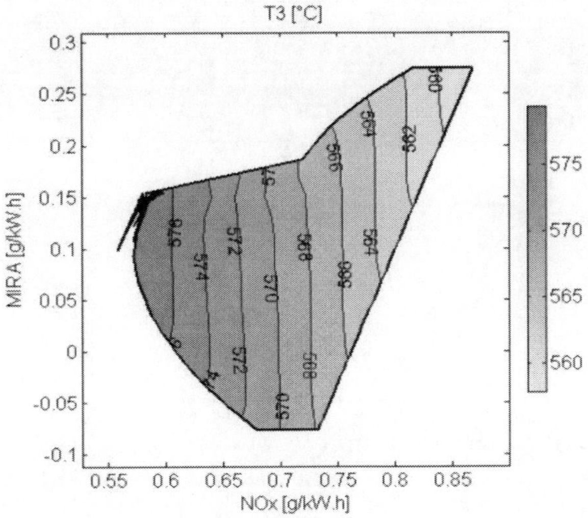

Figure 22. CONFIG III – T3.

EGR versus BSFC

The improvements obtained by increasing the EGR rate - a large reduction of specific emissions of NOx - were the resultants of delayed ignition associated with increased EGR rates. However, increased EGR rates reduce the thermal efficiency of the engine. This effect and the pumping work of the exhaust gases are primarily responsible for the increase in fuel consumption, Jacobs et al. (2003). The increase in pumping losses is the result of the need to increase the differential pressure between exhaust/intake manifold for sufficient EGR rate to be achieved. In Figure 23, the tendency of the engine can be observed when the EGR rate was increased from 2.5 to 12%, in CONFIG I, there was an increase in specific fuel consumption of 5.8%. In CONFIG II the engine presented a similar tendency when EGR rate was increased. For an EGR rate of nearly 15% a BSFC value of approximately 200 g/kW.h was obtained; while an EGR rate of 18% leads to BSFC value of approximately 210 g/kW.h. The improvement in BSFC are justified by changes in air flow due to two-stage turbo used in this assembly.

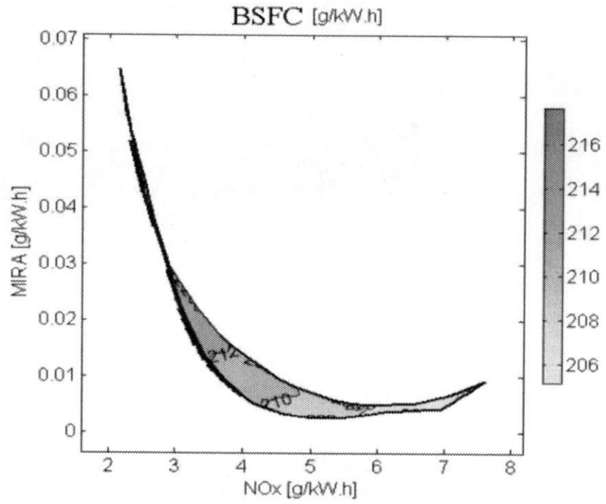

Figura 23. BSFC – CONFIG I.

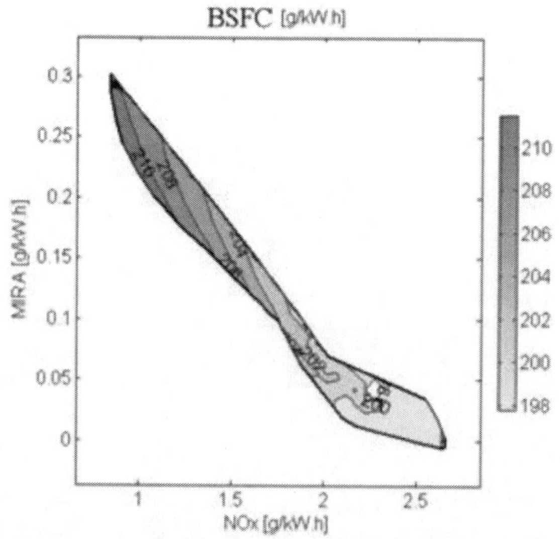

Figure 24. BSFC – CONFIG II.

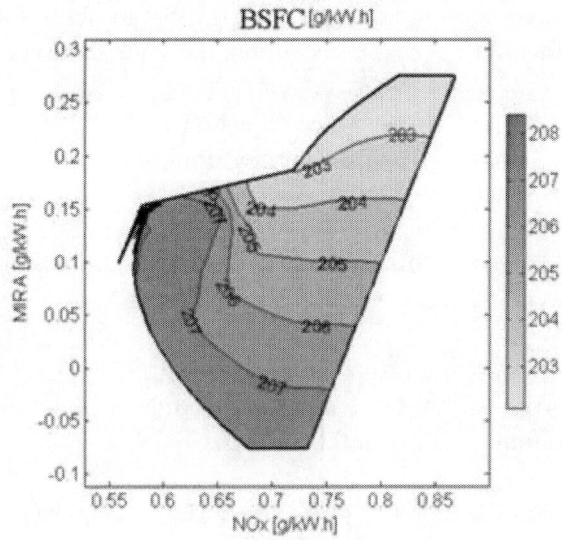

Figure 25. BSFC – CONFIG III.

Final Considerations

The main idea of this chapter was to describe a detailed study on the application and effect of EGR applied in a commercial diesel engine.

The operative testing condition chosen was the ESC steady-state mode. This condition is considered critical and decisive for the definition of EGR strategy.

The engine chosen for the study was the ACTEON 4.12 model. This engine is certified for Euro III with an EGR system already installed. This system was initially tested in order to verify the maximum potential of this configuration to reduce NOx emission. With this configuration it was possible to observe that:

- EGR maximum rate possible was 12%;
- Minimum NOx obtained was 2.18 g / kW h;
- BSFC rises with the increase of EGR;
- Variable of greatest influence in the NOx reduction was, as expected, the opening of the EGR valve.

Thus, in the follow-on test configuration, called CONFIG II, components of the EGR system were exchanged so that an EGR rate of up to 18.2% was obtainable. From this configuration it was possible to observe that:

- NOx emission was reduced to a minimum value of 0.88 g / kW.h;
- BSFC increased with an increase of the EGR rate, but CONFIG II resulted in a lower BSFC compared to CONFIG I;
- PM increased reaching values of 0.289 g / kW.h;

Comparing the first two configurations, it was possible to observe that while increasing the EGR rate was very favorable for NOx reduction, the same does not occur for particulate matter emission.

Thus, still pursuing the goal of a greater reduction of NOx, and also the reduction of particulate matter without compromising engine performance; changes were made to enable higher EGR rates that culminated in CONFIG III.

Before that, however, a preliminary study was conducted to obtain the best post-injection point. This strategy was aimed at PM reduction. In CONFIG III it was possible to achieve an EGR rate of up to 28%. With this configuration it was observed:

- Emission values of minimum NOx at 0.62g / kW.h;
- BSFC increases with the EGR rate increase, but remained within the desirable one;
- PM can reach minimum values of 0.056.

These results motivated the global test of thirteen ESC points. The need was identified for a post-combustion treatment to reduce particulate matter. By necessity with respect to vehicle architecture, the filter will be at a point well downstream of the engine exhaust which requires the application of another post-treatment system, the DOC (diesel oxidation catalyst) which is installed before the particulate filter in order to increase the temperature of exhaust gas and provide system regeneration. Once DOC use was defined (with the DOC having the primary goal of reducing emissions of HC and CO, with an efficiency of 90%), the main focus of the calibration to obtain the lowest values of NOx and fuel consumption. With the DoE technique, optimizations were performed in the remaining 12 points of the ESC cycle and results are shown in Table 6.

Table 6. ESC emission cycles testing results – 13 Points

Pollutant	Euro VI limits g/kWh	Results g/kWh	Post-treatment efficiency, %	Final results
NOx	0.4	0.5	99*	0.45
HC	0.13	0.78	90*	0.078
CO	1.500	1.560	90*	0.156
MIRA	0.01	0.115	99*	0.001

* experimental data.

CONCLUSION

The purpose of this chapter was the study of the application of EGR in diesel engines in order to reduce emissions without compromising the engine performance. As a general conclusion, it can be seen that to control emissions require the application of more than one mitigation strategy and a series of components. Tests were carried out at ever increasing EGR rates. NOx emissions generally decreased with EGR rate increase but the same does not occur with Particulate Matter (PM). Hence a post-injection strategy was implemented allied with a two-stage turbocharger system. The promising results achieved with ESC Point 2 motivated a subsequent, thorough testing of the ESC. The need to apply post-treatment filters was identified after these tests. In this way, emission limits were reduced to values that are generally compatible with or exceed Euro VI requirements.

ACKNOWLEDGMENTS

The authors are grateful to MWM International for their support in this work.

REFERENCES

Arcoumanis C., Bae C-S., Nagwaney A. and Whitelaw J.H., Effect of EGR on Combustion Development in a 1.9L DI Diesel Optical Engine, SAE Transactions, *Journal of Engines,* vol. 104, sec. 3., pp1491-1515, (Paper No. 950850), 1996 (won Arch T. Colwell Award SAE)

Argachoy, C., Aplicação da Técnica de Interferométrica a Laser e do Princípio da Máxima Entropia na Análise de Injetores de Motores Diesel, Dissertação de Mestrado, Instituto Tecnológico de Aeronáutica, 2001.

Benajes, J., Molina, S., García, J.M., Influence of Pre- and Post-Injection on the Performance and Pollutant Emissions in a HD Diesel Engine, Universidad Politécnica de Valencia, *SAE Paper* 2001-01-0526, 2001.

Blauwens, J.; Smets, B.; Peeters J., Mechanism of Prompt NO Formation in Hydrocarbon Flames, *Proceedings of the Combustion Institute*, 16, 1055-1064, 1976.

Calcote, H. F., Mechanisms of Soot Nucleation in Flames – A Critical Review. *Combustion and Flame,* 42, p. 215-242, 1981.

Chomiak, J. Combustion, a Studying Teory, Fact ans Application. Monteraux: *Gordon and Reach Science Publishers,* 1990.

Durigon A.; Krioukov, V. G.; Costa, V. J. Modelagem Matemática da Formação de Fuligem em Mecanismos de Reações com Poliacetilenos. *Lages, DC: UNIPLAC,* 2002.

Fukuda, M., Yamane, K., Neichi, T., Ikegami, M., Reduction of Nitrogen Oxides of Diesel Engines by Exhaust-Gas-Selective Recirculation, *Proceedings of the Fourth International Symposium,* 1998.

Glassman, I. and Sidebotham, G. W., Flame Temperature and Fuel Concentration Effects on Soot Formation in Inverse Diffusion Flames. *Combustion and Flame,* 90, p. 269-283, 1992.

Hawley, J. G., Wallace, F. J., Cox, A., Horrocks, R.W., Bird, G.L., University of Bath e Ford Motor Company Ltda, Reduction of Steady State NOx Levels from an Automotive Diesel Engine Using Optimized VGT/EGR Schedules, *SAE Technical Paper* 1999-01-0835, 1999.

Hayhurst, A.N.; Vince, I.M., Nitric Oxide Formation from N_2 in Flames: The Importance of Prompt NO, *Progress in Energy and Combustion Science*, 6, 35-51, 1980.

Jacobs, T., Assanis, D., Filipi, Z., The Impact of Exhaust Gas Recirculation on Performance and Emissions of a Heavy-Duty Diesel Engine, Automotive Research Center, The University of Michigan, *SAE Paper* 2003-01-1068, 2003.

Khair, M.K., Technical and Synergistic Approaches Towards the 21st Century Diesel Engine, *SAE Technical Paper* 972687, 1997.

Kitamura, Y., Mohammadi A., Ishiyama T., Shioji M., Fundamental Investigation of NOx Formation in Diesel Combustion Under Supercharged and EGR Conditions, *SAE Technical Paper* 2005-01-0364, 2005.

Ladommatos, N., Abdelhalim, S. M., Zhao, H., Hu, Z., The Dilution, Chemical, and Thermal Effects of Exhaust Gas Recirculation on Diesel Engine Emissions-Part 1: Effect of Reducing Inlet Charge Oxygen, *SAE Technical Paper* 961165, 1996 (a)

Ladommatos, N., Abdelhalim, S. M., Zhao, H., Hu, Z., The Dilution, Chemical, and Thermal Effects of Exhaust Gas Recirculation on Diesel Engine Emissions--Part 2: Effects of Carbon Dioxide, *SAE Technical Paper* 961167, 1996 (b)

Ladommatos, N., Abdelhalim, S. M., Zhao, H., Hu, Z., The Dilution, Chemical, and Thermal Effects on Exhaust Gas Recirculation on Diesel Engine Emissions--Part 3: Effects of Water Vapor, *SAE Technical Paper* 971659, 1997 (a)

Ladommatos, N., Abdelhalim, S. M., Zhao, H., Hu, Z., The Dilution, Chemical, and Thermal Effects of Exhaust Gas Recirculation on Diesel Engine Emissions--Part 4: Effects of Carbon Dioxide and Water Vapor, *SAE Technical Paper* 971660, 1997 (b).

Lourenço A.A. M, Martins C.A., Lacava P. T. and Ferreira M.A., Environmental Engineering Science. May 2013, 30(5): 221-231. doi:10.1089/ees.2012.0008.

Mellow, J.P., Mellor, A.M., NOx Emissions from Direct Injection Diesel Engines with Water/Steam Dilution, *SAE Technical Paper* 1999-01-0836, 1999.

MIRA Report No. 1965/10, Nuneaton Warwickshire, UK 1965, AG Dodd and Z. Holubecki.

Miyauchi, T.; Mori, Y.; Imamura, A., A Study of Nitric Oxide Formation in Fuel-Rich Hydrocarbon Flames: Role of Cyanide Species, H, OH and O, *Proceedings of the Combustion Institute*, 16, 1073-1082, 1976.

Richter, H.; Howard J.B.; Formation of polycyclic aromatic hydrocarbons and their growth to soot—a review of chemical reaction pathways. *Progress in Energy and Combustion Science,* V. 26, p. 565 – 608, 2000.

Squaiella L.L.F., Effects of exhaust gas recirculation to control NOx emissions in Diesel engines, Dissertação (Mestrado Profissional), UNICAMP, 2010. http://www.biblioteca digital.unicamp.br/document/?code=000771845.

Squaiella L. L. F., Martins C. A. and Lacava P. T., Strategies for emission control in diesel engine to meet Euro VI, *Fuel* 104 (2013) 183–193.

Stratakis G.A. and Stamatelos A.M., Thermogravimetric analysis of soot emitted by a modern diesel engine run on catalyst-doped fuel, *Combustion and Flame V.* 132, 157–169, 2003.

Wagner, R.M., Green, J.B., Storey, Jr., J.M, Daw, C.S., Extending Exhaust Gas Recirculation Limits in Diesel Engines, AWMA 93rd Annual Conference and Exposition, Salt Lake City, UT, 2000.

Woods, M., Kamo, R., and Bryzik, W., "High Pressure Fuel Injection for High Power Density Diesel Engines," SAE *Technical Paper* 2000-01-1186, 2000, doi:10.4271/2000-01-1186.

Appendix A

Diesel Engine Curve

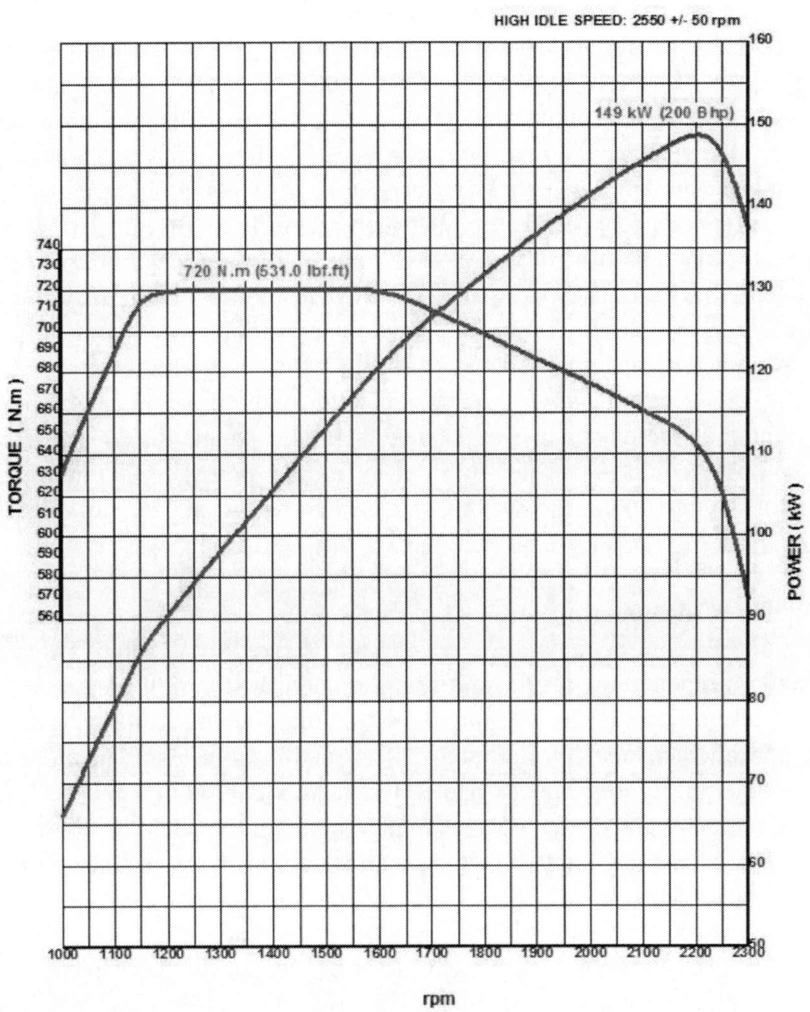

Appendix B

DoE Application in the Settings Configurations

In this appendix the methodology used will be described to define the different operating conditions to which the engine of this study was submitted. The technique applied is based on DoE (Design of Experiments). The first concept of experimental design have been applied in agriculture in 1920 in England. Their application in the industry began in 1930 with the development of the response surface methodology, but it was in 1970 with the techniques developed by the engineer and statistician Genichi Taguchi that the concept of using statistics to improve the quality of the product had its greatest expansion.

The three basic principles for the Design of Experiments are: randomization, replication and blocks. Randomization is important to ensure that the tests are carried out randomly, not following a predetermined sequence, thus avoiding vices in the process or system. The replication principle is different from repetition, because repetition normally occurs in a sequence, that is, the measurements occur one after another. However replication happens in non-regular intervals and it is important in estimating the experimental error because it is possible to determine whether the observed difference is a function of a statistical difference or of a measurement noise. The replicas reflect the sources of variation either between the various combinations of an experiment or between themselves. Blocks, is the technique of experiment design which is used to increase the accuracy between the tests performed by varying the interest factors. They are generally used to reduce or eliminate the variability transmitted by factors which may influence the response variables in an undesirable manner, such as noise variables.

DoE allows the experimenter to evaluate the behavior of a system or process when changes in the input variables of a test are made deliberately so that the consequences of these changes can be observed and identified in order to find the point of maximum yield for two significant factors by analyzing the surface chart. The process of DoE follows some steps, such as:

- Definition of the goal.
- Definition of control factors, X, which can be fixed or variable. Fixed factors are those that influence the results and shall be controlled during the entire experiment, without any variation. Variables are those that must be changed during an experiment in a controlled manner by defining the top and bottom levels in order to evaluate their impact on the response variable. The factors can be quantitative or qualitative depending on their origin. For example, magnitudes such as pressure and fuel injection start are quantitative factors, while types of the piston or turbo compressor are qualitative ones.
- Noise Variable, Z, during the test, are variables that are hardly controlled by the researcher. They are characterized by their ability to affect the response variables, however they are not considered as a factor. Their effects can be minimized by keeping the constants during the tests and taking into account their effects during analysis of response variables, for example the temperature and humidity of the intake air.
- Response Variables Y.

Figure B.1 shows the basic diagram of a DoE.

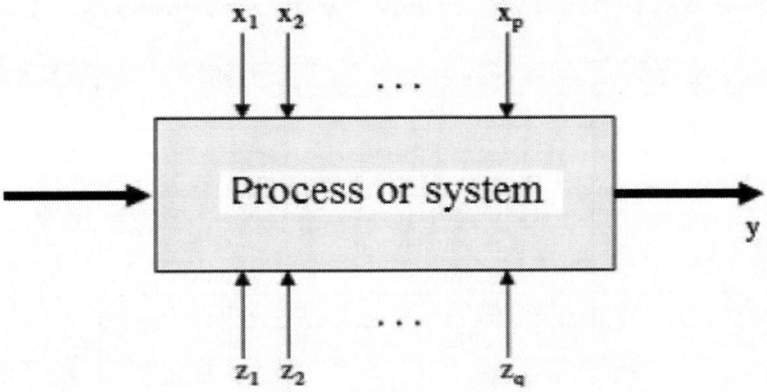

Figure B1. DoE structure.

Application of DoE for CONFIG I

Considering the application of the the DoE technique initially for CONFIG I. Table B.1 provides the objectives for CONFIG I of the rated motor.

Table B.1. Definition of objectives for the CONFIG I

01. Torque	646.8	Nm	-
02. Rotation	1380	rpm	-
03. Smoke	3	FSN	Max.
04. NO_x specific	0.5	g/kWh	Max.
05. CO specific	1.000	g/kWh	Max.
06. HC specific	0.17	g/kWh	Max.
07. MIRA	0.27	g/kWh	Max.
08. Pz	190	bar	Max.
09. Noise of combustion	-	dBa	Min.
10. BSFC observed	-	g/kWh	Min.
11. T3	760	°C	Max.

Table B.2 presents the variables Y which are expected as an answer to the tests. Such variables can be defined as limiting (Y3 and Y4) which shall not exceed the values set in terms of security or variables to be minimized to their minimum value with a defined objective. The temperature, T3, for example, should not exceed 760 °C, and is therefore a limiting variable. Yet, the specific NOx is expected to be minimized with a target value of 0.5g/kWh.

As it is a configuration engine which has already been used for other applications the boundary value for the point and injection pressure was kept the same while varying only the position of the EGR valve and also the turbo VNT vane position, in order to check its potential altogether to achieve the lowest values of NOx. Table B.3 is the definition of the factors for the tests with this engine configuration. Note that the position of the EGR valve (X1) and the position of Vanes (X2) are mixed while the point conditions and injection pressure are kept as fixed values.

Table B.2. Definition of response variables for the CONFIG I

Variables Y	Description	Type	Limit
Y1	Smoke	Minimize	3
Y2	BSFC observed	Minimize	-
Y3	T3	Max. Limit	760
Y4	Pz	Max. Limit	190
Y5	Noise of combustion	Minimize	-
Y6	NO_x specific	Minimize	0.5
Y7	CO specific	Minimize	1
Y8	HC specific	Minimize	0.17
Y9	MIRA	Minimize	0.27

Table B.3. Definition of factors for the CONFIG I

Type: variable factor						
Variables X	Description	Minimum	Nominal	Maximum	Unity	Step
X1	EGR valve position	1000	1140	1280	mV	140
X2	Vanes position	67	70	73	pwm	3
Type: boundary condition						
Variables Z	Description	Minimum	Nominal	Maximum	Unity	Step
Z1	T22	25	30	35	°C	5
Z2	P1	10	15	20	mbar	5
Z3	P4	20	30	40	mbar	10
Z4	Diesel pressure feeding	0.6	0.7	0.8	bar	0.1
Z5	T1	13	15	17	°C	2
Z6	Rotation	1370	1380	1390	rpm	10
Z7	Torque	641.8	647	651.8	Nm	5
Z8	Water Temperature (exit)	90	95	100	°C	5
Z9	Battery Tension	14.2	14.4	14.6	mV	0.2
Z10	Diesel temperature	38	40	42	°C	2

Table B.4. Definition of factors for the CONFIG I

Type: variable factor						
Variables X	Description	Minimum	Nominal	Maximum	Unity	Step
X1	EGR valve position	1000	1140	1280	mV	140
X2	Vanes position	67	70	73	pwm	3
Type: boundary condition						
Variables Z	Description	Minimum	Nominal	Maximum	Unity	Step
Z1	T22	25	30	35	°C	5
Z2	P1	10	15	20	mbar	5
Z3	P4	20	30	40	mbar	10
Z4	Diesel pressure feeding	0.6	0.7	0.8	bar	0.1
Z5	T1	13	15	17	°C	2
Z6	Rotation	1370	1380	1390	rpm	10
Z7	Torque	641.8	647	651.8	Nm	5
Z8	Water Temperature (exit)	90	95	100	°C	5
Z9	Battery Tension	14.2	14.4	14.6	mV	0.2
Z10	Diesel temperature	38	40	42	°C	2

For the results of the tests to be validated, it is necessary to define the boundary conditions (Z1 to Z10) where the maximum and the minimum of parameters that can directly influence the results of engine performance, are provided. Table B4 is of the same description.

With the definition of fixed and variable parameters, following the "Taguchi" method the DoE in Table B.5 describes the 13 points that define a complete mapping of the engine within the limits set in Table B.5, showing the maximum potential of this set reaching the lowest values of emissions and fuel consumption.

Table B.5. Stabilization parameters for the acquisition of emission results. Point 2 of the ESC, CONFIG I

Boundary conditions				Combination number	Factors	
					Positions	
	Rotation	1380	rpm		EGR valves	Vanes
	Power	93	kW		mV	pwm
	Load	92	kgf	1	1000	73
	Torque	646.8	Nm	2	1280	73
Pressures				3	1140	73
	P1	12	mbar	4	1000	67
	P4	30	mbar	5	1140	70
	Diesel	0.7	Bar	6	1140	70
Temperatures				7	1280	67
	T1	15	°C	8	1140	70
	T22	30	°C	9	1140	67
	Water	95	°C	10	1140	70
	Diesel	40	°C			

Application of DoE for CONFIG II

Following the framework to perform DoE for CONFIG II of the rated engine the new goal of torque is presented in Table B6. Note that this proposal provided an improvement in performance in the engine allowing an increase in torque of 11.31 %, from 646.8 (CONFIG I) to 720 Nm.

Table B.6. Definition of objectives for testing in the CONFIG II

01. Torque	720	Nm	-
02. Rotation	1380	Rpm	-
03. Smoke	3	FSN	Máx.
04. NO_x specific	0.5	g/kWh	Máx.
05. CO specific	1.000	g/kWh	Máx.
06. HC specific	0.17	g/kWh	Máx.
07. MIRA	0.27	g/kWh	Máx.
08. Pz	190	Bar	Máx.
09. Combustion noise	-	dBa	Mín.
10. BSFC observed	-	g/kWh	Mín.
11. T3	760	°C	Máx.

The factors that were changed in this assessment were the positions of the EGR valve that will work more open than in the CONFIG I and the pinch valve of the intake air. The turbocharger will not be a factor, because the relief valve was kept fully closed in order to obtain its maximum performance. In table B7 the fixed and variable parameters (noise) for the evaluation of the second engine configuration, are provided. It can be observed that, in relation to proposal 1, the EGR valve will operate in a more open setting and that the increased restrictions on admission (P1) and exhaust (P4) pressure were caused by the increase in air flow obtained with the two-stage turbocharger. In table B8, after the definition of the central point, 13 new points for the mapping of certain variables are defined using the DoE method of Taguchi from Table B7.

Table B.7. Definition of factors for the CONFIG II

Type: variable factor						
Variables X	Description	Minimum	Nominal	Maximum	Unity	Step
X1	EGR valve position	2000	3000	4000	mV	1000
X2	Pinch valve position adm.	0	42	84	pwm	42
Type: boundary condition						
Variables Z	Description	Minimum	Nominal	Maximum	Unity	Step
Z1	T22	25	30	35	°C	5
Z2	P1	10	15	20	mbar	5
Z3	P4	45	55	65	mbar	10
Z4	Diesel pressure feeding	0.6	0.7	0.8	bar	0.1
Z5	T1	13	15	17	°C	2
Z6	Rotation	1370	1380	1390	rpm	10
Z7	Torque	715	720	725	Nm	5
Z8	Water Temperature (exit)	90	95	100	°C	5
Z9	Battery Tension	14.2	14.4	14.6	mV	0.2
Z10	Diesel temperature	38	40	42	°C	2

Table B.8. Stabilization parameters for the acquisition of emission results. Point 2 of ESC, CONFIG II

Boundary conditions				Combination number	Load kgf	Torque Nm	Factors Positions	
	Rotation	1380	rpm				EGR valves	Vanes
	Power	104	kW				mV	PWM
				1	102.3	720.1	4000	5
Pressures				2	102.4	721.0	4000	84
	P1	190	mbar	3	102.5	721.6	4000	42
	P4	45	mbar	4	102.4	720.9	2000	5
	Diesel	0.7	Bar	5	102.6	722.3	3000	42
				6	102.4	721.2	3000	42
				7	103.0	725.5	2000	84
Temperatures				8	102.3	720.6	3000	42
	T1	15	°C	9	102.4	721.3	2000	42
	T22	30	°C	10	102.6	722.8	3000	42
	Water	95	°C	11	102.6	722.8	3000	42
	Diesel	40	°C	12	102.5	722.2	3000	5
				13	102.8	724.7	3000	84

Application of DoE for CONFIG III

Along with the post-injection parameter changes of engine parameters were evaluated such as the pressure and the point of injection and EGR valve position. In Table B.9 the three parameters are illustrated as selected to perform the DoE. The pressure and main injection point ranged from the reference values adopted as the basis of configurations 1 and 2: in this case zero. With the definition of the three factors, following the methodology of DoE, 15 combinations were obtained for the mapping of CONFIG III. See Table B10.

Table B.9. Definition of factors for the CONFIG III

Type: variable factor						
Variables X	Description	Minimum	Nominal	Maximum	Unity	Step
X1	Injection pressure	-100	0	100	bar	100
X2	Injection point	-2	0	2	º Dpms	2
X3	EGR valve position	3000	3500	4000	mV	500
Type: boundary condition						
Variables Z	Description	Minimum	Nominal	Maximum	Unity	Step
Z1	T22	25	30	35	°C	5
Z2	P1	10	15	20	mbar	5
Z3	P4	60	70	80	mbar	10
Z4	Diesel pressure feeding	0.6	0.7	0.8	bar	0.1
Z5	T1	13	15	17	°C	2
Z6	Rotation	1370	1380	1390	rpm	10
Z7	Torque	715	720	725	Nm	5
Z8	Water Temperature (exit)	95	100	105	°C	5
Z9	Battery Tension	14.2	14.4	14.6	mV	0.2
Z10	Diesel temperature	38	40	42	°C	2

Table B.10. Stabilization parameters for the acquisition of emission results. Point 2 of ESC, CONFIG III

				Nº.	Load kgf	Torque Nm	Factors		
Boundary conditions							EGR valves	Pinch valve	Combination
	Rotation	1380	rpm				bar	ºDpms	mV
	Power	104	kW	1	102.9	722.9	-100	2	3500
				2	102.5	719.8	0	0	3500
				3	102.4	719.2	100	2	3500
Pressures				4	102.5	719.8	100	0	3000
	P1	250	mbar	5	102.5	719.7	0	2	3000
	P4	2000	mbar	6	102.6	720.6	-100	0	4000
	Diesel	0.7	Bar	7	102.3	718.5	0	-2	4000
				8	102.5	719.6	100	0	4000
				9	102.7	721.0	0	2	4000
Temperatures				10	102.6	720.5	-100	0	3000
	T1	15	ºC	11	102.5	720.1	-100	-2	3500
	T22	30	ºC	12	102.5	720.0	0	-2	3000
	Water	100	ºC	13	102.4	719.2	0	0	3500
	Diesel	40	ºC	14	102.3	718.6	100	-2	3500
				15	102.3	718.4	0	0	3500

NOMENCLATURE

ATDC	after top dead center
BDC	bottom dead center
BSFC	brake specific fuel consumption
BHP	brake horsepower
CI	compression ignition engine
DI	direct injection
DoE	design of experiments
DOC	diesel oxidation catalyst
DPF	diesel particulate filter
IDI	indirect injection
EGR	exhaust gas recirculation
EPA	environmental protection agency
ESC	european stationary cycle
ETC	european transient cycle
ELR	european load response
FSI	fraction solid inorganic
MIRA	motor industry research association
P22	after cooler air pressure
PM	particulate matter
VNT	variable nozzle turbocharger
MAF	mass air flow
P3	exhaust pressure, before turbine
T3	Temperatura dos gases de escape no coletor antes da turbina
SI	spark ignition engine
EURO	european emission regulations
rpm	rotations per minute
SOF	solid organic fraction
t	time
T	temperature
[ ]	concentrations in $mol.cm^{-1}$

In: Automotive Exhaust Emissions and Energy Recovery
Editor: Apostolos Pesiridis
ISBN: 978-1-63321-493-4

Chapter 4

Particulate Matter Emissions during Transient Diesel Engine Operation with Various Diesel/Biofuel Blends (Biodiesel, Ethanol and N-Butanol)

Evangelos G. Giakoumis*
School of Mechanical Engineering, National Technical Univ. of Athens, Greece

Abstract

The transient operation of turbocharged diesel engines is particularly demanding in terms of engine response, systems reliability and exhaust emissions. It is a situation encountered continuously during the daily driving schedule of passenger, light and heavy-duty vehicles, differentiating radically the engine operation from the respective steady-state conditions. Typically, it results in poor driveability and overshoot in the emitted pollutants (most notably particulate matter).

On the other hand, depleting reserves and growing prices of crude oil, as well as steadily stricter emission regulations and greenhouse gas concerns, have sparked the research to develop alternative fuel sources, with the emphasis placed on biofuels. At the moment, biodiesel is considered the primary alternative fuel for compression ignition engines since it possesses similar properties to diesel fuel and succeeds in substantially reducing the amount of all emitted carbonaceous pollutants (particulate matter (PM), CO, HC). In parallel, the research has also focused on various bio-alcohols (in particular ethanol and n-butanol); although alcohols were initially considered as alternative fuels for gasoline engines, they too have demonstrated significant capacity in particulate matter emission mitigation from diesel-engined vehicles.

The present chapter surveys the literature concerning the effects of biofuel/diesel blends on the PM emissions from diesel engines operating under transient conditions, i.e. acceleration, load increase, starting and in the combined form of transient/driving cycles.

* Corresponding author: School of Mechanical Engineering, National Technical Univ. of Athens, Zografou Campus, 15780, Athens, Greece. Email: vgiakms@central.ntua.gr.

Three very promising biofuels are covered in this review, namely biodiesel, ethanol and n-butanol. The main mechanisms of PM emissions during transients are identified and discussed with respect to the fundamental aspects of transient operation and the differing properties of each biofuel relative to the reference diesel oil.

Lastly, based on the published studies up to date, a statistical analysis of all PM emission data is conducted, and the cumulative trends are presented for the purpose of quantifying each biofuel blend's benefits on the emitted PM during driving cycles, and for various engine types and engine technologies. From this statistical analysis, the superiority of the alcohols' PM reducing capability relative to the biodiesel blends is concluded. It is believed that such statistical analyses can prove useful for future planning and accommodate decision making.

Keywords: Biodiesel, Diesel engine, Driving cycle, Ethanol, n-Butanol, Particulate matter, Transient operation, Turbocharger lag

INTRODUCTION

The diesel engine has for many decades now assumed a leading role in both the medium and medium-large transport sector. Major factors are its superior fuel efficiency over its spark ignition counterpart, its reliability as well as its inherent capability to operate turbocharged and hence produce high torque and power. Nonetheless, discrepancies in the form of exhaust smokiness and combustion noise radiation delayed its infiltration and broad acceptance in the highly competitive passenger car market. Historically, the majority of the research and published studies on diesel engine operation has focused on the steady-state performance. However, only a very small fraction of a vehicle's operating pattern is true steady-state. In fact, the greater part of the daily driving schedules of passenger cars, buses, trucks and non-road engines involves transient operation in the form of changing (engine) speed and/or loading/fueling conditions (Rakopoulos and Giakoumis, 2009). During the transient operation of turbocharged diesel engines, combustion air-supply is delayed compared to fueling (mostly originating in the turbocharger lag), a fact that negatively affects torque build-up and vehicle driveability. As a result of this delay in the response between air-supply and fueling, it is primarily the particulate matter emissions that peak considerably above their acceptable, steady-state values; the latter is experienced as a cloud of black smoke coming out of the exhausts of diesel-engined vehicles (Hagena et al., 2006; Rakopoulos and Giakoumis, 2006; 2009).

On the other hand, during the last decades depleting reserves and growing prices of crude oil, as well as gradually stricter emission regulations and greenhouse gas concerns, have led to a substantial effort to develop alternative fuel sources. Among those, biofuels have assumed a primary role on account of their renewability, which provides an ad hoc advantage as regards the reduction of carbon dioxide (CO_2) (Hansen et al., 2009). The term biofuel refers to any fuel that is derived from biomass, such as sugars, vegetable oils, animal fats, etc. Biofuels made from agricultural products reduce the dependence on oil imports, support local agricultural industries and offer benefits in terms of sustainability and reduced

PM emissions. What is equally important, from an economic point of view, is that they are more evenly distributed than fossil or nuclear resources, since they can be produced domestically (Agarwal, 2007; Demirbas, 2007; Komninos and Rakopoulos, 2012; Giakoumis et al., 2012).

There are numerous biofuels that have been produced and researched so far, ranging from various vegetable oils up to bio-dimethylether or bio-hydrogen. At the moment, it is *biodiesel* that is considered the primary alternative fuel for compression ignition (CI) engines, since it possesses similar properties to diesel fuel and can be blended with diesel practically at any proportion without modifications in the existing distribution infrastructure (Graboski and McCormick, 1998). On the other hand, it is true that bio-alcohols, particularly *ethanol* and *n-butanol*, were initially considered as substitute fuels for spark ignition engines. Nevertheless, they have proven promising for CI engines too, since their high oxygen content has succeeded in a considerable mitigation of particulate matter emission (higher than biodiesel (Giakoumis et al., 2013)), while at the same time exhibiting considerable greenhouse gas emission savings; the latter have been reported to range from 32% (for the case of wheat feedstock) up to 87% (wheat straw feedstock) (EU Directive 2009/28/EC). However, owing to their completely different physico-chemical properties (that resemble those of gasoline rather than diesel), alcohols are usually blended with diesel fuel at small ratios, e.g. 10 or 15% v/v.

The use of biodiesel and alcohols during *steady-state* operation has been researched heavily during the last decades, as has been documented in various review and research papers (e.g. Graboski and McCormick, 1998; Hansen et al., 2005; McCormick et al., 2005; Agarwal, 2007; Lapuerta et al., 2008; Giakoumis et al., 2012; 2013). The target of the present work is to expand on these, and review the literature regarding the impacts of diesel-biofuel blends on the exhaust emissions of CI engines under the very critical *transient* conditions encountered in the every-day engine/vehicle operation i.e., acceleration, load increase, starting and in the collective form of transient cycles. Three very promising biofuels are covered in this chapter, namely biodiesel, ethanol and n-butanol. The US Environmental Protection Agency (EPA), in its national emissions inventory, estimated that diesel vehicles emitted only 7% of total on-road unburned hydrocarbons (HC) and only 5% of on-road carbon monoxide (CO) in 2000, while at the same time being responsible for 60% of the on-road PM emissions (US EPA Report, 2003). This fact combined with the well known finding that PM notoriously peak during each turbocharged diesel engine transient event to values considerably higher than during steady-steady operation (Rakopoulos and Giakoumis, 2009), prove the high significance of the particulate matter emissions in the daily operation of diesel-engined vehicles. As a result, PM will be in the focal point of the analysis that follows.

The usual approach when analyzing alternative fuel impacts on exhaust emissions is by discussing the differing physical and chemical properties of the various blends against those of the reference fuel. Consequently, the composition and properties of biodiesel and bio-alcohols, together with their combustion and PM emission formation mechanisms, will form the basis for the interpretation of the experimental findings. What is equally important is that emphasis will also be placed on the specific attributes and discrepancies encountered during transients too (most notably in the form of turbocharger lag), which may enhance or alleviate the differences observed between the various biofuel blends and the neat diesel operation.

FUNDAMENTAL ASPECTS OF DIESEL ENGINE TRANSIENT OPERATION AND TRANSIENT CYCLES

As already pointed out in the Introduction, the transient operation of a turbocharged diesel engine can prove particularly demanding in terms of engine response (i.e. vehicle driveability), emissions and systems reliability, *irrespective of the fuel used*. It is a very often encountered condition that drastically differentiates the engine operation from the respective steady-state conditions (at least for turbocharged diesel engines, which are the vast majority of compression ignition engines in use). Since the analysis that follows will focus not only on the biofuels chemical and physical specificities relative to the neat diesel fuel but also on the impacts of transient discrepancies, it is considered essential to provide an initial background on transient diesel engine operation. Hence, in this section a typical *discrete* transient event as well as a transient cycle will be discussed briefly, in order to highlight some important aspects of transient conditions, most notably turbocharger lag; a thorough discussion of the subject is provided in Rakopoulos and Giakoumis (2009).

Figure 1 illustrates an acceleration of a medium-duty turbocharged diesel engine commencing from a low engine speed and low load (corresponding to 1st gear engaged in the gearbox). The response of four engine and turbocharger operating parameters as well as the development of smoke opacity[1] are demonstrated in this figure. Such a test, typical in the every-day operation of passenger vehicles and trucks, is quite demanding for both the engine and the turbocharger since the latter accelerates from practically zero boost; hence, it is a good example to pinpoint the peculiarities of transient operation.

As is made obvious from Figure 1, the fuel pump rack responds almost instantly to the fueling increase command and shifts to its maximum position. In fact, the fuel pump rack responds in two stages; initially, a rapid shift to a first peak position is observed, followed by a smoother movement to the maximum position. This behavior highlights in the most explicit way the fuel limiter operating principle; the latter does not allow sharp fueling increases when the compressor boost pressure is still low in order to limit intolerable smoke emissions. Indeed, boost pressure remains practically unchanged during the early transient cycles. Obviously, the increased exhaust gas power produced by the engine is not capable of instantly rising the turbine power output, largely owing to the turbocharger inertia, so that the compressor operating point moves rather slowly towards the direction of increased boost pressure and air-mass flow-rate; during this period (12 engine cycles or 1.17 s for the particular engine), known as *turbocharger lag*, the engine is practically running in naturally aspirated mode or with very limited boost. Turbocharger lag is enhanced by the unfavorable turbocharger compressor characteristics at low loads and speeds (since boost pressure depends strongly on turbocharger rotational speed, with high pressure ratios achieved only when the turbocharger shaft speed has increased). As a result of this slow reaction, the air–fuel equivalence ratio during the early cycles of an acceleration assumes very low values (even lower than unity), deteriorating combustion and leading to slow engine torque and speed response and long recovery period (Watson and Janota, 1982; Rakopoulos and Giakoumis, 2009; Rakopoulos et al., 2010a).

---

[1] It should be noted at this point that although smoke opacity is not among the regulated emissions, it is often used by researchers as a surrogate for the legislated particulate matter, which are considerably more difficult to measure instantaneously.

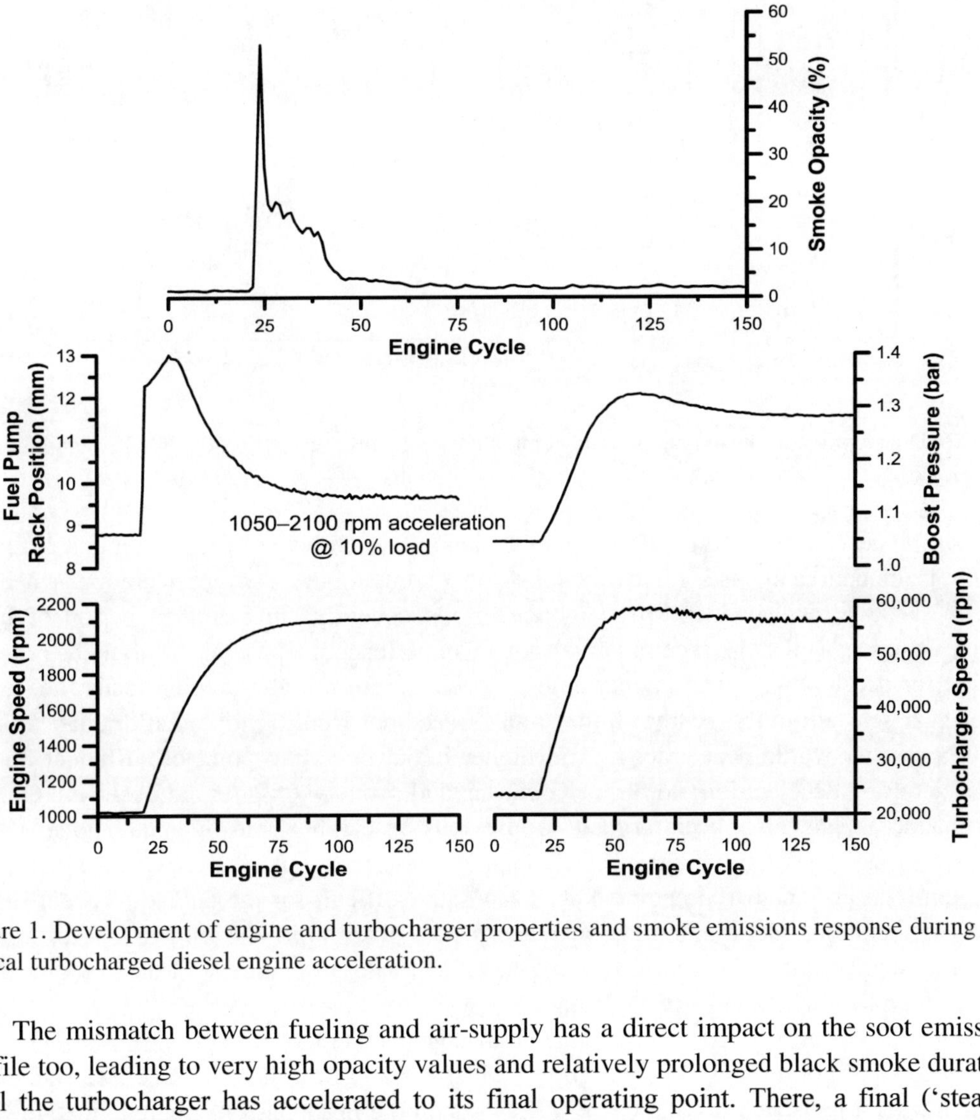

Figure 1. Development of engine and turbocharger properties and smoke emissions response during a typical turbocharged diesel engine acceleration.

The mismatch between fueling and air-supply has a direct impact on the soot emission profile too, leading to very high opacity values and relatively prolonged black smoke duration until the turbocharger has accelerated to its final operating point. There, a final ('steady-state') smoke value is assumed (after approx. 75 engine cycles), of much lower magnitude compared to the previously experienced peak value. For the overshoot in smoke emissions observed in Figure 1 therefore, the main cause is the instantaneous lack of air due to the turbocharger lag. It has also been argued (Hagena et al., 2006) that the latter phenomenon is further aided by the initial sharp increase in ignition delay during the early transient cycles of the acceleration. Rapid increases in fuel injection pressure upon the onset of each instantaneous transient cause the penetration of the liquid fuel jet within the combustion chamber to increase. Since the initial higher-pressure fuel jets are injected into an air environment that is practically unchanged from the previous steady-state conditions, the higher-momentum fuel jet is not accompanied by equally enhanced gas motion. Thus, liquid fuel impingement on the cool combustion walls increases, lowering the rate of mixture preparation and enhancing the heterogeneity of the mixture. Moreover, the subsequent harder combustion course prolongs combustion and reduces the available time for soot oxidation.

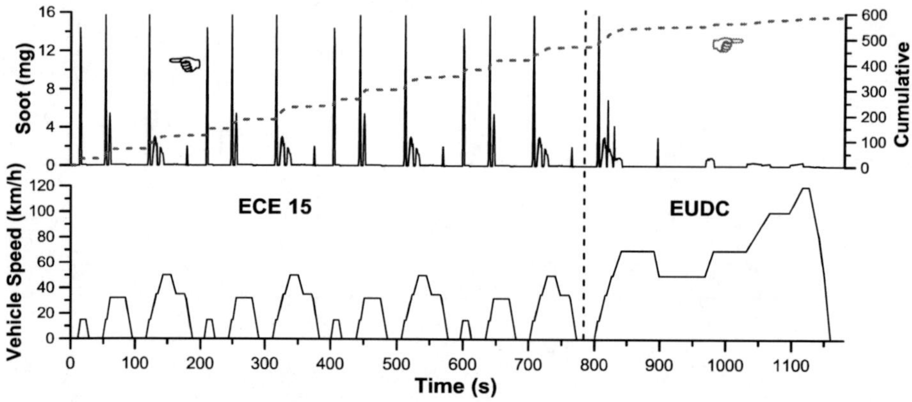

Figure 2. Development of instantaneous and cumulative soot emissions during the NEDC highlighting the peak during vehicle accelerations (adapted from Giakoumis and Lioutas (2011)).

For the acceleration depicted in Figure 1, this mismatch between fueling and air-supply is very prominent and lasts relatively long too owing to the hard acceleration schedule (approx. 1100 rpm demanded speed increase); consequently, an extended period of high opacity values as well as high cumulative soot are experienced. However, it should be pointed out that for accelerations from higher engine speeds or from higher engine loads (i.e., with higher gear selected in the gearbox), the smoke overshoot would not be that dramatic, since the turbocharger would accelerate from a higher initial operating point of sufficient boost. Likewise during deceleration, no turbocharger lag and soot spike issues are observed. On the other hand, as has also been argued in previous research (Watson and Janota, 1982; Rakopoulos and Giakoumis, 2006; 2009), load increase transients (more typical for heavy-duty, non-road and industrial engines) are even more difficult for the engine to cope, since, unlike acceleration, the initial load increase causes a drop in the engine speed, which constitutes another major burden for the engine to deal with; it might even lead to engine stall if a very high loading is instantly applied.

Both acceleration and load-change transients, together with (cold) starting, are experienced during transient cycles. Figure 2 illustrates a typical diesel engine PM emissions profile during a legislated driving cycle; it focuses on the instantaneous and cumulative soot during the European passenger car and light-duty NEDC cycle (Giakoumis and Lioutas, 2011). More data regarding the transient/driving cycles discussed in this chapter are available in the Appendix.

As is made obvious in Figure 2, the points in the cycle where the most abrupt accelerations (or load increases for heavy-duty engine transient cycles) are experienced lead also to a considerable overshoot in soot emissions. Net soot production (fueling as well, hence CO_2 emissions) is mainly dependent on the engine load. As the load increases, more fuel is injected into the cylinders, increasing the temperatures in the fuel-rich zones. Moreover, the duration of diffusion combustion is increased favoring soot formation, whereas the remaining time after combustion as well as the availability of oxygen – both of which enhance the soot oxidation process – decrease; thus, the production of soot is favored. During each acceleration in the cycle, the above mechanism is remarkably enhanced by the locally very low values of air–fuel ratios experienced during turbocharger lag. It is obvious that the steeper or the more frequent the accelerations (or the load increases) the higher and the more frequent the turbocharger lag discrepancies, subsequently leading to higher soot spikes and cumulative emissions.

Various measures have been proposed in the literature in order to cope with these transient discrepancies that affect both the vehicle driveability and the emitted PM (and NO_x), e.g. *smaller turbocharger moment of inertia* (coupled with a waste-gate valve to avoid extreme pressure ratios and in-cylinder peak pressures at high speeds), *combined supercharging* (mechanical compressor in series with the turbocharger), *two-stage series or sequential turbocharging, electrically assisted turbocharger* etc. (Rakopoulos and Giakoumis, 2009).

PHYSICAL AND CHEMICAL PROPERTIES OF BIOFUELS

Irrespective of the biofuel used, spray properties may be altered with respect to normal diesel operation owing to differences in the physical and chemical properties such as molecular structure, cetane number, latent heat of vaporization, viscosity, surface tension, bulk modulus of elasticity, and boiling point. All these, in turn, affect the injection timing, the ignition delay, as well as the balance between premixed and diffusion combustion, shifting the emission pattern to lower or higher emission values depending on other specific conditions (blending percentage, injection system and engine technology in general, transient schedule and exhaust gas after-treatment). Hence, in this section a brief description of the key physical and chemical properties of each one of the surveyed biofuels will be provided; this will aid the analysis of the individual combustion and emission mechanisms, in relation to the reference diesel fuel that will follow.

Biodiesel

Owing to their high viscosity (10–20 times higher than that of diesel fuel) the use of vegetable oils in diesel engines is not very popular, and is usually limited to small blending ratios (up to 20%) in order to avoid problems such as injector coking, carbon deposits and piston oil ring sticking. The transesterification of oils, on the other hand, yields biodiesel, which is characterized by much smaller viscosity (closer to that of the conventional diesel fuel) and at the same time higher cetane number. Biodiesel produced by transesterification of vegetable oils, animal fats or recycled cooking oils, consists of long-chain alkyl esters, which contain two oxygen atoms per molecule (Graboski and McCormick, 1998; Agarwal, 2007; Demirbas, 2005). The transesterification reaction proceeds with catalyst (base or acid) or without catalyst by using primary or secondary monohydric aliphatic alcohols, where the glycerol–based triesters (or triacyl glycerides) that make up the fats and oils are converted into monoesters yielding free glycerol as a byproduct:

$$\text{Triglycerides} + \text{Monohydric alcohol} \rightarrow \text{Mono-alkyl esters} + \text{Glycerol}$$

The substitution of normal diesel fuels with soybean methyl ester (SME) is already a reality in certain activities in the US, while its substitution with rapeseed methyl ester (RME) comprises a commercial activity in many countries in Central Europe (Austria, Germany, France). In other countries with warmer climates, as for example in Asia, palm oil methyl ester (PME) is the corresponding substitute. Other popular biodiesels are sunflower (SuME), cottonseed (CME), waste cooking (WCME) and tallow (TME) methyl ester, collectively known as fatty acid methyl esters (FAME), owing to the extensive use of methanol in the esterification process. However, issues such as increase of the corresponding food prices and biodiversity have sparked the research into biodiesels from non-edible feedstock (e.g. algae, jatropha and karanja).

The chemical composition of biodiesel is dependent upon the length and degree of unsaturation of the fatty acid alkyl chains. With few exceptions, the carboxylic (fatty) acids are all straight-chain compounds ranging in size from 3−18 carbons. Acids may be saturated (cannot chemically add hydrogen), which means that they contain only single bonds, or unsaturated (can be hydrogenated), which means that they contain at least one double bond (Graboski and McCormick, 1998). For example, RME consists mainly of mono-unsaturated fatty acids (of which, a high percentage of the C18:1 oleic one), whereas SME, safflower and SuME are primarily high in poly-unsaturated (C18:2 linoleic) fatty acids. Beef tallow, on the other hand, is primarily composed of saturated C16:0 palmitic and C18:0 stearic acids, with the remainder being mono-unsaturated oleic acid (Hoekman et al., 2012; Giakoumis, 2013). As is made obvious, since the composition of the originating oils/fats varies considerably, it is expected that the physical and chemical properties of biodiesel will differ (sometimes substantially) from place to place, influenced primarily by the oil or fat used but also affected by the exact transesterification process. To this aim, both the European Union (EU) and the US have issued specifications that should be met by the methyl esters intended for use in compression ignition engines (EN 14214:2008 and ASTM D6751 respectively).

Due to its chemical structure, which usually contains a high percentage of unsaturated fatty acids, biodiesel is more prone to oxidation compared to conventional diesel fuel, particularly for long term storage. Hence, additives such as anti-oxidants are usually required. It is the more saturated biodiesels (PME or TME) that exhibit better oxidative behavior but still inferior to mineral diesel. Moreover, biodiesel has worse cold-flow properties (pour point and cloud point) than petrodiesel, requiring cold-flow improvers. In this case, it is the more saturated esters that suffer from poor low-temperature attributes. Biodiesel also shows increased dilution and polymerization of the engine sump oil, thus requiring more frequent oil changes. For all the above reasons, it is generally accepted that blends of standard diesel fuel with up to 20% (by volume) biodiesels can be safely used in existing diesel engines without any modifications, but there are concerns about the use of higher percentage blends that can limit the durability of various components.

Apart from the above-mentioned physico-chemical specifications, another major technical barrier associated with the use of biodiesel is its higher production cost (largely owing to the cost of the feedstock); a notable exception being when the feedstock used is waste cooking oil.

Alcohols

Alcohols are defined by the presence of a hydroxyl group (–OH) attached to one of the carbon atoms. *Ethanol*, in particular, (or ethyl alcohol C_2H_5OH) is a biomass based renewable fuel (bio-ethanol), which can be produced from corn, sugar cane, sugar beets, barley, and from (non-food) agricultural residues such as straw, feedstock and waste woods; the process is relatively easy and of low cost and is known as alcoholic fermentation (Hansen et al., 2005; Agarwal, 2007).

Because of its high octane number, ethanol was initially and primarily intended as a substitute for gasoline in spark-ignition engines. Nonetheless, it has been considered as a suitable fuel for compression ignition engines too, mainly in the form of blends with diesel fuel. It was its high capacity in PM emission reduction (originating in the high oxygen content of the corresponding fuel blend) that motivated the blending of ethanol with diesel fuel in CI engines (Meiring et al., 1983; Ecklund et al., 1984; Hansen et al., 2005; Agarwal, 2007; Rakopoulos et al., 2007; 2008).

In any case, there are several critical issues to consider with the use of ethanol in the diesel fuel. While anhydrous ethanol is soluble in gasoline, its miscibility in diesel fuel is problematic. This forms a critical drawback, since the lack of miscibility may cause phase separation between diesel fuel and ethanol, with serious consequences on the engine operation. To overcome this important drawback, additives in the form of emulsifiers or co-solvents are usually applied especially at low temperatures (below 10°C). Moreover, ethanol possesses lower flash point and lower viscosity than diesel fuel. Ethanol addition in the diesel fuel reduces the lubricity of the blend and creates potential wear problems in fuel pumps, particularly during starting. This is valid primarily in rotary and distributor-type pumps but also in modern common-rail systems that employ a fuel-based lubrication. Ethanol is also characterized by corrosiveness and a much lower cetane number that reduces the cetane level of the diesel/ethanol blend, thus requiring the use of cetane enhancing additives to improve ignition delay and mitigate cyclic irregularity (McCormick and Parish, 2001; Hansen et al., 2005).

In view of the serious disadvantages when using ethanol in diesel engines, a very challenging alcohol competitor has emerged in the form of *butanol* (butyl alcohol). Butanol is a biomass-based renewable fuel too that can be produced by alcoholic fermentation of sugar beet, sugar cane, corn, wheat (bio-butanol), although butanol produced from fossil fuels also exists. Butanol ($CH_3(CH_2)_3OH$) has a 4-carbon structure and is a higher-chain alcohol than ethanol, as the carbon atoms can either form a straight chain or a branched structure, thus resulting in different properties. Consequently, it exists as different isomers depending on the location of the hydroxyl group (–OH) and carbon chain structure, with butanol production from biomass tending to yield mainly straight chain molecules. 1-butanol, better known as n-butanol (normal butanol), has a straight-chain structure with the hydroxyl group (–OH) at the terminal carbon (Jin et al., 2011).

N-butanol is of particular interest as a renewable biofuel as it is less hydrophilic, and possesses higher energy content, higher cetane number, higher viscosity, lower vapor pressure, higher flash point and higher miscibility than ethanol, making it more preferable than ethanol for blending with diesel fuel. Therefore, the problems associated with ethanol mentioned in the previous paragraphs are solved to a considerable extent when using n-butanol, which is also less corrosive. However, at the moment, its production rate by ABE (acetone-butanol-ethanol) fermentation is much lower than that of the yeast ethanol fermentation process, a fact explaining the much more vigorous research on ethanol compared with n-butanol during the last decades, particularly after the petroleum crisis in the 1970s.

The literature concerning the use of n-butanol/diesel fuel blends in diesel engines and its effects on their steady-state performance and (exhaust) emissions has revealed the beneficial effects of using various blends of n-butanol with diesel fuel on smoke and CO emissions at various loads, however at the expense of higher NO_x and HC emissions. As was the case with ethanol, it is the high oxygen content of n-butanol that leads to enhanced in-cylinder soot oxidation, which is in turn responsible for the decrease in smoke emissions (Hansen et al., 2005; Rakopoulos et al., 2010b; Yao et al., 2010; Giakoumis et al., 2013).

Comparison of the Physical and Chemical Properties between the Various Biofuels and Conventional Diesel

Table 1 summarizes the most important physical and chemical properties of the three biofuels covered in this chapter against those of the reference diesel fuel. The following comments can be made with respect to the values demonstrated in this table.

1. All three biofuels are characterized (by default) by high oxygen content that leads to proportionally lower energy density (lower heating value). Thus, more fuel needs to be injected in order to achieve the same engine power output. Further, the inbound oxygen increases the air–fuel equivalence ratio and so lowers the exhaust gas temperatures (provided that the injection system retains its diesel-fuel calibration). The above arguments are more pronounced for ethanol that possesses the higher oxygen content.
2. All three biofuels have almost zero (biodiesel) or even zero (alcohols) natural sulfur content, which is considered a soot precursor. However, this advantage seems to fade away gradually owing to the continuous desulfurization of the reference diesel fuel.
3. Both alcohols exhibit a very low cetane number (CN) (and high octane number, accordingly); CN represents the ignitability of the fuel, with lower CN values leading to longer ignition delay. The increase in the premixed-phase of combustion originating in the longer ignition delay period of the alcohol blends results also in a proportionately higher amount of fuel burned under constant volume conditions, which entails higher cycle efficiency but also elevated combustion noise radiation (Rakopoulos et al., 2010a). The ignitability issues associated with the use of alcohols in diesel engines are more prominent during cold starting. On the other hand, biodiesels (in particular those methyl esters derived from saturated feedstocks) have higher CN values than diesel fuel. Higher cetane numbers promote faster auto-ignition of the fuel, and often lead to lower NO_x emissions (although no unambiguous trend has been established), particularly during low-load (i.e. premixed-controlled) engine operation (Graboski and McCormick, 1998; Lapuerta et al., 2008).
4. Both alcohols have lower density than petrodiesel, which means that volumetrically-operating fuel pumps inject smaller mass of alcohol than conventional diesel fuel. On the contrary, biodiesel is characterized by higher density, which leads to higher amount of injected fuel; this in turn will affect the air–fuel ratio, hence the local gas temperatures and NO_x emissions, as long as the engine retains its diesel-fuel calibration. The density of biodiesel exhibits a strong correlation with the degree of

unsaturation, with the density increasing with the increase in the number of double bonds. This means that the more unsaturated the originating oil (e.g. soybean or rapeseed or sunflower compared to tallow or coconut), the higher the density of the derived methyl ester.

5. Vegetable oils have high viscosity (one order of magnitude higher than the diesel fuel), which means that they cannot be used safely as fuels in a compression ignition engine, at least not without prior heating. The transesterification process, on the other hand, reduces considerably the viscosity of the FAME to levels comparable to (but still higher than) that of petrodiesel. The higher viscosity of biodiesel in relation to petrodiesel, leads to less accurate operation of the fuel injectors, poorer atomization of the fuel spray, increase in the Sauter mean diameter of the fuel droplets and of the break-up time; viscosity decreases with increasing unsaturation of the methyl ester.
6. A very important drawback of alcohols with respect to diesel (and biodiesel) is their low flash point, which is a measure of the temperature to which a fuel must be heated such that the mixture of vapor and air above the fuel can be ignited. Hence, ethanol (and to a smaller extent n-butanol) is less safe than diesel fuel in that respect, whereas biodiesel is much safer, with flash point values higher than 100°C.
7. Both alcohols (particularly ethanol) have higher heat of vaporization than diesel. Thus, larger amount of heat is needed to evaporate the liquid alcohol, which eventually leads to smaller amount of heat remaining for the increase of gas temperature.

Table 1. Primary physical and chemical properties of biodiesel, ethanol and n-butanol in comparison to the low-sulfur automotive diesel fuel (Hansen et al., 2005; Jin et al., 2011; Giakoumis, 2013)

	Low-sulfur automotive diesel fuel	Biodiesel	Ethanol C_2H_5OH	n-Butanol C_4H_9OH
Oxygen content (% weight)	0	10–12	34.7	21.6
Lower heating value (kJ/kg)	~43,000	36,500–39,500	26,800	33,000
Sulfur content (ppm)	<50 <10 for ultra low-sulfur diesel fuel	<10	0	0
Cetane number	~50	46–65	5–8	17
Octane number	~30	40–15	108	96
Density/15°C (kg/m^3)	820–850	870–890	789	810
Kinematic viscosity/40°C (cSt)	2–3.5	3.5–6.2	1.20	2.5
Flash point (°C)	50–90	120–180	13	35
Molecular weight (kg/kmol)	170	290 *	46	74
Stoichiometric air–fuel ratio	~15	12.5 *	9	11.2
Latent heat of evaporation (kJ/kg)	265	230 *	900	585
Boiling temperature (°C)	180–360	345 *	78	118
Bulk modulus of elasticity (bar)	16,000	17,500 *	13,200	15,000

* average values.

Lastly, all three biofuels do not contain any aromatic or poly-aromatic hydrocarbons, which act as soot precursors. Based on the physical and chemical properties alone, as these are documented in Table 1, n-butanol seems more appropriate than ethanol to be used in a diesel engine, although its lower oxygen content will lead to smaller PM emission benefits than ethanol when the same v/v ratio is applied (Giakoumis et al., 2013). Both alcohols, however, fall short of biodiesel, whose physico-chemical properties resemble more closely those of the conventional diesel fuel.

EFFECTS OF BIOFUEL BLENDS ON PM EMISSIONS DURING TRANSIENT OPERATION

Diesel particulates consist mainly of combustion generated carbonaceous material (soot) on which some organic and inorganic compounds have been absorbed; the organic ones arise mainly from unburned fuel and lubricating oil. The health effects of inhaling particulate matter include asthma, lung cancer, cardiovascular issues, respiratory diseases, birth defects, and even premature death. Contrary to the other regulated pollutants (carbon monoxide CO, unburned hydrocarbons HC and nitrogen oxides NO_x), particulates exist in solid rather than gaseous form. Further, they are distributed over a wide size range, typically from 20 nm to 10 μm, a fact that renders them respirable. A usual classification of the PM is in soluble organic fraction (SOF) and insoluble or dry fraction; the latter is often used as an estimation of soot. The soot percentage in the particulate matter from diesel exhaust varies, but is typically higher than 40–50%. Other particulate matter constituents are: unburned or partially burned fuel and lubricant oil, bound water, wear metals and fuel-derived sulfate.

Soot, in particular, or elemental or black carbon, is formed from unburned fuel that nucleates from the vapor phase to a solid phase in fuel-rich regions at elevated temperatures. Hydrocarbons or other available molecules may condense on, or be absorbed by soot depending on the surrounding conditions. The evolution of liquid- or vapor-phase hydrocarbons to solid soot particles, and possibly back to gas-phase products, involves six commonly identified processes, namely: pyrolysis, nucleation, surface growth, coagulation, agglomeration and oxidation (Tree and Svensson, 2007). Soot is responsible for the black color of the emitted exhaust from diesel-engined vehicles, and is substantially increased during turbocharged diesel engines accelerations (typical in passenger cars and trucks/buses) or load increase transients (typical in stationary or marine applications) (Rakopoulos and Giakoumis, 2009).

Most of the surveyed in this chapter works have dealt with engine emissions during transient/driving cycles. Such cycles require sophisticated experimental facilities (a fully automated test-bed with electronically controlled motoring and dissipating (chassis) dynamometer, exhaust gas analyzers, dilution tunnels, etc.) in order to be accurately reproduced. Figure 3 demonstrates graphically the PM experimental procedure during such a transient cycle measurement. Particulates are usually collected on filters that are weighed before and after the test; these filters must be conditioned with respect to temperature and humidity prior to weighing. On the other hand, when discrete transient tests are investigated, a high-response emission analyzer is required capable of continuous (smoke opacity) measurement in order to be able to capture the short-term emissions spikes.

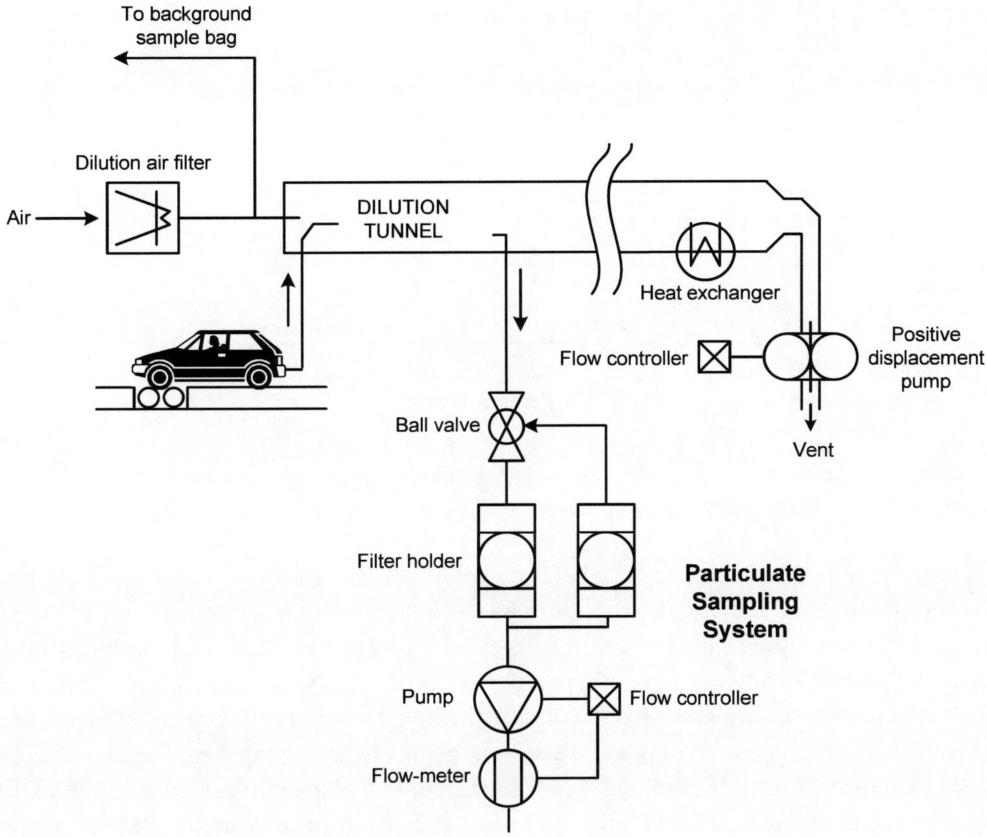

Figure 3. Typical constant volume sampling (CVS) configuration with positive displacement pump for PM emission measurements of diesel-engined vehicles.

Biodiesel Effects

The extended research on biodiesel combustion during steady-state diesel engine operation has revealed that the methyl esters exhibit significant capacity in mitigating the amount of the emitted PM (Graboski and McCormick, 1998; Agarwal, 2007; Hansen et al., 2009). In general, similar results have been reported during transient conditions too for both passenger cars and heavy-duty diesel engines (Giakoumis et al., 2012). A representative result is provided in Figure 4 that illustrates smoke opacity development during typical (fully warmed-up) acceleration of a medium-duty turbocharged diesel engine; smoke measurement was accomplished applying a high response analyzer particularly suited to transient experimentation (Rakopoulos et al., 2010a). Clearly, smoke was found to reduce with biodiesel addition in the fuel blend as was also the period of unacceptable smoky operation, and this was largely attributed to the extra fuel-bound oxygen that was available for combustion. Even higher reduction is expected, and has actually been documented, for higher biodiesel blend ratios (Giakoumis et al., 2012) (usually, the biodiesel blend ratio is identified with the letter B and a number that shows the amount of biodiesel volume in the fuel blend, e.g. B30 meaning 30% biodiesel and 70% diesel fuel per volume (v/v)).

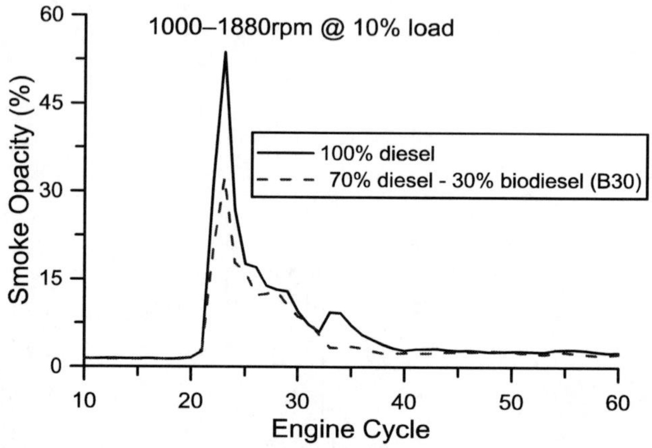

Figure 4. Smoke opacity during a 1000–1880 rpm acceleration at low load for a medium-duty turbocharged diesel engine using neat diesel and a B30 blend.

Similar results have been reported during load increase transient events (Armas et al., 2006), and for the vast majority of the transient cycles, as is representatively documented in Figure 5. The latter figure summarizes results from various American and European heavy-duty and light-duty cycles as have been reported in the literature for various engine and vehicle types (Graboski et al., 1996; Lujan et al., 2009; Kooter et al., 2011; Pelkmans et al., 2011); collective results concerning all the driving cycles tested so far regarding the effects of biodiesel combustion on the emitted PM compared to the neat diesel operation will be presented and discussed later in the text (Figures 9 and 10). In fact, with only few exceptions, the vast majority (90%) of all individual experimental measurements so far have reported decreasing PM emissions when biodiesel was added into the fuel (Giakoumis, 2012). It is also noteworthy mentioning that PM emissions were usually reduced proportionally to the oxygen content of the fuel blend, as is also demonstrated in Figure 5.

Despite the rather clear behavior of biodiesel blends on PM emissions during fully warmed-up transients, during *cold starting* the respective performance of biodiesel has been reported to exhibit an adverse trend compared to the one illustrated in Figures 4 and 5, i.e. an increase in the amount of the emitted PM has been documented. This behavior has been primarily attributed to the higher initial boiling point of biodiesel and its higher viscosity with respect to conventional diesel (Table 1). Specifically, the higher boiling point is responsible for more difficult fuel evaporation, whereas the higher viscosity reduces the rate of spray atomization, both of which prove particularly influential at low ambient temperatures. As a result, biodiesel combustion during starting leads to worse fuel–air mixing and, thus, more intense soot formation at low temperatures; the latter increase was measured, for example, up to 80% for an automotive, HSDI diesel engine, when comparing B100 with neat diesel operation (Armas et al., 2006), and is expected to increase the higher the biodiesel fuel blend ratio. Even more pronounced are the negative effects of cold starting on CO and HC, originating also in the diesel oxidation catalyst's lower efficiency with biodiesel blends (Giakoumis et al., 2012). Nonetheless, later in the warm-up phase, when the engine assumes its normal operating temperature, the advantages of biodiesel combustion prevail, and soot (CO and HC too) emissions decrease compared to the neat diesel fuel as already illustrated in Figures 4 and 5.

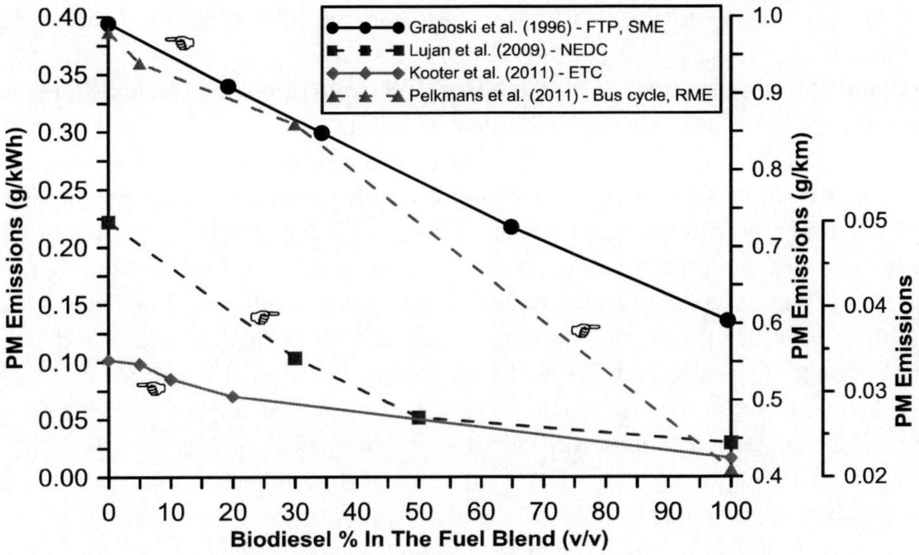

Figure 5. Particulate matter emissions during various driving cycles with biodiesel blends.

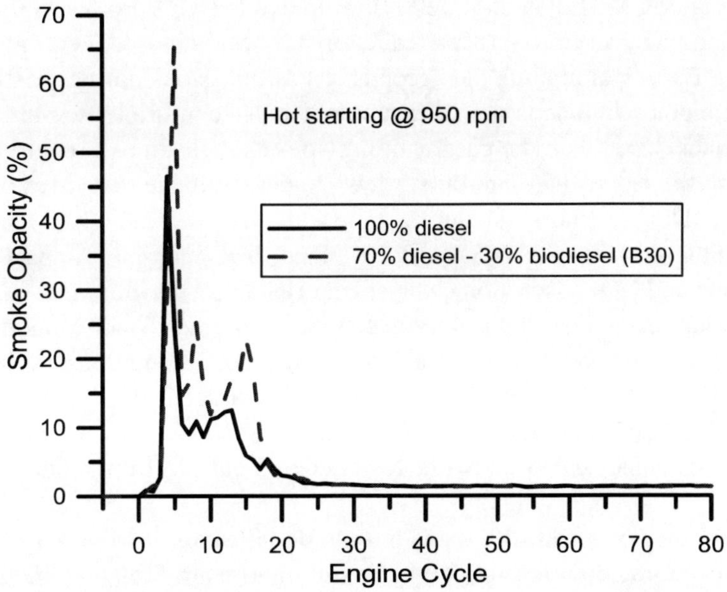

Figure 6. Development of smoke opacity during hot starting of a medium-duty turbocharged diesel engine for neat diesel and B30 operation

The peculiar cold-starting PM emissions with biodiesel combustion have been confirmed by Rakopoulos et al. (2011) during various hot-starting tests performed on an engine dynamometer for a medium-duty turbocharged bus/truck engine, as is demonstrated in Figure 6; it is obvious in this figure that starting not only increased the peak in the smoke opacity (from 45 to 65%) but also the duration of unacceptable black smoke emissions. Likewise during transient cycles, Graboski et al. (1996) measured higher PM emissions during the cold-started runs of the heavy-duty FTP compared to the hot tests, but in either case, a significant benefit relative to the neat diesel operation was established for each biodiesel blend tested.

Regarding the composition of the emitted PM from biodiesel combustion, and in particular its soluble organic fraction (SOF), most of the studies conducted so far have concluded that, despite the sometimes substantial PM decrease, biodiesel-fueled engines produce a higher fraction of SOF in their exhausted PM than when conventional diesel fuel is used; a typical summary of composite transient particulate composition is provided in Table 2 for a heavy-duty diesel engine running on the FTP transient cycle for two biodiesel blends and the nominal diesel fuel. SOF emissions generally increased with increasing biodiesel content, although it has been claimed that the exact percentage is engine dependent. Moreover, the lube-oil portion of the SOF was found to be unaffected by the biodiesel content. Although the exact mechanism is not absolutely clear, this higher SOF percentage of biodiesel has been attributed to its lower volatility (higher boiling point) as well as to heavy fuel-related organic compounds that remain intact through combustion (Sharp et al., 2000; Knothe et al., 2006; Karavalakis et al., 2007).

Interestingly, application of a modern, high-efficiency exhaust gas after-treatment system (e.g. in the form of a diesel particulate filter - DPF) being capable of reducing soot emissions to negligible levels, can practically downgrade to a large extent the biodiesel beneficial effects on PM (Lujan et al., 2009). Moreover, since Euro 5 specifications are even stricter regarding PM than Euro 4 (80% reduction has been imposed, i.e. from 25 to 5 mg/km, from Euro 4 to Euro 5 as regards the acceptable PM emissions during the NEDC), it seems that the use of biodiesel alone is not sufficient to achieve conformity to the current standards (as could, perhaps, have been the case with previous specifications), requiring a DPF in any case. Regarding the latter, there are only a few studies available that investigated the effects of biodiesel on DPF regeneration during transients. Tatur et al. (2008) found that under normal operating conditions, a SME B20 blend had marginal impact on both DPF regeneration rate and NO_x adsorption catalyst lean-rich cycle development for a passenger car running on various American, light-duty cycles. On the other hand, Muncrief et al. (2008), concluded that a diesel vehicle operating under low-load conditions (as the refuse truck studied) with neat cottonseed or soybean derived biodiesel had relatively cool exhaust to adequately oxidize the accumulated soot on the DPF with NO_2. It was speculated, however, that this might be counter-balanced by the lower engine-out PM from biodiesel combustion.

Table 2. Summary of composite transient particulate composition from a heavy-duty diesel engine during the FTP cycle (adapted from Sharp et al. (2000))

Fuel	DOC	Volatile Organic Fraction (g/kW h)				Non-volatile (g/kW h)	
		Total	Oil	Fuel & Other	PM	SO_3+H_2O	Soot
Diesel	No	0.0039	0.023	0.016	0.137	0.013	0.084
B20	No	0.0044	0.024	0.020	0.118	0.009	0.064
B100	No	0.0043	0.024	0.019	0.070	0.001	0.024
Diesel	Yes	0.0027	0.009	0.017	0.101	0.001	0.071
B20	Yes	0.0024	0.008	0.016	0.079	0.004	0.051
B100	Yes	0.0020	0.009	0.011	0.040	0.003	0.017

Alcohol Effects

Historically, ethanol was the first alcohol to be investigated in diesel engines blended into the diesel fuel, whereas research on the combustion and emissions of n-butanol blends with diesel fuel appeared recently. An initial investigation of ethanol/diesel fuel blend effects during transients was conducted by Armas et al. (2007), who studied discrete transient schedules (cold starting and load increase) of a passenger car engine running on E10 fuel blend (i.e. 10% ethanol/90% diesel fuel) without a stabilizing agent. This ethanol/diesel blend was proven successful in reducing both the peak and the final smoke opacity values throughout the transient event. Since the inlet air flows were measured for each fuel to behave similarly during the transient test, it was concluded that the reduced opacity values of E10 could be primarily attributed to the high oxygen content of ethanol (34.7% according to Table 1), and also to its lack of sulfur and aromatics that are well known to promote the creation of soot precursors. Greater reductions in smoke opacity should be expected if higher ethanol blends are added in the fuel, as revealed, for example, the results reached by Merritt et al. (2005) during the FTP smoke test for 3 non-road engines (Figure 7).

Ethanol addition in the fuel blend during cold starting, on the other hand, has proven more difficult compared to the reference diesel operation (Merritt et al., 2005; Armas et al., 2007), with the engine experiencing cranking for several seconds before firing. This is not surprising since the low ethanol cetane number is expected to worsen the engine behavior (spray penetration, wall impingement and fuel vaporization) the lower the coolant and cylinder wall temperatures. During cold starting, Armas et al. (2012) reported substantial smoke opacity increases when again 10% v/v ethanol (or 16% v/v n-butanol) was added in the fuel blend. A possible explanation provided was the fact that during cold starting the opacity detected was not only soot but rather contained high concentrations of hydrocarbons too. In contrast, ethanol addition in the fuel blend has been reported beneficial during hot starting (Armas et al., 2012), i.e. when the engine was fully warmed up.

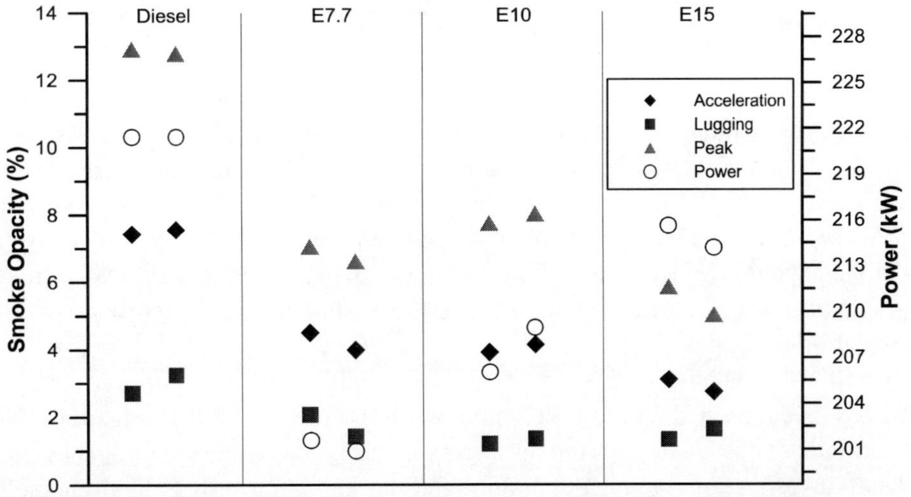

Figure 7. Cumulative smoke opacity and power results for a 8.1 L non-road engine running on three different ethanol/diesel blends during two runs of the FTP smoke test (adapted from Merritt et al. (2005)).

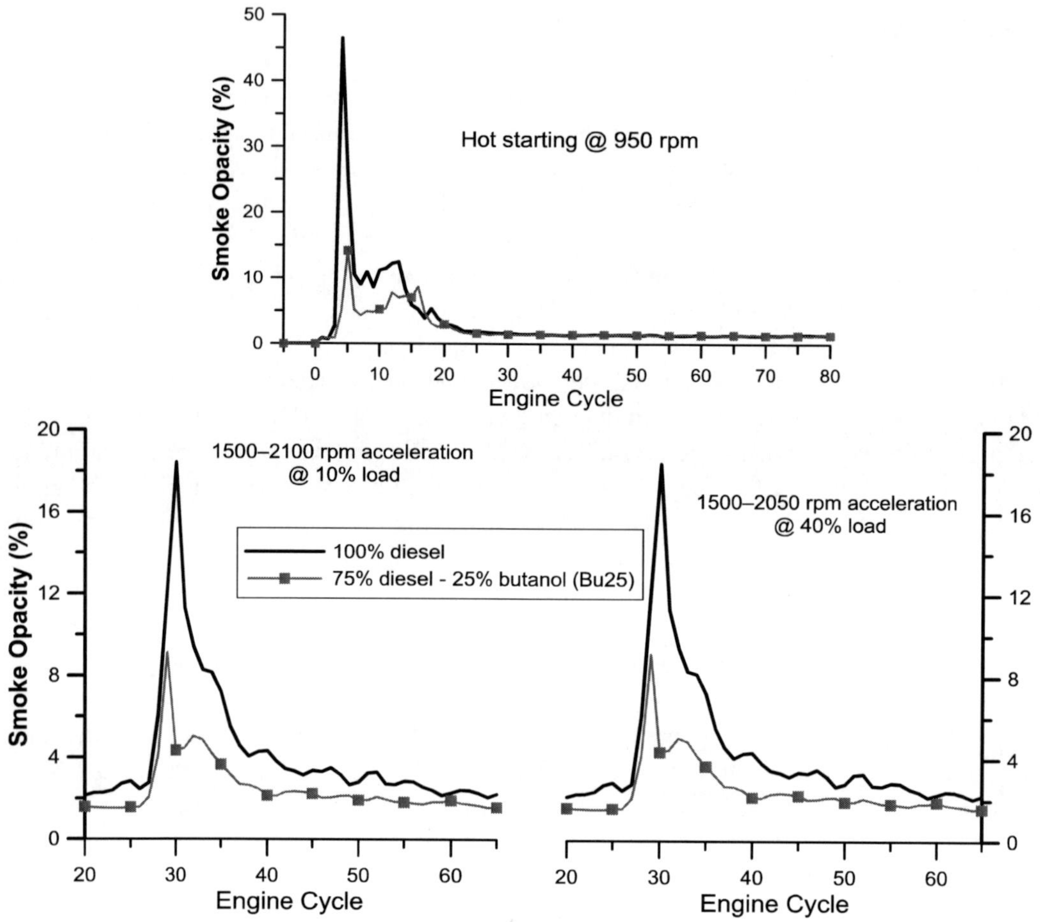

Figure 8. Development of smoke opacity during two accelerations and one hot starting event of a turbocharged diesel engine for neat diesel and a 25% n-butanol/75% diesel (Bu25) fuel blend (adapted from Rakopoulos et al. (2010b, 2011))

In parallel to the ethanol studies, Kozak (2011) and Miers et al. (2008) reported results from transient cycles studies with various n-butanol/diesel blends. The former investigated a 10% n-butanol/90% diesel (Bu10) fuel blend during the European passenger car NEDC. The blend was found capable of reducing PM emissions up to 21% with respect to the reference operation while, interestingly, maintaining the amount of emitted NO_x and CO_2 (g/km), however at the expense of increasing CO and HC emissions. Miers et al. (2008) applied even higher n-butanol ratios in the fuel blend (20 and 40%) studying various American cold and hot-started light-duty cycles.

An even more fundamental investigation into the effects of n-butanol on diesel engine emissions has been carried out by Rakopoulos et al. (2010b; 2011) regarding discrete transient schedules, in particular during acceleration and hot starting of a turbocharged diesel engine. These investigations revealed that during all accelerations tested, the n-butanol/diesel fuel blend emitted lower smoke than the respective neat diesel fuel operation (but with higher amounts of NO); figure 8 is a collective example of the results reached in these two studies. Its lower two diagrams illustrate the response of PM emissions from a medium-duty, turbocharged diesel engine during two medium-to-high speed accelerations, when running on neat diesel fuel and a blend of 25% n-butanol/75% diesel fuel (Bu25); the latter blend succeeded in decreasing soot opacity nearly 50% relative to the conventional diesel operation. Even better results (69% reduction), was reported during hot starting transients, as is demonstrated in the upper sub-diagram of Figure 8. Further, whereas soot opacity exceeded the 10% value for 10 engine cycles (1.9 s) for the neat diesel operation, for the Bu25 blend the respective period was 'only' 3 engine cycles (0.5 s) during the hot starting test. It was argued that it is most probably the lower viscosity of n-butanol and its higher volatility compared to diesel fuel that were responsible for the reduction in the soot production during hot starting, a finding confirmed also by Armas et al. (2012).

As was also the case with ethanol blends, a rough engine operation was noticed by Miers et al. (2008), during the butanol/diesel cold-started runs, particularly for high blending ratios (40% v/v), accompanied by reduced fuel economy that was attributed to increased misfire. Advanced injection techniques, with one pilot and one main injection, might prove beneficial in this case.

Emission Mechanisms

Based on the research conducted so far, and the introductory comments made in the previous two sub-sections, the beneficial effects of biofuel blends on smoke opacity values/PM emissions during transients can be primarily attributed to the following factors and mechanisms that are both fuel and transient operation-related.

- The most influential factor that is responsible for reduced soot emissions with biofuels is their inherent oxygen concentration that aids considerably the soot oxidation process. Soot formation caused by high temperature decomposition mainly takes place in the fuel-rich zone at high temperatures and pressures, within the core region of each fuel spray, effectively preventing carbon atoms from participating in the soot–precursor reactions. If the fuel is partially oxygenated, as is the case with biofuels, it possesses the ability to reduce the locally fuel-rich regions and limit soot nucleation early in the formation process, thus reducing PM emissions and smoke opacity.
- Further, and this is a mechanism related with the fundamentals of transient operation, the formation of soot is strongly dependent on engine load, with higher loads (e.g. cruising portions of the driving cycle) promoting higher temperature, longer duration of diffusion combustion (where particles are mostly formed) and lower overall oxygen availability (i.e. lower air–fuel equivalence ratio). The locally very low values of air–fuel equivalence ratio experienced during turbocharger lag at the onset of each acceleration and load increase, enhance the above mechanism, which is more pronounced the higher the engine rating, i.e., the higher the full-fueling to no-fueling difference. The excess oxygen of the biofuel blend, however, aids in maintaining these air–fuel equivalence ratio discrepancies during turbocharger lag (where soot is primarily produced (Rakopoulos and Giakoumis, 2009)) milder relative to the neat diesel transient operation, provided that the engine has retained its diesel-fuel calibration (as is the case with all the experimental measurements carried out so far).

This mechanism, however, is expected to act unfavorably as regards the production of NO_x (Giakoumis et al., 2012).

- The absence of aromatic (primarily) and sulfur compounds in the biofuels' molecule act in support of lower soot emissions, since these compounds are well known to promote soot, through their function as soot precursors. Nonetheless, this advantage of biofuels was more prominent during the previous decades, while nowadays it gradually faints following the continuous desulfurization of the base diesel fuel.
- Another contributing factor (as regards biodiesel) comes from the fact that combustion with methyl ester blends shifts to more controlled mode, which means that there is more time available after diffusion combustion for soot oxidation at a high temperature environment.
- Biodiesel and alcohols are also characterized by lower air–fuel equivalence ratios (Table 1) compared to the reference diesel fuel, a fact that is expected to reduce the possibility of fuel-rich regions in the non-uniform fuel–air mixture (Giakoumis et al., 2012).
- A further crucial factor is located in the biofuels' lower heating value. Bannister et al. (2010) found that the increase of the biodiesel percentage in the fuel blend led to an approximately linear reduction in the maximum tractive force at any given speed owing to the lower energy density of methyl esters. Consequently, when an engine that has been calibrated for neat diesel operation runs on biodiesel or any other biofuel such as alcohols, more fuel is required to achieve the demanded torque/vehicle speed. Similar conclusions can be derived from the power curves in Figure 7 as regards ethanol/diesel blends combustion. As a result, the ECU strategy may dictate an earlier start of injection and, more importantly, a decrease in the exhaust gas recirculation rate, since fueling is an input to the EGR control. Both measures result in elevated gas temperatures inside the cylinder that promote soot oxidation (again at the expense of NO_x). Sze et al. (2007) actually quantified this EGR effect. They found that the lower calorific value of biodiesel affected 'throttle' position (up to 3.3% relative to the petroleum diesel operation), and subsequently other relevant engine parameters, such as the fuel rail pressure, VGT position and boost pressure, and ultimately decreased the EGR rate up to 10.5% (depending on the investigated cycle) for the cases examined, which included biodiesel blends, but similar results are expected for alcohol/diesel blends.
- One specific factor that enhances the PM emission reduction with biodiesel blends in particular, is the different structure of the soot particles formed during the methyl ester combustion. An important factor affecting the DPF regeneration rate is the oxidative reactivity of particulate matter. Boehman et al. (2005) found that the use of a B20 blend during steady-state operation can significantly lower the balance point temperature (BPT; is the DPF inlet temperature at which the rate of particle oxidation approximately equals the rate of particle collection). They presented results showing that it is not the increased availability of NO_2 that is responsible for the decrease in BPT but rather the inherent differences in soot reactivity for different fuels, and specifically for B20, the more highly disordered soot nanostructure, such that the soot is more reactive or it is reactive at lower temperatures. The same research group further concluded that the more reactive toward oxidation behavior of biodiesel soot

derives from enhanced incorporation in the soot of surface oxygen functionality. These results were confirmed by Williams et al. (2006), again during steady-state operation; they showed that on average, the BPT is 45°C and 112°C lower, respectively, for B20 blends and neat biodiesel than for 2007 certification diesel fuel. Filter regeneration rate measurements indicated that biodiesel causes a significant increase in regeneration rate, even when B5 blends are employed. Overall, their results suggested significant benefits from the use of biodiesel blends in engines equipped with DPFs.

- Despite the significant positive effect of biodiesel on PM emissions, it has been reported in the literature that the biodiesel-bound oxygen may be actually under-utilized. This is due to the fact that methyl esters undergo decarboxylation, which yields a CO_2 molecule directly from the ester (Szybist et al., 2007). Thus, the oxygen in the biodiesel blend is used less effectively to remove carbon from the pool of soot precursors compared, for example, with ethanol or ether/diesel blends. As will be discussed in the next section (Figure 13) this is also reflected in the different PM emission benefit profile with different biofuel blends, i.e. it will be shown that combustion of different biofuels results in different PM benefits over the reference diesel fuel operation, even if the oxygen content in the fuel blend remains the same. This obviously implies that there are other decisive, and maybe not absolutely clear and understood, factors, apart from the oxygen content, that differentiate one biofuel's effects from the other and define the respective PM emission behavior. One such factor is probably located in the specific biofuel's molecular structure. For example, Shudo et al. (2009) found that when blending ethanol to PME, the flame region with high luminosity and high temperature shrank. This suggests that the ethanol blending increased premixed combustion and reduced the region with high local fuel-air equivalence ratio, which leads to the soot formation. Likewise, Botero et al. (2012) reported that the addition of ethanol to diesel fuel delayed the onset of the yellow luminosity of the flame, because of its volatility and thereby preferential gasification. This indicates a corresponding reduction in soot formation during the early stage of burning, when the flame size and thereby soot content are large, which was not evident during their experiments with biodiesel. The fact that alcohols possess lower cetane number and higher heat of evaporation means that the ignition delay period increases when an alcohol is added into the fuel blend, increasing accordingly the amount of fuel burned during the premixed phase of combustion. Consequently, a reduction in the diffusion phase is experienced, where the majority of soot is normally produced.

Statistical Analysis of the Measured Data

Motivation and Methodology

In 2002, the US Environmental Protection Agency (EPA) published a comprehensive analysis of biodiesel impacts on exhaust emissions, with the available at that time emissions data collected and analyzed in order to quantify the effects of biodiesel blends on all regulated pollutants. Although that work was primarily intended for the US market, and at the time the vast majority of the investigations concerned heavy-duty engines running on SME blends, the calculated best-fit curves have since been used by researchers all over the world in order to demonstrate, or even predict, the expected emission benefit or penalty when biodiesel is added into the fuel blend. Recently, an update of these emission predictions was accomplished (Giakoumis, 2012), in order to broaden the engine, vehicle and biodiesel feedstock database, and include recent results from newer production engines equipped with common rail injection systems, EGR control and modern antipollution technology running on various driving cycles that also incorporate cold starting.

Moreover, a similar procedure was adopted for the first time for ethanol and n-butanol/diesel combustion studies (Giakoumis et al., 2013)). Overall, 67 papers, published in international Journals, well established Conferences and final Reports issued by renowned research centers, all dealing with transient/driving cycles experimentation were gathered regarding biodiesel blends combustion (studies up to the end of 2011), and 20 studies concerning alcohol-diesel blends. A complete list of the papers/reports included in the two databases is available in (Giakoumis, 2012; Giakoumis et al., 2013). All of the studies concerned four-stroke, direct injection engines running on various chassis or engine dynamometer transient cycles, most notably the American heavy-duty engine-dynamometer FTP, the European light-duty chassis dynamometer NEDC and the American heavy-duty chassis-dynamometer UDDS.

For those cases where more than one sets of measurements were available (typically in final reports), average data were used, taking into account possible diesel fuel drift effects. If the vehicle was fitted with a diesel particulate filter (DPF), PM emission data were only collected if they were available upstream of the DPF. As regards the (many) cases with a diesel oxidation catalyst (DOC), engine-out but also DOC-out data have been included in the database (sometimes both were available for the same investigation). On the other hand, some markedly 'extreme' data were only taken into account if the researchers had confirmed in their study the validity of these results with a second or third run. Data were not taken into account if the researchers had used additives, e.g. cetane improvers in the (biodiesel) blend; on the contrary, a small percentage of additives is, in any case required for ethanol/diesel blends to avoid separation of the mixture.

For the estimation of the best-fit curves, a regression analysis was chosen in order to be able to also provide the coefficient of determination (R^2) that indicates the degree of data variability. In particular, a quadratic best-fit curve was selected for most of the cases examined; the latter is believed to be a good combination between simplicity and satisfactory regression capability. There is another critical point that should be taken into account for the discussion of the results in the next two sub-sections. *All available measurements in the database with various biofuel blends combustion concern engines that were initially calibrated for the use of conventional diesel fuel, and did not undergo any kind of modification prior to the biofuel blend operation.* More details about the procedure, the variability of the data, the derived best-fit curves, as well as an expansion of the analysis to the other regulated pollutants (CO, HC, and NO_x) can be found in (Giakoumis, 2012; Giakoumis et al., 2012; 2013).

RESULTS

Biodiesel

The statistical analysis of all PM emission measurements during transients with various biodiesel/diesel blends yields the quadratic best-fit curve presented in Figure 9. In this figure, all available individual measurements are also provided, from which the variability of the results can be assessed. Roughly 30% of the experimentations focused on the American FTP heavy-duty, transient cycle, and another one quarter on the European passenger car NEDC. The most investigated biodiesel blends are B20 and B100. Further, the most popular methyl ester in the research so far is soy-derived; it has been the subject in half of the studies and 40% of all investigated biodiesels during transients (Giakoumis, 2012; Giakoumis et al., 2013).

Overall, although a rather high degree of variation is observed in Figure 9, and documented in the R^2 value (due to the fact that data from all kinds of cycles, engines, and originating oils has been included), a clear decreasing trend of the PM emissions with rising biodiesel blend ratios is established. Interestingly, although a considerable amount of new data has been included in the current statistical analysis, the earlier EPA best-fit curve (discontinuous red line in Figure 9) is only slightly affected for blend ratios up to B50. One of the main reasons for the PM differences observed for higher than B50 blend ratios is probably located in the fuel sulfur content. During the 90s, road commercial No 2 diesel in the US (or its equivalent in the EU) contained up to 500 ppm sulfur in contrast to less than 50 ppm nowadays, or even less than 15 ppm for ultra-low sulfur diesel. This fact has accordingly lowered the respective diesel PM reference level and reduced the differences with biodiesel.

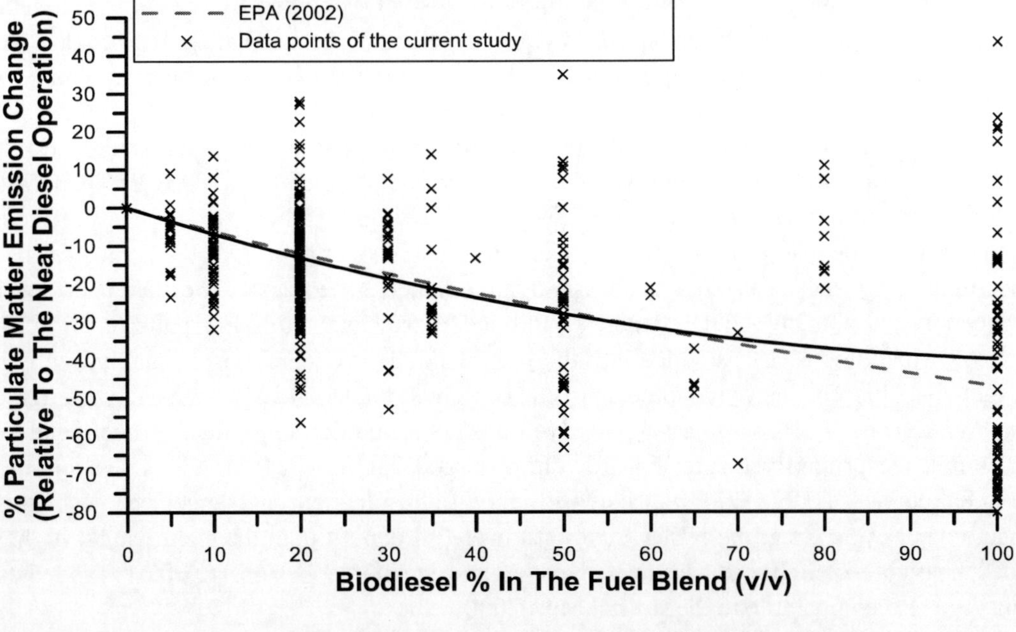

Figure 9. Collective results of PM reduction when using various biodiesel-diesel blends.

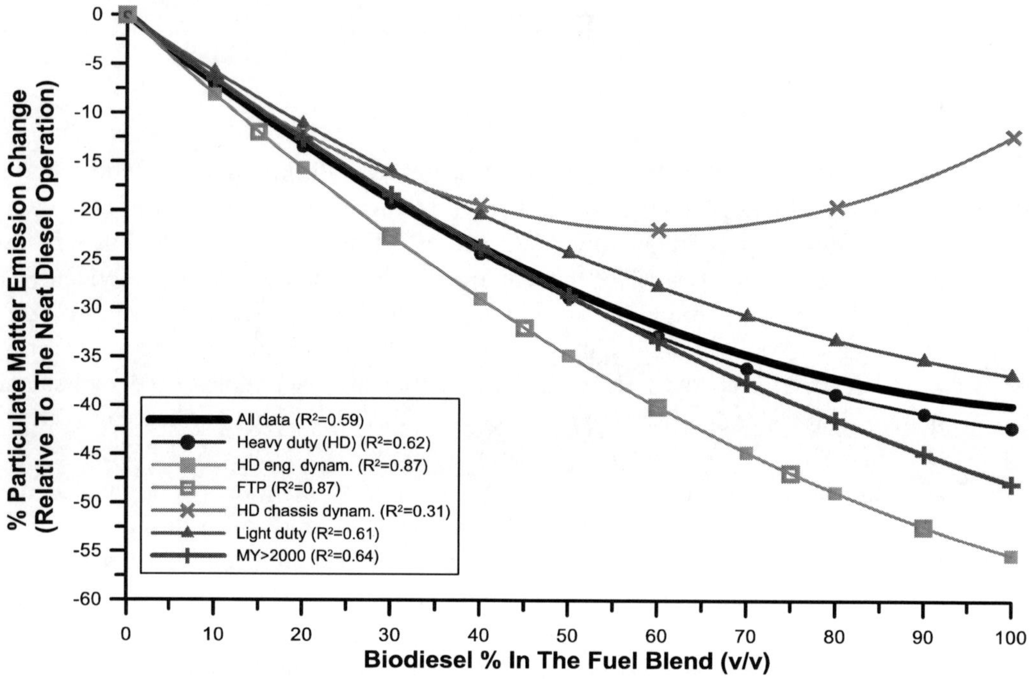

Figure 10. Collective best-fit PM results for various transient cycles and engine types.

An expansion of the overall results is demonstrated in Figure 10 that provides the effects of engine type, engine model year and driving cycle schedule on the PM emissions from biodiesel-blended engines. The heavy-duty engine dynamometer and the FTP curves in Figure 10 practically coincide, since very few data are available from non-FTP, heavy-duty, engine-dynamometer cycles. It is not surprising that these data correlate very well with the earlier EPA trend-line in Figure 9, which almost exclusively included FTP results. In general, the FTP data were found the most cohesive (coefficient of determination R^2=0.87). Since these FTP tests were conducted applying a rather narrow range of engines running mainly on SME blends, it is not surprising that the results exhibit minimal disparity.

On the other hand, inclusion of heavy-duty, chassis-dynamometer data (mainly results from the American heavy-duty chassis-dynamometer UDDS cycle) shifts the PM benefit to much lower values, whereas R^2 drops to 0.62 if all heavy-duty PM emissions (irrespective of dynamometer schedule employed) are taken under consideration (R^2=0.31 for the data concerning only the heavy-duty, chassis-dynamometer schedules). The latter effect is produced because many UDDS studies exhibited a ‘reverse’ emission pattern, with PM increases when biodiesel was added into the fuel blend. This occurs most probably owing to the fact that this chassis-dynamometer cycle is lighter loaded than the respective engine-dynamometer one (FTP), with less aggressive transient schedules and milder turbocharger lag phases that are primarily responsible for turbocharged engines PM emissions (Rakopoulos and Giakoumis, 2006; 2009). Likewise, the lightly-loaded passenger car, chassis-dynamometer cycles (e.g. the NEDC) exhibit lower PM benefit than the significantly higher-loaded heavy-duty engine dynamometer cycles and smaller cohesion (R^2=0.61), and these results are also comparable to the overall MY>2000 values.

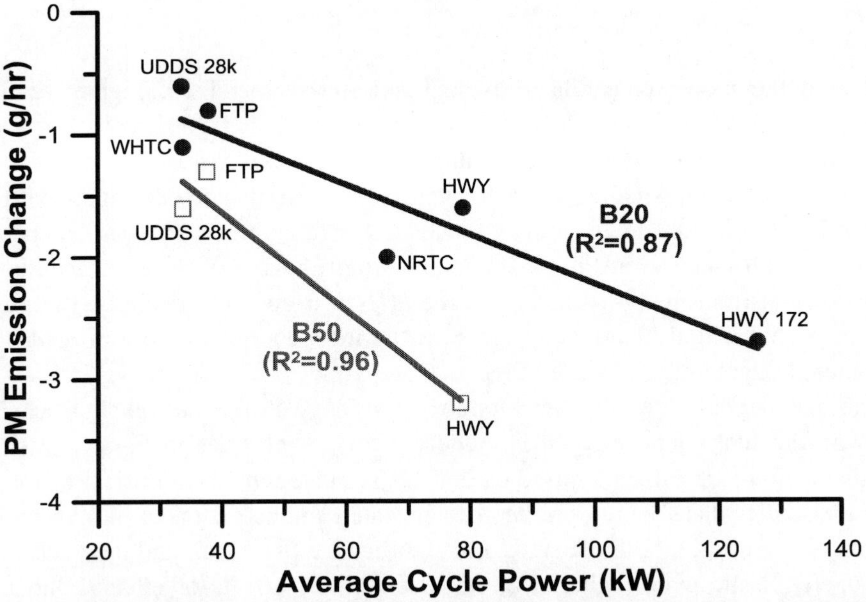

Figure 11. Effect of average cycle power on PM emissions for a 2006 medium-duty diesel engine running on various transient cycles (adapted from Sze et al. (2007)).

An interesting investigation of the transient cycle effects on the biodiesel PM emission benefit has been provided by Sze et al. (2007), who studied a 2006 MY engine running on 7 different heavy-duty engine and chassis-dynamometer cycles. A value of 0.87 (for B20 blends) and 0.96 (for B50) was established for the coefficient of determination of PM emissions with respect to the average power cycle, as it is demonstrated in Figure 11. It seems then that the higher the average cycle power (i.e. the more aggressive the cycle), the higher the benefit from biodiesel-blended fuels on PM emissions relative to the neat diesel oil. Obviously, the higher the loading or the aggressiveness of a cycle, the lower the air–fuel equivalence ratios or the harsher and the more frequent the turbocharger lag phases induced respectively; both lead to more intense soot production, where the beneficial effects of biodiesel (most importantly its increased oxygen content) prevail over the neat diesel operation. Moreover, high loadings and abrupt accelerations both dictate lower EGR rates through the usually applied ECU calibration, which again act in favor of lower PM emission rates (at the expense of NO_x).

It is also worth mentioning that contrary to NO_x emissions that have been found to correlate satisfactorily with the degree of unsaturation of the methyl ester used (Giakoumis. 2012; Giakoumis et al., 2013) (with higher feedstock saturation leading to lower NO_x emissions), no such correlation could be established between biodiesel feedstock and transient emissions. This finding suggests that the positive effects of biodiesel on PM are most probably not associated with specific 'internal' physical or chemical attributes of the fuel, such as, for example, the number of double bonds or the iodine value, but rather depend on the oxygen content. Hence, methyl esters with different molecular structure or chain length that have the same oxygen content, succeed equally well in reducing the emitted PM.

Alcohols

Further to the biodiesel results provided in Figures 9 and 10, Figure 12 illustrates graphically the respective alcohol/diesel blend overall results as regards both PM and smoke changes relative to the neat diesel operation during transients.

Owing to the much smaller range of research on transient emissions with various alcohol/diesel blends so far, the database contains rather few observations (in comparison, the respective biodiesel database in Figure 9 contained more than 400 different measurements). In order to demonstrate in the same figure the effects from both alcohols, the per weight oxygen content of the fuel blend was chosen as the independent variable instead of the per volume content in the biodiesel-related Figures 9 and 10.

As was also the case with biodiesel, a fairly high degree of statistical significance of the PM data with the fuel-bound oxygen is evident in the graph-lines in Figure 12 (R^2=0.85), implying that from the arguments raised earlier this is indeed the most critical factor for each biofuel. If the studies that focused on the non-legislated smoke are taken into account (upper sub-diagram of Figure 12), both the statistical significance (R^2=0.95) and the benefit over the reference operation are even higher. Not surprisingly, the effects of ethanol and n-butanol were found almost identical as regards the PM emission benefit, provided that the oxygen content in the fuel blend was the same.

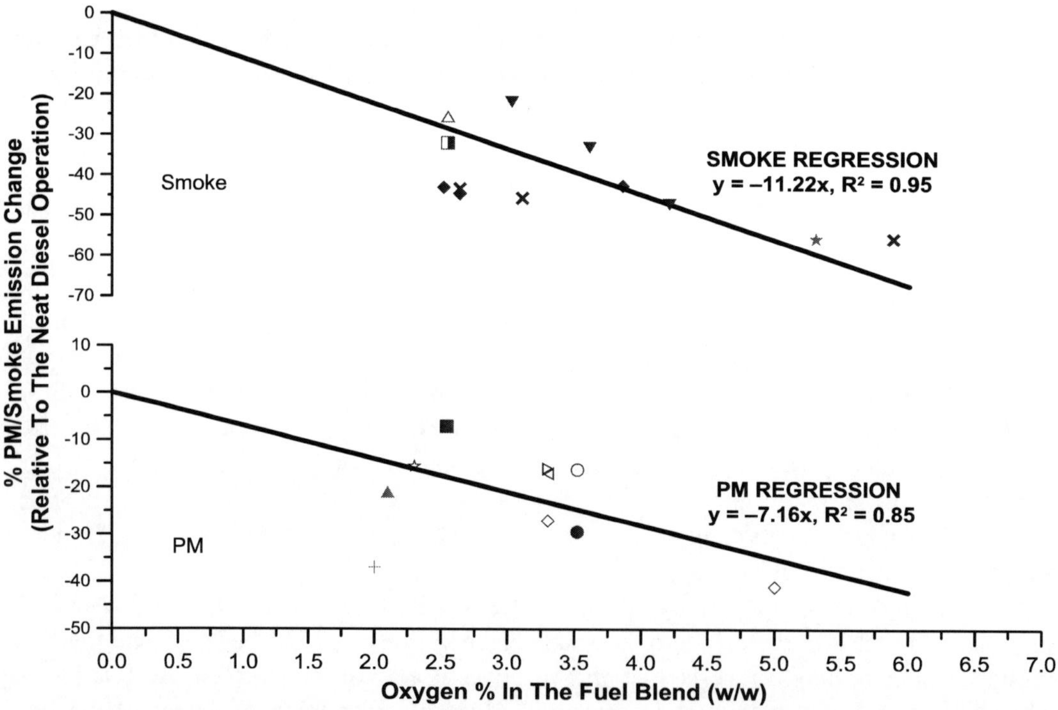

Figure 12. Collective PM (lower sub-diagram) and smoke (upper sub-diagram) emission benefits from ethanol and n-butanol/diesel fuel blends combustion during transient schedules.

Comparison between Biodiesel and Alcohol PM Reduction Results

Lastly, Figure 13 compares the results from Figures 9 and 12, providing further insight into the relevant phenomena and supporting the arguments made earlier regarding each biofuel's specific effects on the transient PM emissions.

In Figure 13, the overall transient cycle best-fit curves from biodiesel and alcohol/diesel blends combustion are directly compared. The per weight oxygen content (w/w) was chosen again as the independent variable in order to be able to compare results from different biofuels. From Figure 13 it can be concluded that an alcohol/diesel blend with 3.3% w/w oxygen content (roughly 10% ethanol (E10) or 16% n-butanol (Bu16) v/v in the final blend, assuming zero oxygen for the neat diesel fuel) is expected to decrease PM by 24%. Likewise, a 5% w/w oxygen content (approximately 15% ethanol (E15) or 24% n-butanol (Bu24) v/v) reduces the amount of the emitted PM relative to the neat diesel fuel operation by 36%. In contrast, the respective benefit from biodiesel combustion assumes a comparable value of 36% when a B70 blend is applied, which corresponds to 7.8% w/w oxygen content.

These very interesting findings prove that when comparing different biofuels, it is not the amount of oxygen alone that is responsible for higher or more efficient in-cylinder soot oxidation rate. Alternatively, the fuel-bound oxygen does not influence the PM emission mitigation in the same manner for all biofuels. There are other, inherent in the molecule of each (oxygenated) fuel, contributing parameters (e.g. flame structure and velocity or the balance between premixed and diffusion combustion) that ultimately differentiate one biofuel from the other and produce the final benefit over the reference diesel operation (Barrienteros et al., 2013). Hence, it is the attributes inherent in the combustion of alcohols, most notably their propensity towards premixed-flame combustion combined with the hydroxyl group (–OH), that differentiate the PM emission benefit over biodiesel combustion, increasing the capacity for reduced exhaust gas smokiness over the reference diesel fuel operation. On the other hand, as was argued earlier in this chapter, since the methyl esters during biodiesel blends combustion undergo decarboxylation that yields a CO_2 molecule directly from the ester, the respective biodiesel benefit is smaller than the one from the alcohol combustion (Szybist et al., 2007). Even greater PM reduction capability has been reported from ether/diesel blends combustion (Liotta and Montalvo, 1993; Rakopoulos et al., 2012).

Particle Size Distribution

Historically, diesel particulate matter legislation has been based on the emitted particle mass (g/km or g/kWh). Nonetheless, particle size distribution has gained increased attention recently in terms of air quality, as it is believed that the toxicity increases as the particle size decreases. In fact, a correlation between elevated ambient particulate matter concentration and hospital admissions has been suggested (Lave and Seskin, 1973). To this aim, the EU has legislated an intermediate Euro 5b specification level that also includes a particle number limit of the order of $6x10^{11}$/km over the NEDC for both passenger cars (category M) and light-duty vehicles (categories N_1 and N_2), applicable from September 2011.

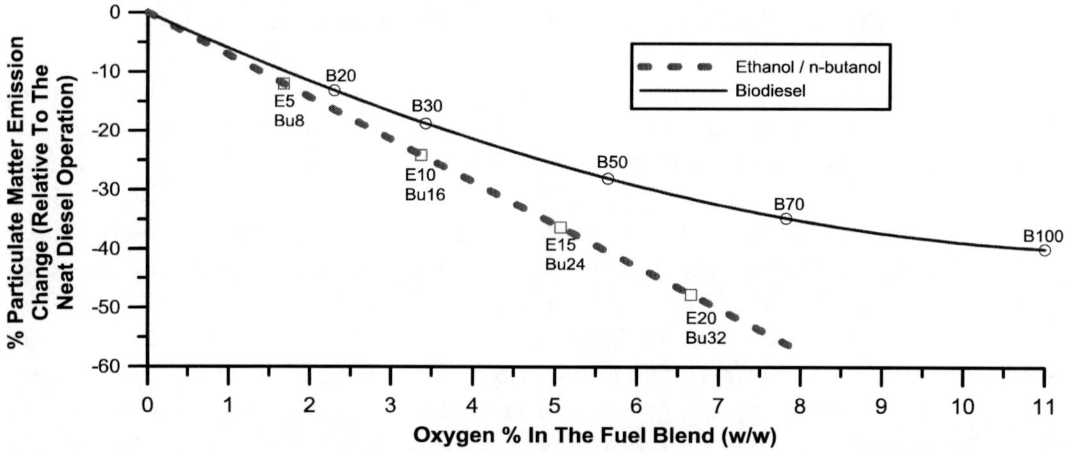

Figure 13. Comparison in the PM emission benefit over the neat diesel operation during driving cycles between ethanol or n-butanol/diesel blends (discontinuous red line) and biodiesel blends (solid blue line).

Unlike mass, particle number is not conserved. Within the exhaust pipe of a diesel engine, the continuously changing conditions may lead to particulates nucleation (resulting in an increase in both the particle number and mass concentration), condensation and coagulation (resulting in a reduction in the number concentration and increase in particulate size).

Most of the particle mass exists in the so-called accumulation mode in the 0.1–0.3 μm diameter range. This is where the carbonaceous agglomerates and associated adsorbed materials reside. The nuclei mode typically consists of particles in the 0.005–0.05 μm diameter range. This mode usually consists of volatile organic hydrocarbon and sulfur compounds that form during exhaust dilution and cooling, and may also contain solid carbon and metal compounds. By number, more than 90% of the particles emitted by a diesel engine fall into the nuclei mode, which, however, contains only 1–20% of the particle mass. The coarse mode contains 5–20% of the particle mass; it consists of accumulation mode particles that have been deposited on cylinder and exhaust system surfaces and later re-entrained (Kittelson, 1998).

Among the major factors that have been found to influence both particle number concentration and size distribution is the type of fuel used, with the ultra-low sulfur fuel being preferable for overall decreased PM emissions, although an increase in the nanoparticles concentration has been documented. The same holds true for higher fuel injection pressures or lower EGR rates. Whereas diesel particulate filters can reduce particle number concentration by 1 to 2 orders of magnitude, mainly in the accumulation mode, concern has been expressed that the current DPF technology may be allowing nanoparticles to pass through, or even producing them at high engine loads during the fast oxidation process, e.g., through oxidative fragmentation of soot aggregates.

Recent steady-state experimentation has, likewise, indicated that the decrease of the emitted PM from biodiesel use is usually associated with an increase in the number of the more toxic nanoparticles (Krahl et al., 2002; Tsolakis et al., 2006) although the trend is not unambiguous (Lapuerta et al., 2008; Giakoumis et al., 2012). In other words, it seems that a trade-off may exist between particle number distribution and mass, with lower aerodynamic diameter particles favored with decreasing PM mass emissions. This fact rather suggests that biodiesel (or other PM-reducing techniques, such as the DPF) actually decompose larger (visible) particles into many nanoparticles that are non-visible (hence, the 'whiter' image of the exhaust gas) but may be more (or equally) hazardous. Luckily, typical residence time in the atmosphere for 10 nm (nano) particles is only about 15 min.

As with almost all aspects of transient operation, particle concentration and size/mass distribution differentiates from the respective steady-state conditions, with changing load and engine speed affecting strongly both the number and size distribution of the emitted particles. For example, Fontaras et al. (2009) found that the use of B100 SME increased total particle number up to 200% compared with the neat diesel operation during the NEDC, and a B50 blend behaved similarly. This increase in the particle number arose from a general shift towards smaller-diameter (nano)particles and was evident for all transient cycles examined. Possible causes for this behavior were argued to be the higher viscosity of biodiesel, the increase of the SOF, the increase of the injection pressure and the advance in the injection timing with rising biodiesel percentage in the fuel blend. Similar results were reached by Tinsdale et al. (2010), who reported a 25% increase in the nucleation mode particle number over the NEDC for a B30 blend compared with the neat diesel case. Likewise, Chien et al. (2009), using MOUDI and nano-MOUDI devices, concluded that as the (waste-cooking derived) biodiesel percentage increased, the ultra-fine and nanoparticles number increased too; in fact, when neat biodiesel was used, nanoparticles dominated the size distribution.

Similarly to the steady-state tests, contradicting results have been reported during transients too by some researches, who measured decreasing number of nanoparticles with biodiesel addition in the fuel blend. One such example was reported by Macor et al. (2011), who found that the total number of emitted particles decreased for both tested vehicles between 10–20% over the examined driving cycles (NEDC, Artemis) relative to the neat diesel case. A possible explanation that has been argued by some researchers in such cases is the absence of sulfur in the biodiesel. The respective size distribution profiles from the same study revealed that the larger-diameter particles (for all fuel blends) were produced during those sections that include frequent accelerations or high power. This behavior can be explained by the fact that generally the trend is towards larger particles with increasing load, whereas nanoparticles are favored mainly at idling conditions (Rakopoulos and Giakoumis, 2009). As the load increases and more fuel is injected, the formation of larger particles is favored, owing to: a) the longer diffusion combustion duration, b) the higher combustion temperatures, and c) the reduced oxidation rate of the soot in the expansion stroke, since there is less time available after the end of the diffusion combustion and also lower oxygen availability.

An expansion of the previous results was provided by Fontaras et al. (2010) who also included biodiesel feedstock effects on the particle number concentration investigation. It was found that the effect of biodiesel on total particle number, including volatile and semi-volatile particles, was variable. Although reductions were observed over the low-power (sections of the) cycles, and for sunflower and unused frying methyl esters, the PME and the very popular in Europe RME blends were associated with up to 3 times higher particle number emissions than the base diesel fuel over the Artemis Motorway test.

As regards alcohol/diesel blends effects on the emitted particle size distribution, on the other hand, the only available result during transients has been reported by Armas et al. (2012) as regards discrete transient events; no clear picture could be established however, as to the effects of ethanol or n-butanol addition in the fuel blend on the emitted particles distribution.

Conclusion

An assessment of the literature concerning PM emissions from diesel engines and diesel-engined vehicles when running on various biofuel/diesel blends during transient conditions was conducted. Three very popular biofuels were covered in the investigation of this chapter, namely biodiesel, ethanol and n-butanol. The investigation covered all types of transient operation, from discrete transients (such as accelerations, load increases and starting) up to the more complicated cases of engine or vehicular transient cycles. The primary mechanisms of PM exhaust emissions were identified and discussed; further, they were inter-related with the inherent discrepancies observed during transients, most notably turbocharger lag. The most important of the conclusions derived are summarized below:

- Confirming the principal observations during steady-state operation, for the majority of the transients a decreasing trend in PM emissions is established when the biodiesel or the alcohol percentage in the fuel blend increases, at least for fully warmed-up conditions. During starting, however, biodiesel combustion exhibits an adverse PM behavior.
- It is primarily the increased oxygen concentration in the biofuels blend, which aids the soot oxidation process (most prominently during the critical turbocharger lag cycles) that has been identified as the key contributor for the observed PM emission mitigation. The absence of aromatic compounds and sulfur, and the lower stoichiometric air–fuel ratio aid also in the general decreasing PM emission trend. In particular for biodiesel, its specific soot structure has also been found to contribute to the lower emitted PM.
- Owing to the lower calorific value of biofuels, when an engine that has been calibrated for neat diesel operation runs on biodiesel or alcohol or n-butanol fuel blends, a lower EGR rate is established, therefore contributing towards a decrease in PM (and an increase in NO_x emissions) in relation to the conventional diesel operation.
- Biodiesel-blended engines produce a higher fraction of soluble organic fraction in their exhausted particulate matter during transients than when petroleum-based diesel fuel is used, even when the total particulate emissions are lowered.

- There is no unanimous trend with regards to the effects of biodiesel on particle number concentration and distribution, as is also the result during steady-state conditions; a slight majority (in the limited number of transient investigations so far) suggests that an increase in the number of nanoparticles is observed with increasing biodiesel blend.
- Based on a large amount of published data from four-stroke engines during the last 25–30 years, correlations were reached for quantification of biodiesel and alcohol benefits on the emitted particulate matter or smoke relative to the neat diesel operation, providing useful results for future planning. From this statistical analysis it was concluded that when a fuel blend with a specific (per weight) oxygen content is used, the PM emission benefit over the reference diesel operation is higher with ethanol or n-butanol than with biodiesel. The latter behavior most probably originates in the propensity of the alcohols towards premixed-flame combustion combined with their hydroxyl group (–OH).
- Irrespective of driving cycle type, the biodiesel impact on PM emissions appears to be related to the fuel consumption or the average cycle load. In particular, the PM benefit with biodiesel appears to increase for more aggressive cycles/driving patterns that incorporate steeper and more frequent accelerations or load increase events, hence harsher turbocharger lag phases that also provoke lower EGR rates. On the other hand, no correlation has been established between biodiesel feedstock and transient PM emissions. This suggests that the positive effects of biodiesel on PM are most probably not associated with physical or chemical attributes of the fuel but rather depend on the inherent oxygen concentration.

APPENDIX. TRANSIENT CYCLE DATA

A transient test cycle is a sequence of test points, each with a defined vehicle speed to be followed by the vehicle under study, or with a defined rotational speed/torque to be followed by the engine under transient conditions; these test points are divided in time steps, usually seconds, during which acceleration is assumed constant. Transient cycles are employed for type approval of new engines/vehicles, and are usually characterized by long duration (up to 30 minutes). In order for the exhaust emission measurements to be representative of real engine operation, transient test cycles usually incorporate:

- Cold or hot starting,
- Frequent accelerations and decelerations,
- Changes of load,
- Idling conditions typical of urban driving,
- Sub-urban or rural driving schedule, and
- Motorway driving.

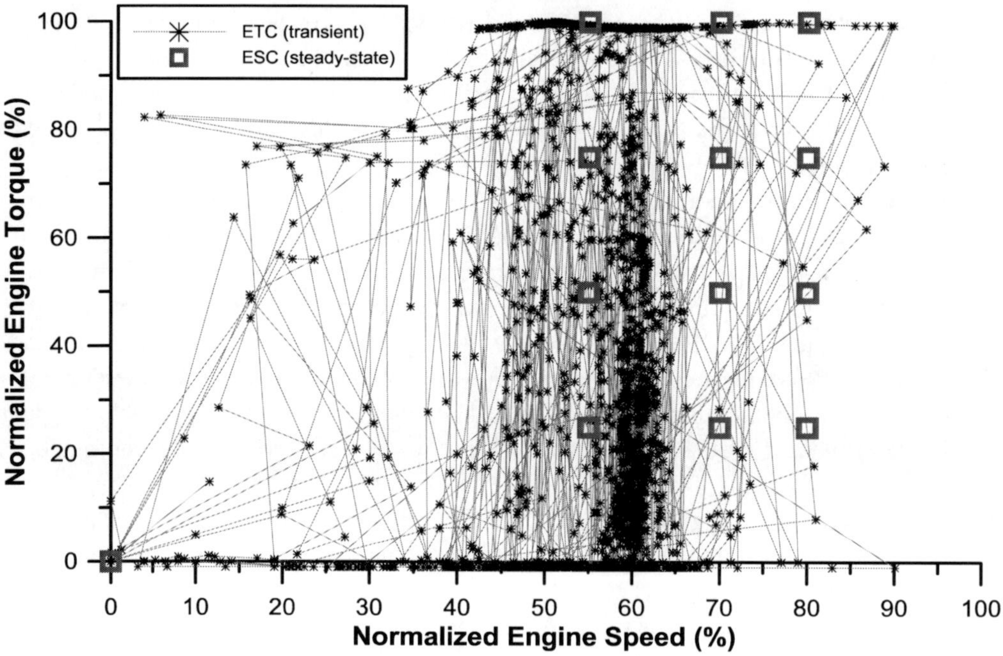

Figure A1. Comparison between the transient (ETC) and the steady-state (ESC) cycles for heavy-duty engines certification in the EU.

By applying a transient cycle for the testing of new vehicles, the complete engine operating range is tested and not just the maximum or some specific power/torque operating points. Figure A1 demonstrates in an unambiguous way the significant dissimilarities encountered when the engine is certified using a transient cycle, in this case the ETC (used for the certification of heavy-duty engines in the EU) compared to its steady-state counterpart, the ESC. It is made obvious that during a transient cycle, the serious discrepancies that are experienced during abrupt transients are taken into account.

Passenger cars and light-duty vehicles usually undergo a vehicle speed vs. time test cycle on a chassis dynamometer and the results are expressed in g/km. Since vehicle testing is much more difficult for heavy-duty or non-road vehicles, the exhaust emission certification procedure for the latter usually makes use of an engine rather than a vehicle cycle; this is realized on an engine test bed, where the engine under study follows a prescribed engine speed/torque vs. time procedure and the results are usually expressed in g/kWh. Figures A2 (engine dynamometer) and A3 (chassis dynamometer) demonstrate the time pattern of the most important transient cycles discussed in this chapter.

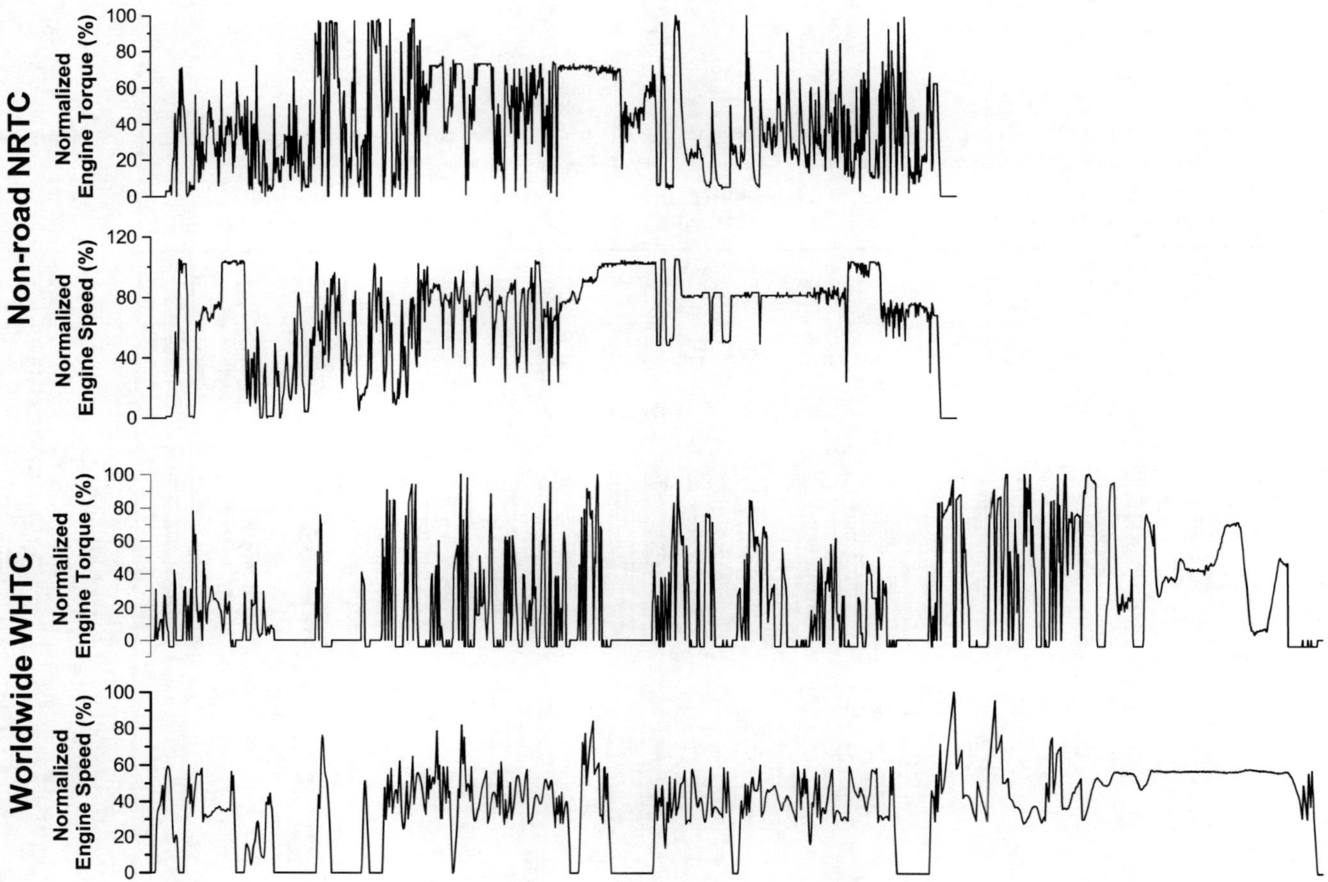

Figure A2. (Continued)

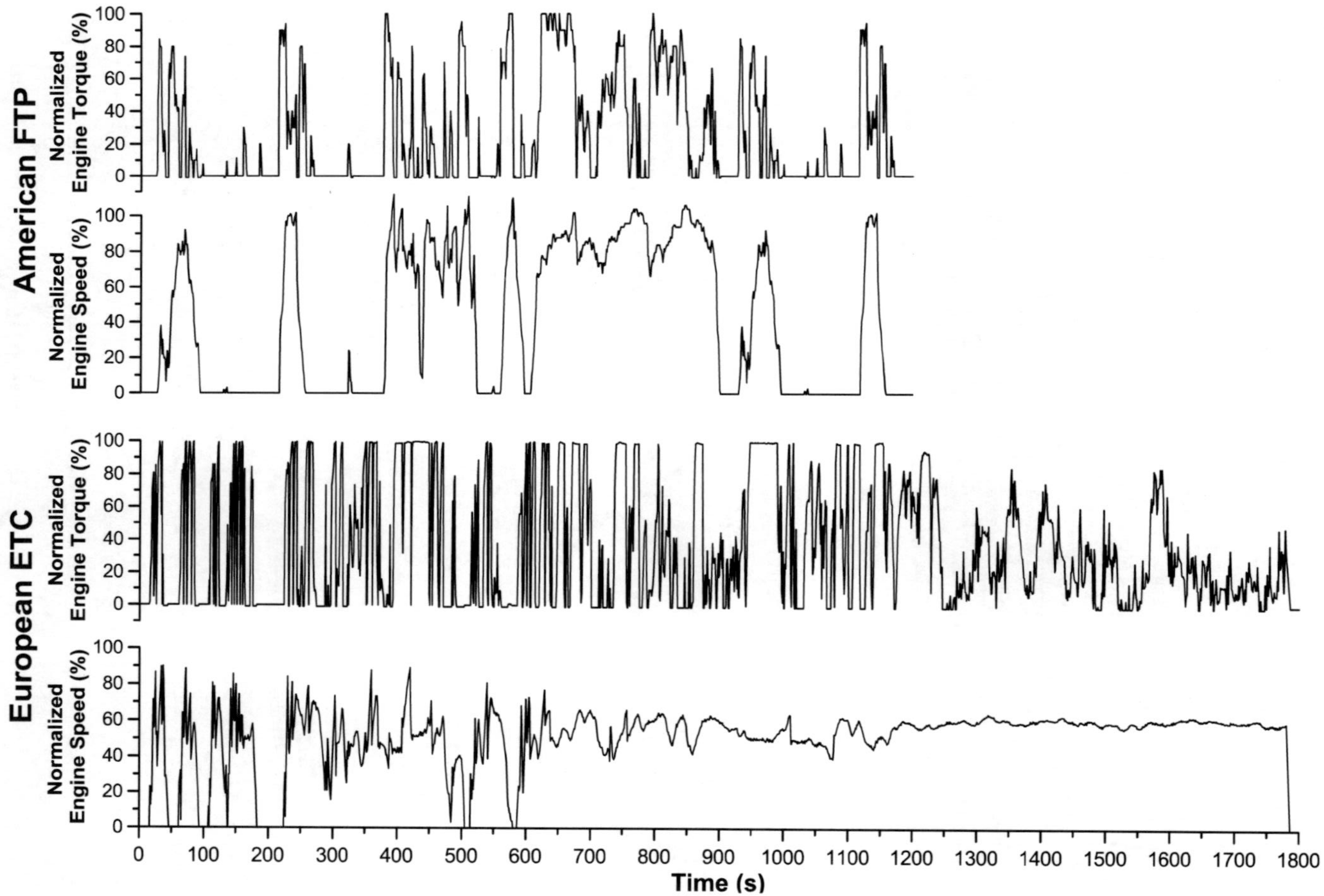

Figure A2. Normalized engine speed and torque values for four engine-dynamometer transient cycles (three heavy-duty and one non-road) mentioned in this chapter.

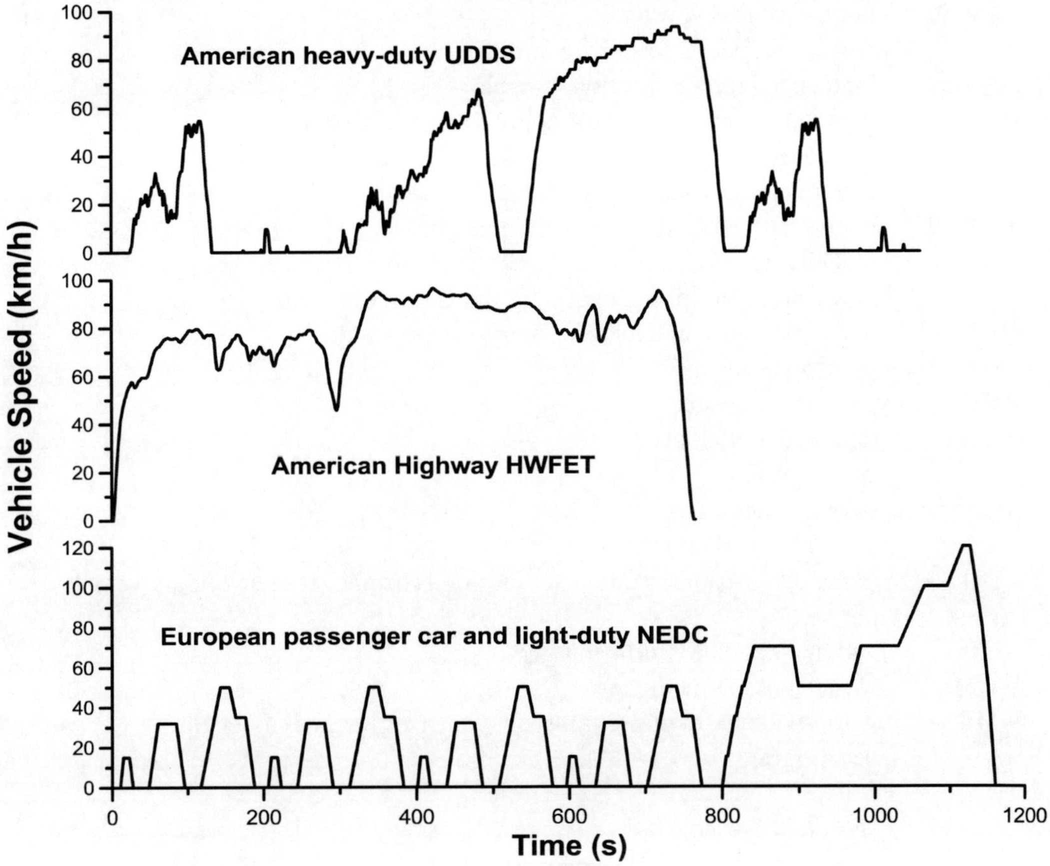

Figure A3. Vehicle speed vs. time for various light-duty and heavy-duty chassis-dynamometer driving cycles mentioned in this chapter.

NOMENCLATURE

Abbreviations

Bx	x% biodiesel / (100-x)% diesel v/v
CN	cetane number
CPC	condensation particle counter
CVS	constant volume sampling
DI	direct injection
DOC	diesel oxidation catalyst
DPF	diesel particulate filter
ECU	engine control unit
EGR	exhaust gas recirculation
EPA	environmental protection agency
ETC	European transient cycle (for heavy-duty engines)
EU	European Union

FAME	fatty acid methyl ester
FTP	American federal test procedure (for heavy-duty engines)
HSDI	high-speed direct injection (engine)
HWFET	American highway fuel economy test cycle (HWY)
LD	light duty
LHV	lower heating value
MOUDI	micro-orifice uniform deposit impactor
MY	model year
NEDC	new European driving cycle (for light-duty vehicles)
NRTC	non-road transient cycle (for non-road engines)
PM	particulate matter
PME	palm methyl ester
RME	rapeseed methyl ester
SI	spark ignition
SME	soybean methyl ester
SuME	sunflower methyl ester
UDDS	American urban dynamometer driving schedule (for heavy-duty vehicles)
v/v	per volume
VGT	variable geometry turbocharger
WCME	waste cooking methyl ester
WHTC	world-wide harmonized transient cycle (for heavy-duty engines)
w/w	per weight

REFERENCES

Agarwal, A. K. (2007). Biofuels (alcohols and biodiesel) applications as fuels in internal combustion engines. *Progr. Energy Combust. Sci.*, 32, 233−271.

Armas, O., Hernandez, J. J. & Cardenas, M. D. (2006). Reduction of diesel smoke opacity from vegetable oil methyl ester during transient operation. *Fuel*, *85*, 2427–2438.

Armas, O., Cardenas, M. D. & Mata, C. (2007). Smoke opacity and NO_x emissions from a bioethanol/diesel blend during engine transient operation. SAE paper 2007-24-0131.

Armas, O., Garcia-Contreras, R. & Ramos, A. (2012). Pollutant emissions from engine starting with ethanol and butanol diesel blends. *Fuel Process. Technol.*, *100*, 63−72.

Bannister, C. D., Hawley, J. G., Ali, H. M., Chuck, C. J., Price, P. et al. (2010). The impact of biodiesel blend ratio on vehicle performance and emissions. *IMechE*, *224* (Part D), 405–421.

Barrienteros, E., Lapuerta, M. & Boehman, A. L. (2013). Group additivity in soot formation for the example of C-5 oxygenated hydrocarbon fuels. *Combustion Flame*, *160*, 1484-1498.

Boehman, A. L., Song, J. H. & Alam, M. (2005). Impact of biodiesel blending on diesel soot and the regeneration of particulate filters. *Energy Fuels*, *19*, 1857–1864.

Botero, M. L., Huang, Y., Zhu, D. L., Molina, A. & Law, C. K. (2012). Synergistic combustion of droplets of ethanol, diesel and biodiesel mixtures. *Fuel*, *94*, 342–347.

Chien, S.-M., Huang, Y.-J., Chunag, S.-C. & Yang, H.-H. (2009). Effects of biodiesel blending on particulate and polycyclic aromatic hydrocarbon emissions in nano/ultrafine/fine/coarse ranges from diesel engine. *Aerosol Air Quality Res.*, *9*, 18–31.

Demirbas, A. (2005). Biodiesel production from vegetable oils via catalytic and non-catalytic supercritical methanol transesterification methods. *Progr. Energy Combust. Sci.*, *31*, 466−487.

Demirbas, A. (2007). Progress and recent trends in biofuels. *Progr. Energy Combust. Sci.*, *33*, 1−18.

Directive 2009/28/EC of the European Parliament and of the Council on the promotion of the use of energy from renewable sources, 2009.

Ecklund, E. E., Bechtold, R. L., Timbario, T. J. & McCallum, P. W. (1984). State-of-the-art report on the use of alcohols in diesel engines. SAE paper 840118.

Fontaras, G., Karavalakis, G., Kousoulidou, M., Tzamkiozis, T., Ntziachristos, L., Bakeas, E., Stournas, S. & Samaras, Z. (2009). Effects of biodiesel on passenger car fuel consumption, regulated and non-regulated pollutant emissions over legislated and real-world driving cycles. *Fuel*, *88*, 1608–1617.

Fontaras, G., Kousoulidou, M., Karavalakis, G., Tzamkiozis, T., Pistikopoulos, P., Ntziachristos, L., Bakeas, E., Stournas, S. & Samaras, Z. (2010). Effects of low concentration biodiesel blend application on modern passenger cars, Part 1: Feedstock impact on regulated pollutants, fuel consumption and particle emissions. *Environ. Pollution*, *158*, 1451–1460.

Giakoumis, E. G. (2012). A statistical investigation of biodiesel effects on regulated exhaust emissions during transient cycles. *Appl. Energy*, *98*, 273-291.

Giakoumis, E. G. (2013). A statistical investigation of biodiesel physical and chemical properties, and their correlation with the degree of unsaturation. *Renew. Energy*, *50*, 858–878.

Giakoumis, E. G. & Lioutas, S. C. (2011). Diesel-engined vehicle nitric oxide and soot emissions during the European light-duty driving cycle using a transient mapping approach. *Transport Res.*, *15* (Part D), 134−143.

Giakoumis, E. G., Rakopoulos, C. D., Dimaratos, A. M. & Rakopoulos, D. C. (2012). Exhaust emissions of diesel engines operating under transient conditions with biodiesel fuel blends. *Progr. Energy Combust. Sci.*, *38*, 691–705.

Giakoumis, E. G., Rakopoulos, C. D., Dimaratos, A. M. & Rakopoulos, D. C. (2013). Exhaust emissions with ethanol or n-butanol diesel fuel blends during transient operation. *Renew. Sustain. Energy Rev.*, *17*, 170–190.

Graboski, M. S. & McCormick, R. L. (1998). Combustion of fat and vegetable oil derived fuels in diesel engines. *Progr. Energy Combust. Sci.*, *24*, 125−164.

Graboski, M. S., Ross, J. D. & McCormick, R. L. (1996). Transient emissions from No. 2 diesel and biodiesel blends in a DDC Series 60 engine. SAE paper 961166.

Hagena, J. R., Filipi, Z. S. & Assanis, D. N. (2006). Transient diesel emissions: analysis of engine operation during a tip-in. SAE paper 2006-01-1151.

Hansen, A. C., Zhang, Q. & Lyne, P. W. L. (2005). Ethanol/diesel fuel blends - a review. *Biores. Technol.*, *96*, 277−285.

Hansen, A. C., Kyritsis, D. C. & Lee, C. F. (2009). Characteristics of biofuels and renewable fuel standards. In: Biomass to biofuels - Strategies for global industries, Vertes A.A., Blaschek, H.P., Yukawa, H., Qureshi, N. (eds). John Wiley, New York.

Hoekman, S. K., Broch A., Robbins C., Ceniceros E. & Natarajan M. (2012). Review of biodiesel composition, properties and specifications. *Renew. Sustain. Energy Rev.*, *16*, 143–169.

Jin, C., Yao, M., Liu, H., Lee, C.-F. & Ji, J. (2011). Progress in the production and application of n-butanol as a biofuel. *Renew. Sustain. Energy Rev.*, *15*, 4080−4106.

Karavalakis, G., Tzirakis, E., Zannikos, F., Stournas, S., Bakeas, E., Arapaki, N. & Spanos, A. (2007). Diesel/Soy methyl ester blends emissions profile from a passenger vehicle operated on the European and the Athens driving cycles. SAE paper 2007-01-4043.

Kittelson, D. B. (1998). Engines and nanoparticles: a review. *J. Aerosol Sci.*, *29*, 575–588.

Knothe, G., Sharp, C. A. & Ryan, T. W. (2006). Exhaust emissions of biodiesel, petrodiesel, neat methyl esters, and alkanes in a new technology engine. *Energy Fuels*, *20*, 403–408.

Komninos, N. P. & Rakopoulos, C. D. (2012). Modeling HCCI combustion of biofuels: A review. *Renew. Sustain. Energy Rev.*, *16*, 1588–1610.

Kooter, I. M., van Vugt, M., Jedynska, A. D., Tromp, P., et al. (2011). Toxicological characterization of diesel engine emissions using biodiesel and a closed soot filter. *Atmos. Environ.*, *45*, 1574–1580.

Kozak, M. J. (2011). Exhaust emissions from a diesel passenger car fuelled with a diesel fuel-butanol blend. SAE Paper 2011-28-0017.

Krahl, J., Buenger, J., Schroeder, O., Munack, A. & Knothe, G. (2002). Exhaust emissions and health effects of particulate matter from agricultural tractors operating on rapeseed oil methyl esters. *J. Am. Oil Chem. Soc.*, *79*, 717–724.

Lapuerta, M., Armas, O. & Rodriguez-Fernandez, J. (2008). Effect of biodiesel fuels on diesel engine emissions. *Progr. Energy Combust. Sci.*, *34*, 198−223.

Lave, L. B. & Seskin, E. P. (1973). An analysis of the association between U. S. mortality and air pollution. *J. Amer. Statistical Assoc.*, *68*, 284−290.

Liotta, F. J. & Montalvo, D. M. (1993). The effect of oxygenated fuels on emissions from a modern heavy-duty diesel engine. SAE paper 932734.

Lujan, J. M., Bermudez, V., Tormos, B. & Pla, B. (2009). Comparative analysis of a DI diesel engine fuelled with biodiesel blends during the European MVEG-A cycle: Performance and emissions (II). *Biomass Bioenergy*, *33*, 948–956.

Macor, A., Avella, F. & Faedo, D. (2011). Effects of 30% v/v biodiesel/diesel fuel blend on regulated and unregulated pollutant emissions from diesel engines. *Appl. Energy*, *88*, 4989−5001.

McCormick, R. L. & Parish, R. (2001). Technical barriers to the use of ethanol in diesel fuel. National Renewable Energy Laboratory Report, NREL/MP-540-32674.

McCormick, R. L., Tennant, C. J., Hayes, R. R., Black, S., Ireland, J. et al. (2005). Regulated emissions from biodiesel tested in heavy-duty engines meeting 2004 emission standards. SAE Paper 2005-01-2200.

Meiring, P., Allan, R. S., Hansen, A. C. & Lyne, P. W. L. (1983). Tractor performance and durability with ethanol–diesel fuel. *Trans. ASAE*, *26*, 59–62.

Merritt, P. M., Ulmet, V., McCormick, R. L., Mitchell, W. E. & Baumgard, K. J. (2005). Regulated and unregulated exhaust emissions comparison for three Tier II non-road diesel engines operating on ethanol/diesel blends. SAE paper 2005-01-2193.

Miers, S. A., Carlson, R. W., McConnell, S. S., Ng, H. K., Wallner, T. & Esper, J. L. (2008). Drive cycle analysis of butanol/diesel blends in a light-duty vehicle. SAE paper 2008-01-2381.

Muncrief, R. L., Rooks, C. W., Cruz, M. & Harold, M. P. (2008). Combining biodiesel and exhaust gas recirculation for reduction in NOx and particulate emissions. *Energy Fuels*, *22*, 1285−1296.

Pelkmans, L., Lenaers, G., Bruyninx, J., Scheepers, K. & De Vlieger, I. (2011). Impact of biofuels blends on the emission of modern vehicles. *IMechE*, *225* (Part D), 1204–1220.

Rakopoulos, C. D. & Giakoumis, E. G. (2006). Review of thermodynamic diesel engine simulation under transient operating conditions. SAE paper 2006-01-0884.

Rakopoulos, C. D. & Giakoumis, E. G. (2009). Diesel engine transient operation. Springer, London.

Rakopoulos, C. D., Antonopoulos, K. A. & Rakopoulos, D. C. (2007). Experimental heat release analysis and emissions of a HSDI diesel engine fueled with ethanol/diesel fuel blends. *Energy*, *32*, 1791−1808.

Rakopoulos, D. C., Rakopoulos, C. D., Kakaras, E. C. & Giakoumis, E. G. (2008). Effects of ethanol/diesel fuel blends on the performance and exhaust emissions of heavy duty DI diesel engine. *Energy Convers. Manage.*, *49*, 3155−3162.

Rakopoulos, C. D., Dimaratos, A. M., Giakoumis, E. G. & Peckham, M. S. (2010a). Experimental assessment of turbocharged diesel engine transient emissions during acceleration, load change and starting. SAE paper 2010-01-1287.

Rakopoulos, C. D., Dimaratos, A. M., Giakoumis, E. G. & Rakopoulos, D. C. (2010b). Investigating the emissions during acceleration of a turbocharged diesel engine operating with biodiesel or n-butanol diesel fuel blends. *Energy*, *35*, 5173−5184.

Rakopoulos, C. D., Dimaratos, A. M., Giakoumis, E. G. & Rakopoulos, D. C. (2011). Study of turbocharged diesel engine operation, pollutant emissions and combustion noise radiation during starting with biodiesel or n-butanol diesel fuel blends. *Appl. Energy*, *88*, 3905–3916.

Rakopoulos, D. C., Rakopoulos, C. D., Giakoumis, E. G. & Dimaratos, A. M. (2012). Characteristics of performance and emissions in high-speed direct injection diesel engine fueled with diethylether/diesel fuel blends. *Energy*, *43*, 214−224.

Sharp, C. A., Howell, S. A. & Jobe, J. (2000). The effect of biodiesel fuels on transient emissions from modern diesel engines - Part I: Regulated emissions and performance. SAE paper 2000-01-1967.

Shudo, T., Nakajima, T. & Hiraga, K. (2009). Simultaneous reduction in cloud point, smoke, and NOx emissions by blending bioethanol into biodiesel fuels and exhaust gas recirculation. *Int. J. Engine Res.*, *10*, 15−26.

Sze, C., Whinihan, J. K., Olson, B. A., Schenk, C. R. & Sobotowski, R. A. (2007). Impact of test cycle and biodiesel concentration on emissions. SAE paper 2007-01-4040.

Szybist, J. P., Song, J., Alam, M. & Boehman, A. L. (2007). Biodiesel combustion, emissions and emission control. *Fuel Process. Technol.*, *88*, 679–791.

Tatur, M., Nanjundaswamy, H., Tomazic, D. & Thornton, M. (2008). Effects of biodiesel operation on light-duty Tier 2 engine and emission control systems. SAE paper 2008-01-0080.

Tinsdale, M., Price, P. & Chen, R. (2010). The impact of biodiesel on particle number, size and mass emissions from a Euro 4 diesel vehicle. SAE paper 2010-01-0796.

Tree, D. R., Svensson, K. I. (2007). Soot processes in compression ignition engines. *Progr. Energy Combust. Sci.*, *33*, 272–309.

Tsolakis, A. (2006). Effects on particle size distribution from the diesel engine operating on RME-biodiesel with EGR. *Energy Fuels*, *20*, 1418–1424.

US Environmental Protection Agency, (2002). A comprehensive analysis of biodiesel impacts on exhaust emissions. Draft Technical Report. EPA 420-P-02-001, US, EPA, Washington DC, USA.

US Environmental Protection Agency, (2003). National air quality and emissions trends report. EPA 454/R-03-005, US, EPA, Washington DC, USA.

Watson, N. & Janota, M. S. (1982). Turbocharging the internal combustion engine. McMillan, London.

Williams, A., McCormick, R. L., Hayes, R. & Ireland, J. (2006). Biodiesel effects on diesel particle filter performance. National Renewable Energy Laboratory Report, NREL/TP-540-39606.

Yao, M., Wang, H., Zheng, Z. & Yue, Y. (2010). Experimental study of n-butanol additive and multi-injection on HD diesel engine performance and emissions. *Fuel*, *89*, 2191–2201.

In: Automotive Exhaust Emissions and Energy Recovery
Editor: Apostolos Pesiridis
ISBN: 978-1-63321-493-4

Chapter 5

EXHAUST GAS AFTERTREATMENT TECHNOLOGIES AND MODEL BASED OPTIMIZATION

***Dimitrios Karamitros*[1], *Stavros A. Skarlis*[1] *and Grigorios Koltsakis*[2,*]**
[1]Exothermia SA, Thessaloniki, Greece
[2]Aristotle University Thessaloniki, Greece

ABSTRACT

The enforcement of stricter emissions legislation on a global level has given a tremendous impulse to the development of very efficient emission reduction technologies for combustion engines. Despite the impressive improvements in clean combustion technologies using advanced fuel injection and air-path handling, catalytic exhaust after-treatment will be necessary across the whole range of engine sizes and applications in the foreseeable future. In the present chapter the exhaust gas aftertreatment technologies are presented and categorized based on their functions and mission. For the case of Diesel engine the presentation will cover Diesel Oxidation Catalysts, Selective Catalytic Reduction, NOx storage catalyst, Diesel Particulate Filters and Ammonia Slip Catalysts. The applicability and system design challenges will be discussed separately for each application of interest. For the case of the gasoline engine the main emphasis will be given to the 3-way catalyst for passenger cars and secondarily to particulate filtration. For each technology, the main operating principles and the governing physico-chemical phenomena will be described. To facilitate in-depth understanding of the processes, the respective mathematical modeling equations describing the transport and reaction processes will be presented.

Keywords: Exhaust aftertreatment, catalysis, selective catalytic reduction, particulate filtration, NOx storage

---

[*] Aristotle University, Thessaloniki, Greece; Email: Grigoris@auth.gr.

INTRODUCTION

Automotive emission control has a history of more than 35 years. Already since the 1950s, pollutants emissions from the growing vehicle fleet led to deterioration of air quality in urban areas. The city of Los Angeles was one of the most characteristic examples of this situation, since formation of ground level ozone was detected, due to photochemical interactions between volatile organic compounds (VOC) and nitrogen oxides (NO_x), both originating from vehicles tailpipes. Remarkably, during the 1960s, typical levels of major pollutants emissions from cars in the USA were: unburned hydrocarbons (HCs) 15 g/mil, CO 90 g/mile and NO_x 6 g/mile [1]. Nevertheless, it was not until the 1970s when a significant change took place with respect to public awareness on environmental issues, such as energy conservation and greenhouse gases emissions. Paul G. Rogers published in the EPA journal of February 1990 *"Historians of the environmental movement are likely to peg Earth Day 1970 as a key turning point in the American public's consciousness about environmental problems"* [2]. In 1975, the Clean Air Act (CAA) came into force in the USA, which consists in a comprehensive federal law which regulates air emissions from stationary and mobile sources. Main scope of the Act was to determine and achieve national ambient air quality standards in every state, by 1975 in order to address the public health and welfare risks posed by certain widespread air pollutants [3].

First, improvements in engine technology were made in an attempt to comply with emission standards. However, due to increasingly stringent legislation, these modifications could not allow to meet automakers requirements. As an alternative, catalytic systems were proposed in order to accomplish efficient emissions control. Introduction of automotive catalytic converters had an outstandingly positive effect on air quality in urban environments. In the USA, during 1982-1991, despite the fact that vehicle miles travelled increased, CO and VOC emissions were decreased up to 40%, whereas NO_x emissions attenuated up to 20% [1]. The first type of catalytic converter, used in the automotive sector was an oxidation catalyst, for the aftertreatment of CO and HCs. Early oxidation catalysts were based on pelleted platinum, originating from process industry, placed in flat radial flow-like reactors [4]. In order to achieve pollutants conversion under rich conditions, the oxidation catalyst had to be fed with additional air. By 1988, the oxidation catalysts were systematically applied in diesel vehicles and they are still the leading technology for both passenger cars and light duty trucks in Europe as well as for heavy duty diesel engines in the U.S.A. [5].

The first engines operating with aftertreatment systems were using carburetors for fuel dosage. However, this technology could not ensure precise fuel supply, resulting in a randomly altered lean/rich engine operation. In 1976, the integration of an "oxygen storage component" into the oxidation catalyst was proposed, so as to allow the catalytic converter to moderating the intense oscillations in exhaust stoichiometry [6]. This evolution of the oxidation catalyst was annotated as the three-way catalyst, due to the fact that this device was able to simultaneously treat CO, HC in oxygen excess and NO_x under rich conditions.

Nitrogen oxides reduction can be efficiently performed in three way catalysts, under rich conditions, but it is very challenging in excess of oxygen, which is the case of lean burn mobile diesel engines. Currently, the limitation of nitrogen oxides emissions from diesel engines is of significant concern. In 2010, statistics collected within countries-members of the European Environmental Agency showed that almost 40% of overall nitrogen oxides emissions in Europe originate from road transportation [7]. First, the task of controlling NO and NO_2 emissions from mobile lean burn engines was faced by means of proper engine design and/or calibration. Nevertheless, engine technologies such as the exhaust gas recirculation (EGR) did not allow to sufficiently limit in-cylinder NO_x formation. Therefore, in order to meet forthcoming stringent legislation advanced aftertreatment technologies were introduced. The NO_x Storage and Reduction catalyst was initially introduced in the automotive sector by Toyota in the mid-1990s [8,9]. This technology enables to reduce emitted NO and NO_2 through reducing components, such as hydrocarbons CO and H_2. NSR catalysts exhibit high deNO_x activity at low temperature range, whereas their performance is hindered at higher ones. The alternative to this method is the Urea or NH_3–SCR catalyst, which enables the reduction of NO_x through ammonia. The Urea-SCR concept was broadly applied for NO_x abatement from stationary diesel engines and particularly from coal fired boilers in Japanese power plants, already since the late 1970s [10]. SCR introduction, for the reduction of NO_x from the exhaust stream of mobile sources was initially applied to marine engines, when the first SCR systems were installed in 1989 and 1990 on two Korean carriers [11]. Only after 2000, SCR was applied to passenger cars and heavy duty vehicles. In 2006, Mercedes-Benz was the first manufacturer to apply a urea SCR catalyst on a passenger car named E320 BlueTEC [12].

Besides the aftertreatment of CO, HCs and NO_x, particulate emissions from lean burn diesel engines raised significant environmental and public health concerns. Diesel particulate filters have been applied on European passenger cars already at the end of the 1990s, whereas their commercialization in Japan and the USA began only after 2005, due to regulations coming into force. Nowadays, the technology of diesel particulate filter consists in the state of the art for removing particles, existing in diesel engines exhaust stream. Filters technology is currently expanding to the field of gasoline engines, where restrictions in the number of emitted particles will be imposed.

Highly sophisticated modern emission control systems are nowadays computer-controlled units. To meet emission standards over a broad range of engine operating conditions, catalytic systems have to achieve optimal performance, which requires continuous monitoring and control of their operating parameters. On board diagnostics (OBD) can efficiently serve this role. Basic, first generation OBDs were installed in all cars sold in California by 1990 [13], whereas four years later, new OBD systems were introduced, which were able to detect and warn the operator about malfunctions such as engine misfiring and/or catalytic components inappropriate operation.

In this context, the scope of the present chapter is to present a comprehensive overview of modern aftertreatment systems. First, the theoretical background concerning catalytic converters operating principals is discussed. Particular emphasis is given on the mathematical description of governing phenomena. Then, the state of the art of aftertreatment technologies is presented for both diesel and gasoline engines. Operating principles, catalytic formulations and global reaction schemes are of major concern. Finally, OBD systems for each technology are briefly discussed.

CATALYTIC CONVERTER TECHNOLOGIES

Principles and Operation

A catalyst is a substance which has the ability to increase the rate of certain chemical reactions, without itself being consumed or altered. An automotive catalytic converter is a device, placed downstream of the engine, which is used to convert the unwanted emissions into harmless products. The required catalytic material is distributed over a carrier with high specific surface area in order for the mass-transfer characteristics between the gas phase and the catalyst surface to be large enough to allow for maximum accessibility of the reactants to the active material [14].

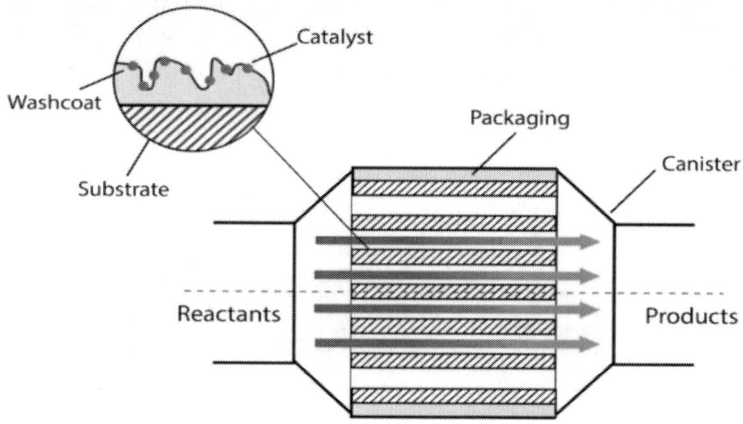

Figure 1. Schematic of a catalytic converter.

The most commonly used catalytic converter design is presented in Figure 1. The catalyst substrate is a ceramic honeycomb monolith, with square cross-section channels in the axial direction, which does not have any catalytic reactivity of its own. The active catalytic material is impregnated into a high porosity carrier, called washcoat, which is coated on the channel walls. The catalyzed substrate is shelled with a packaging material and placed inside a steel canister to form the catalytic converter.

The major advantages of these honeycomb structures, as opposed to pellet-based packed beds used in other industrial chemical processes, include the high geometric surface area (GSA) per unit volume, the low pressure drop and the high durability [15].

The honeycomb substrates are typically made of ceramic or metal. The majority of today's ceramic-supported catalysts are made of cordierite. The ceramic monoliths usually have square cells while other cross-sections are also used, such as round, triangular etc. The metallic substrates are made of thin metal foils, corrugated in a way to form a honeycomb structrure with sinusoidal cells. The number of cells varies depending on the application, with gasoline applications generally having a higher cell density than the diesel ones. The substrate does not serve only as a coating support material. Its geometry and material properties have to be carefully selected as they greatly affect the catalyst's performance parameters such as conversion efficiency, light-off temperature and pressure drop. The coating of the bare monolith is usually a two-step procedure which includes: (1) the washcoat application on the substrate and (2) the impregnation of the applied washcoat with the precious metal catalyst [15]. The washcoat is an inorganic material which serves as a carrier for the precious metal catalyst. The high surface area for the dispersion of the catalytic material, the good thermal stability and ageing resistance are the desired characteristics of washcoat materials. The latter include aluminum and silica oxides, cerium dioxides and zeolites. The selection of washcoat materials depends on the specific application. The most common catalysts, used for the impregnation of the washcoat, are metals or metal oxides such as Pt, Pd, Rd, V_2O_5 etc.

Heterogeneous Chemistry

Automotive catalysts can be classified as heterogeneous catalysts, since the solid catalytic material is in a different phase than the gaseous reactants. The heterogeneous catalytic reactions take place on the surface of the active catalytic sites which are dispersed on the washcoat. The number of reactant molecules converted to products is directly related to the number of available catalytic sites [16], therefore a fine dispersion of the catalytic material over a high surface area washcoat is required to maximize the reaction rate.

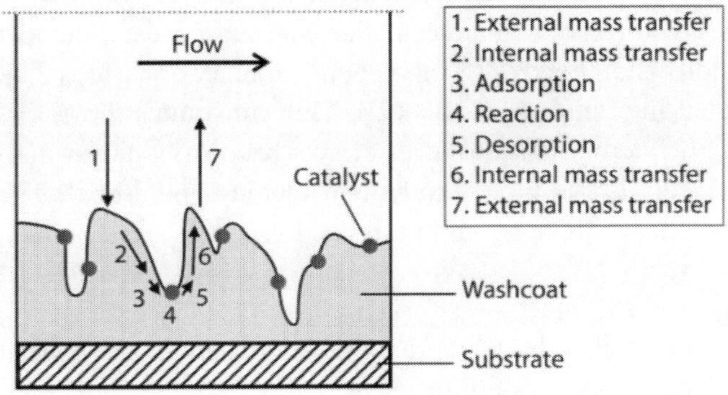

Figure 2. Schematic representation of the main processes involved in the conversion of species in a catalytic converter.

The following elementary steps are usually considered in heterogeneous catalysis [17], as illustrated in Figure 2:

1. *External diffusion:* Transfer of the reactants from the bulk gas to the external surface of the washcoat.
2. *Internal diffusion:* Transfer of the reactants through the porous washcoat towards the active sites.
3. *Adsorption:* Physisorption and chemisorption of the reactants at the catalytic sites.
4. *Surface reaction:* Chemical reaction of the adsorbed species to form adsorbed products.
5. *Desorption:* Release of the adsorbed products.
6. *Internal diffusion:* Transfer of products towards the external washcoat surface.
7. *External diffusion:* Transfer of products to the bulk gas.

The above elementary steps can be grouped into the following three categories which are characterized by different reaction mechanisms and rate expressions [15].

(i) *Chemical reactions* which include the surface reaction (step 4) as well as the adsorption of reactants and desorption of products (steps 3 and 5 respectively).
(ii) *Pore diffusion* which consists of steps 2 and 6 and depends mainly on the microstructural properties of the pore and the nature of the diffusing species.
(iii) Bulk mass transfer which is represented by steps 1 and 7 and is a function of the specific molecules, the flow dynamics and the external surface area of the catalyst.

Fundamentals of Flow-Through Catalyst Modeling

The monolith reactor can be represented by a single channel (1D approach), if uniform inlet flow distribution and negligible heat losses to ambient are assumed. The mathematical model should take into account the 7 main steps that are involved in the conversion of species in a catalytic converter [18] that were described in the previous section.

A common simplifying assumption is to approximate the washcoat with a solid-gas interface ("film" approach") where all the reactions take place [19]. In this case, it is assumed that all catalytic sites are directly available to the gaseous species, neglecting the internal diffusion effects. Pore diffusion (steps 2, 6) and chemical reactions (steps 3, 4, 5) are modeled implicitly and are bulked into an overall rate step. This constitutes the 1D model, which can be extended to 1D+1D if internal mass-transfer effects need to be taken into account. In this case, internal diffusion steps 2 and 6 need to be considered explicitly. A detailed description of the 1D and 1D+1D models will be given in this section.

1D Model

The schematic and the basic geometric properties of a channel cross-section are presented in Figure 3, for the case of a coated monolith.

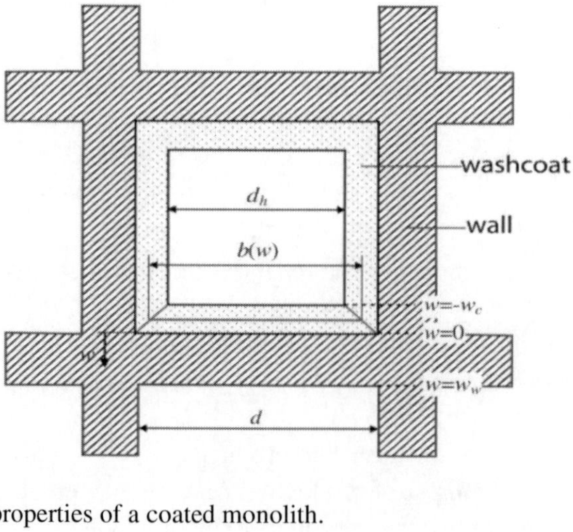

Figure 3. Basic geometric properties of a coated monolith.

The temperature and species concentrations in the channel are computed by solving the quasi-steady state balance equations for the heat eq. 1 and mass eq. 2 transfer. To avoid the solution of the complete boundary layer, the "film" approach is followed which is based on the use of local transfer coefficients [18,19].

$$\rho_g C_{p,g} v_g \frac{\partial T_g}{\partial z} = -h \cdot \left(\frac{S_F}{\varepsilon}\right) \cdot (T_g - T_s) \qquad \text{eq. 1}$$

$$\frac{\partial (v_g y_{g,j})}{\partial z} = -k_j \cdot \left(\frac{S_F}{\varepsilon}\right) \cdot (y_{g,j} - y_{s,j}) \qquad \text{eq. 2}$$

The heat and mass transfer coefficients are calculated according to the following definitions:

$$h = \frac{Nu \cdot \lambda_g}{d_h} \qquad \text{eq. 3}$$

$$k_j = \frac{Sh \cdot D_{mol,j}}{d_h} \qquad \text{eq. 4}$$

The dimensionless Nusselt and Sherwood numbers can be derived from empirically or theoretically derived correlations. The semi-empirical correlation, proposed by Hawthorn [20], is the most commonly used and is applicable to laminar flows in square ducts:

$$Nu = 2.976\left(1 + 0.095 \cdot \frac{Re \cdot Pr \cdot d_h}{z}\right)^{0.45} \qquad \text{eq. 5}$$

$$Sh = 2.976\left(1 + 0.095 \cdot \frac{Re \cdot Sc \cdot d_h}{z}\right)^{0.45} \qquad \text{eq. 6}$$

If the influence of internal diffusion is considered negligible, no transverse concentration gradients in the washcoat layer are considered and only one surface species concentration is defined.

The total surface reaction rate for each species is equal to the external species mass transfer from/to the exhaust gas:

$$\frac{\rho_g}{M_g} k_j \left(\frac{S_F}{\varepsilon}\right) (y_{g,j} - y_{s,j}) = R_j \qquad \text{eq. 7}$$

The reaction rates R_i are calculated based on the applicable theories of heterogeneous catalysis depending on the controlling surface mechanisms for each reaction. The major reactions relevant to each aftertreatment device will be presented in the corresponding sections.

The transient energy balance of the solid phase is written as:

$$\rho_s C_{p,s} \frac{\partial T_s}{\partial t} = \lambda_s \frac{\partial^2 T_s}{\partial z^2} + S \qquad \text{eq. 8}$$

The convective heat transfer H_{conv} due to the gas flow in the channels and the heat release H_{react} by chemical reactions, are included in the source term S:

$$S = H_{\mathrm{conv}} + H_{\mathrm{react}} \qquad \text{eq. 9}$$

$$H_{\mathrm{conv}} = h\left(\frac{S_{\mathrm{F}}}{1-\varepsilon}\right)(T_{\mathrm{g}} - T_{\mathrm{s}}) \qquad \text{eq. 10}$$

$$H_{\mathrm{react}} = \frac{1}{1-\varepsilon}\sum_{\mathrm{k}=1}^{n_{\mathrm{k}}} \Delta H_{\mathrm{k}}\, R_{\mathrm{k}} \qquad \text{eq. 11}$$

The above equations were used to describe the basic 1D catalyst model. The extension of this model to 1D+1D, taking into account the diffusion phenomena in the washcoat layer, will be presented in the following section.

(1D+1D) model

When the internal diffusion effects become important, the simplifying assumption of negligible internal diffusion resistance is not valid and a more detailed model which accounts for the mass transfer both in the gas and the solid phase is required.

The convective mass transfer from the gas to the washcoat surface is now formulated as :

$$\frac{\partial(v_{\mathrm{g}} y_{\mathrm{g},j})}{\partial z} = k_j\left(\frac{S_{\mathrm{F}}}{\varepsilon}\right)\left(y_{\mathrm{s},j}\big|_{w=-w_{\mathrm{c}}} - y_{\mathrm{g},j}\right) \qquad \text{eq. 12}$$

According to the quasi-steady state approach, the transient accumulation terms are neglected and the species balance inside the washcoat layer is written as:

$$-D_{\mathrm{w},j}\frac{\partial^2 y_{\mathrm{s},j}}{\partial w^2} = \sum_{k} n_{j,\mathrm{k}} R_{\mathrm{k}} \qquad \text{eq. 13}$$

The boundary conditions for the washcoat layer are:

$$D_{\mathrm{w},j}\left.\frac{\partial y_{\mathrm{s},j}}{\partial w}\right|_{w=-w_{\mathrm{c}}} = k_j\left(y_{\mathrm{g},j} - y_{\mathrm{s},j}\big|_{w=-w_{\mathrm{c}}}\right) \qquad \text{eq. 14}$$

$$\left.\frac{\partial y_{\mathrm{s},j}}{\partial w}\right|_{w=0} = 0 \qquad \text{eq. 15}$$

where $w = 0$ corresponds to the wall boundary and $w = -w_c$ to the external surface of the washcoat.

Various theoretical models have been proposed in the literature for the estimation of the effective diffusivity $D_{w,j}$. The most commonly used approximations are the mean transport pore model [21] and the random pore model [22].

The mean transport pore model uses the expression:

$$\frac{1}{D_{\mathrm{w},j}} = \frac{\tau}{\varepsilon_{\mathrm{pore}}}\left(\frac{1}{D_{\mathrm{mol},j}} + \frac{1}{D_{\mathrm{knud},j}}\right) \qquad \text{eq. 16}$$

with the Knudsen diffusivity:

$$D_{\mathrm{knud},j} = \frac{d_{\mathrm{pore}}}{3}\sqrt{\frac{8RT}{\pi M_j}} \qquad \text{eq. 17}$$

The porosity ε_{pore} and the mean pore size d_{pore} can be extracted from the microstructural properties of the washcoat while tortuosity τ is an empirical parameter.

The random pore model considers a bi-dispersive washcoat material with two characteristic pore sizes, namely micropores and mesopores. The total diffusivity is given as a function of the respective Knudsen diffusivities:

$$D_{\mathrm{w},j} = \varepsilon_{\mathrm{meso}}^2 D_{\mathrm{knud},j}^{\mathrm{meso}} + \varepsilon_{\mathrm{micro}}^2 D_{\mathrm{knud},j}^{\mathrm{micro}} + \frac{4(\varepsilon_{\mathrm{micro}} - \varepsilon_{\mathrm{meso}}^2)}{\frac{1}{D_{\mathrm{knud},j}^{\mathrm{meso}}} + \frac{\left((1-\varepsilon_{\mathrm{meso}})/\varepsilon_{\mathrm{micro}}\right)^2}{D_{\mathrm{Knud},j}^{\mathrm{micro}}}} \qquad \text{eq. 18}$$

Multi-Dimensional Model Extension

If the non-uniformities in the inlet temperature and flow distribution need to be considered, a 2D or 3D model is required. In this case, the transient energy balance equation is extended to two or three dimensions. The heat conduction equation is formulated in polar coordinates for 2D simulations and in Cartesian coordinates for 3D simulations.

$$\rho_{\mathrm{s}} c_{\mathrm{p,s}} \frac{\partial T_{\mathrm{s}}}{\partial t} = \lambda_{\mathrm{s,z}} \frac{\partial^2 T_{\mathrm{s}}}{\partial z^2} + \lambda_{\mathrm{s,z}} \frac{1}{r}\frac{\partial}{\partial r}\left(r \frac{\partial T_{\mathrm{s}}}{\partial r}\right) + S \qquad \text{eq. 19}$$

$$\rho_{\mathrm{s}} c_{\mathrm{p,s}} \frac{\partial T_{\mathrm{s}}}{\partial t} = \lambda_{\mathrm{s,x}} \frac{\partial^2 T_{\mathrm{s}}}{\partial x^2} + \lambda_{\mathrm{s,y}} \frac{\partial^2 T_{\mathrm{s}}}{\partial y^2} + \lambda_{\mathrm{s,z}} \frac{\partial^2 T_{\mathrm{s}}}{\partial z^2} + S \qquad \text{eq. 20}$$

FILTER TECHNOLOGIES

Wall-Flow Monoliths

Carboneous particles produced due to heterogeneous combustion conditions, are major pollutants of mobile diesel and gasoline engines. Worldwide, forthcoming legislation imposes significant limitations not only on emitted particles mass, but on their respective number as well. Particulate filters have been established to control soot emissions from passenger cars and heavy duty vehicles. Based on the engine type, they can be distinguished in diesel and gasoline particulate filters. Both technologies are discussed in details later.

Nanometer scale particulate matter is very difficult to be captured through inertial filtration mechanisms with reasonable filter structures, characterized by collector sizes in the order of 10-100 microns. On the other hand, exhaust aftertreatment filters can efficiently separate the typically sub-μm particulate matter from the exhaust stream through various transport mechanisms, including Brownian diffusion and to a lesser extent direct interception [23]. As with all filtration processes, the maximization of filtration efficiency and the minimization of pressure drop during operation are major challenges. Both targets can be met by achieving low filtration velocities, which require high surface areas. In this respect, the monolithic honeycomb wall-flow filter has been established as a very efficient filter concept. This is an extruded ceramic structure with a large number of small parallel channels, similar to the common flow-through catalytic converter. Nevertheless, contrary to the latter, the channels of a "wall-flow" monolith are alternatively plugged, forming inlet and outlet channels. The particulate-containing gas entering the inlet channels is hence forced to flow through the porous walls which act as the filtering medium for solid particles (Figure 4).

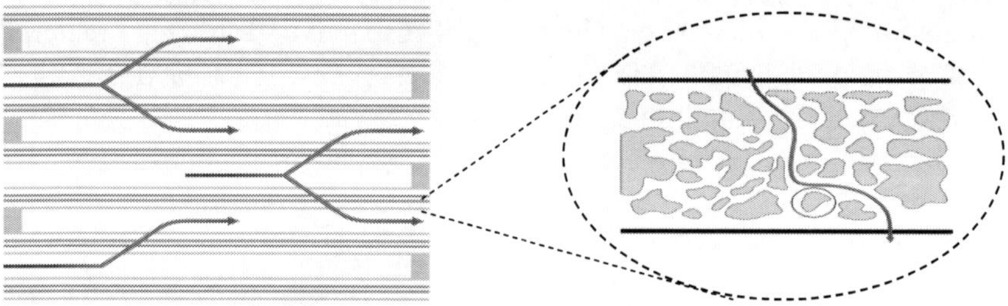

Figure 4. Wall-flow filtration principle in diesel particulate filters.

Irrespectively the powertrain type to be integrated into, the filter materials have to meet a number of requirements, including high filtration efficiency (with respect to both particles mass and number), resistance to heat and mechanical loadings, as well as tolerance to chemical compounds existing in engine exhaust stream. Given the harsh conditions in the engine exhaust system and the durability standards of the automotive industry, only few materials could be used. Cordierite is well-known from its use in exhaust catalytic converters as a low-cost material with excellent thermal and corrosion stability. The porosity and the mean pore size of extruded cordierites can be engineered to achieve high filtration efficiency and reasonable backpressure.

For diesel particulate filter applications, cordierite based substrates are currently widely used, mainly for heavy-duty trucks. On the other hand, in the passenger car sector, silicon carbide (SiC) based materials play a dominant role, due to their higher thermal stability. Other ceramic materials used are Aluminum Titanate, mullite and SiN. In all cases, the channel structure is similar with cell densities in the order of 100 to 300 cells per square inch and wall thickness in the order of 10 to 15 mils. To achieve high filtration efficiencies and reasonably low pressure drop, the wall porosities lie in the range of 45 to 60%, whereas the optimal pore size range for maximum retention of the diesel particles is in the order of 10 to 20 microns. It is not only the porosity and the mean pore size of the wall microstructure that affects the filter performance but also the pore size distribution. Narrower pore size distributions close to the optimum values are more favorable.

Similarly to diesel particulate filters, cordierite is also used for gasoline particulate filters as well. Wall thicknesses may range within 5 – 15 mils, with cell densities in the same order of 200 – 350 cells per square inch [24,25]. Alternatively, metallic foams can be also applied, which can achieve filtration efficiencies as high as 93% [26]. Nevertheless, wall porosity has to be carefully determined, since gasoline engines operating at stoichiometric conditions are sensitive to backpressure oscillations.

Fundamentals of Wall-Flow Filter Modeling

One-dimensional modeling is sufficient in the case where all filter channels behave in the same way. This is true in the case of uniform radial flow distribution at the filter inlet and of negligible heat losses to ambient, conditions that are usually met during normal filter operation. Multi-dimensional models are interesting in the case of high temperature forced regenerations where the inlet thermal and flow field is highly non-uniform and the heat losses cannot be neglected. In the current section, the single-channel 1D modeling approach will be presented, the fundamentals of which have been established in the work of Bissett [27].

A schematic of the front and side view of a filter channel is given in Figure 5. The governing equations for the conservation of mass, momentum, energy and species are given in the following sub-sections [28].

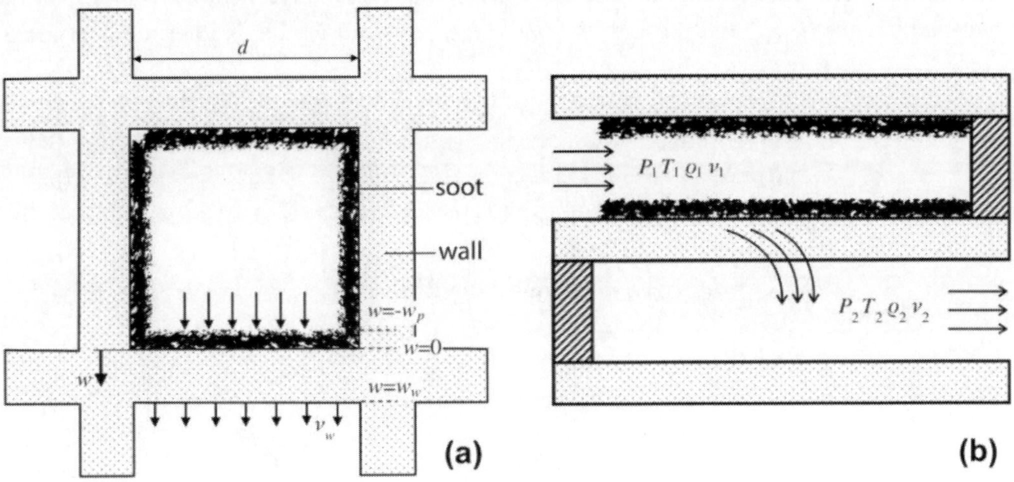

Figure 5. Schematic of a filter channel. Front (a) and side (b) view.

Mass-Momentum Balance

Conservation of mass of channel gas. The mass balance equation for the gas flowing in the inlet and outlet channels is:

$$\frac{\partial}{\partial z}\left(d_i^2 \rho_i v_i\right) = (-1)^i \cdot 4 d \rho_{\mathrm{w}} v_{\mathrm{w}} \qquad \text{eq. 21}$$

Conservation of axial momentum of channel gas. Considering the mass gain/loss through the porous wall and the friction in the axial direction z, the momentum balance of the exhaust gas can be written as:

$$\frac{\partial p_i}{\partial z} + \frac{\partial}{\partial z}\left(\rho_i v_i^2\right) = -\,\alpha_1 \mu v_i / d_i^2 \qquad \text{eq. 22}$$

Energy Balance

Energy balance of the gas phase. The conservation of energy of the channel gas has two expressions, one for the inlet eq. 23 and one for the outlet eq. 24 channel.

$$C_{p,g}\rho_1 v_1\big|_z \frac{\partial T_1}{\partial z} = h_1 \frac{4}{d_1}(T_s - T_1) \qquad \text{eq. 23}$$

$$C_{p,g}\rho_2 v_2\big|_z \frac{\partial T_2}{\partial z} = \left(h_2 + C_{p,g}\rho_w v_w\right)\frac{4}{d}(T_s - T_2) \qquad \text{eq. 24}$$

Energy balance of the solid phase. The temperature field in the filter is described by the transient heat conduction equation with heat sources in axi-symmetric coordinates:

$$\rho_s \cdot C_{p,s}\frac{\partial T_s}{\partial t} = \lambda_{s,z}\frac{\partial^2 T_s}{\partial z^2} + S \qquad \text{eq. 25}$$

In the case of 2D or 3D simulations, equations eq. 19 and eq. 20 can be used respectively.

The source term S includes the contribution of the convective heat transfer of the gas flow in the channel H_{conv} and through the wall H_{wall} as well as the exothermic heat release H_{react}.

$$S = H_{conv} + H_{wall} + H_{react} \qquad \text{eq. 26}$$

$$H_{conv} = h_1 \cdot S_F \cdot (T_1 - T_s) + h_2 \cdot S_F \cdot (T_2 - T_s) \qquad \text{eq. 27}$$

$$H_{wall} = \rho_w \cdot v_w \cdot S_F \cdot C_{p,g} \cdot (T_1 - T_s) \qquad \text{eq. 28}$$

$$H_{react} = S_F \sum_k \left(\int_{-w_p}^{w_w} f_w R_k dw \right) \cdot \Delta H_k \qquad \text{eq. 29}$$

The definition of the geometrical parameter f_w is:

$$f_w = \frac{b(w)}{d} \qquad \text{eq. 30}$$

The width available to the flow $b(w)$ varies in the particulate layer and remains constant in the wall:

$$b(w) = \begin{cases} d + 2w, w < 0 \\ d, w \geq 0 \end{cases} \qquad \text{eq. 31}$$

Species Balance

The importance of species transfer to account for oxygen convection to the reacting soot layer was recognized in [29]. In addition, the solution of the species transfer equation is essential to model the catalyzed filters with respect to pollutant conversion and their interactions with the accumulated soot.

The governing advection-reaction-diffusion equation for the conservation of mass of any species within the soot layer and wall is:

$$v_{\mathrm{w}}\frac{\partial y_j}{\partial w}-D_{\mathrm{w},j}\frac{\partial}{\partial w}\left(f_{\mathrm{w}}\frac{\partial y_j}{\partial w}\right)=\frac{f_{\mathrm{w}}}{c_{\mathrm{m}}}\sum_{\mathrm{k}}n_{j,\mathrm{k}}R_{\mathrm{k}} \quad \text{eq. 32}$$

The calculation of the effective diffusivity $D_{w,j}$ is based on the mean transport pore model, which was previously described by the equations eq. 16 and eq. 17.

The boundary conditions couple the wall with the gas concentrations in the channels. The 'film' approach is used to compute the convective mass transfer from the bulk gas to the wall surface. The mass transfer coefficients $k_{i,j}$ correspond to laminar flow for both inlet and outlet channels:

$$\frac{\partial(v_1 y_{1,j})}{\partial z}=-\frac{4}{d\cdot f_{\mathrm{w}}{}^2}v_{\mathrm{w}}y_{1,j}+\frac{4}{d\cdot f_{\mathrm{w}}}k_{1,j}(y_{1\mathrm{s},j}-y_{1,j}) \quad \text{eq. 33}$$

$$\frac{\partial(v_2 y_{2,j})}{\partial z}=\frac{4}{d\cdot f_{\mathrm{w}}{}^2}v_{\mathrm{w}}y_{2\mathrm{s},j}+\frac{4}{d\cdot f_{\mathrm{w}}}k_{2,j}(y_{2\mathrm{s},j}-y_{2,j}) \quad \text{eq. 34}$$

DIESEL TECHNOLOGIES

Diesel Oxidation Catalyst (DOC)

The diesel oxidation catalyst (DOC) has been an integral part of the diesel exhaust aftetreatment systems since first regulations were introduced for the abatement of gaseous pollutant emissions of diesel engines. Its main function is the oxidation of carbon monoxide (CO) and unburned hydrocarbons (HC). Additional benefits from the presence of a DOC include the necessary temperature increase for the active regeneration of diesel particulate filters (DPFs) and lean NO_x traps (LNTs), as well as the generation of a favourable NO_2/NOx ratio for the efficient operation of the selective catalytic reduction (SCR) and the passive DPF regeneration.

In the 1970s, the early diesel catalysts were virtually the same as those developed for gasoline applications. These catalysts were used to reduce the carbon monoxide emissions and diesel odor but had certain disadvantages such as the increase of sulfate particulate emissions [15]. The introduction of stricter emissions regulations in the 1990s, such as the Euro 2 norm and the emission standards for highway diesel engines, forced the development of a new generation of oxidation catalysts specifically designed for diesel applications. These catalysts were mainly optimized in reducing the emission of particulates rather than the abatement of CO and HC. The reduction in PM emissions was achieved by oxidizing the soluble organic fraction (SOF) of the particulates resulting in a conversion efficiency of 15 to 30% [15]. To avoid the oxidation of sulfur compounds such as the SO_2 and the subsequent formation of sulfate particulates, these catalysts were formulated with a very low Pt loading. This resulted in a low conversion of CO and HC which was acceptable at the time since it was enough to comply with the regulations. The introduction of low sulfur diesel fuel in the market allowed for higher catalytic activity formulations to be used. Moreover, the use of wall-flow monoliths (DPFs) for the control of PM emissions has become standard and consequently the main function of the DOC is the abatement of CO and unburned HCs. Nowadays, the diesel oxidation catalysts are very effective in controlling CO and HC emissions with achieved conversion efficiencies in excess of 90% at sufficiently high temperatures.

The emissions reduction in the DOC occurs through the oxidation of the pollutants on the active catalytic sites. The main chemical reactions are [30] :

$$CO + {}^{1}/_{2}\,O_2 \rightarrow CO_2 \qquad \text{reac. 1}$$

$$C_xH_y + (x + y/4)O_2 \rightarrow xCO_2 + (y/2)H_2O \qquad \text{reac. 2}$$

$$NO + {}^{1}/_{2}\,O_2 \leftrightarrow NO_2 \qquad \text{reac. 3}$$

Carbon monoxide and gas-phase hydrocarbons are oxidized to form carbon dioxide according to reactions (reac. 1) and (reac. 2) respectively. The NO oxidation to NO_2 (reac. 3) is a reversible reaction, limited by thermodynamic equilibrium, which is considered desirable in certain applications. Additionally, the DOC is used to oxidize the SOF compounds and thereby leading in a reduction of the total PM emission mass but with no significant reduction in the number of the emitted particles [31].

$$SOF + O_2 \rightarrow CO_2 + H_2O \qquad \text{reac. 4}$$

An active catalyst will promote the oxidation of all species of a reducing character which may result in the formation of undesirable products. This is the case with the oxidation of sulfur dioxide to sulfur trioxide which can subsequently lead to the formation of sulfuric acid, as described by (reac. 5) and (reac. 6).

$$SO_2 + {}^{1}/_{2}\,O_2 \rightarrow SO_3 \qquad \text{reac. 5}$$

$$SO_3 + H_2O \rightarrow H_2SO_4 \qquad \text{reac. 6}$$

Sulfuric acid is a precursor of liquid sulfuric particles that add to the measured Particulate Matter. These particles are in fact formed as secondary byproducts downstream the emission control system and are therefore not possible to capture in the Particulate Filter. In fact, the presence of the particulate filter removes the bigger solid particles that would otherwise serve as adsorbers of the liquid sulfuric particles. This explains the paradox of particle generation in the case of DOC/DPF systems operating with high fuel sulfur content.

Other reactions, frequently mentioned in the literature, are the oxidation of H_2 and the NO_2 reduction by both CO and HC [30,32].

$$H_2 + {}^{1}/_{2}\,O_2 \rightarrow H_2O \qquad \text{reac. 7}$$

$$CO + NO_2 \rightarrow CO_2 + NO \qquad \text{reac. 8}$$

$$C_xH_y + (2x + y/2)NO_2 \rightarrow xCO_2 + (y/2)H_2O + (2x + y/2)NO \qquad \text{reac. 9}$$

During the cold engine start the catalyst is not able to oxidize the carbon monoxide and the hydrocarbons present in the exhaust gas. In order to minimize the cold-start emissions, zeolites are added to the catalytic washcoat which promote the low temperature adsorption of

HCs, preventing them from escaping to the environment. The adsorbed hydrocarbons are released and oxidized when temperature is sufficiently high and the catalyst has reached its light-off temperature. Therefore, the adsorption/desorption of HCs need to be considered :

$$C_xH_y \leftrightarrow C_xH_y(l) \qquad \text{reac. 10}$$

The main DOC reactions that take place at lean conditions where presented above. In some cases, such as the forced regeneration of a DPF or a LNT, the DOC operates at fuel-rich conditions and some additional reactions occur. These reactions include the water gas shift and steam reforming reactions, the NO reduction by CO and H_2 etc. [33].

The diesel oxidation catalysts are precious metal-based, commonly containing platinum (Pt) supported on a Al_2O_3 carrier. Pt only catalysts are very reactive but they are characterized by high cost and poor thermal durability due to Pt hydrothermal sintering. The addition of palladium (Pd) can be beneficial as both the thermal stability and the light-off temperature of HC oxidation are significantly enhanced when Pt and Pd coexist in the catalyst [34]. The good resistance to thermal ageing is particularly important when fuel injection in the exhaust gas is employed in order to generate the necessary exotherms to actively regenerate the DPF. However, replacing a part of Pt with Pd can potentially lead to a reduction of the NO to NO_2 conversion as the latter is poor on the Pd surface [35]. The NO oxidation towards NO_2 is a desirable reaction that greatly affects the performance of devices commonly found downstream of the DOC, such as the DPF and the SCR. The regeneration of a DPF is enhanced by the presence of NO_2 [36] which reacts with the soot at temperatures as low as 250-300°C, making the high temperature forced regenerations less frequent. In SCR catalysts, the NO_2/NO_x ratio is known to have a significant effect on de-NO_x performance [37]. More specifically, an equimolar NO-NO_2 mixture at the SCR inlet is known to promote the so-called 'fast SCR' reaction and to maximize the achieved NO_x conversion efficiency [38]. It is evident that the various, often conflicting, functional and durability requirements of the DOC are associated with the complete exhaust line configuration. Recent developments in DOC technologies include, among others, the addition of active metals such as Pd and gold (Au) by advanced impregnation techniques and the use of new support materials such as ceria-zirconia and silica-titania [39,40].

Diesel Particulate Filter (DPF)

The Diesel Particulate Filter (DPF) is designed to physically capture the solid fraction of particulate matter (PM) existing in diesel engines exhaust stream. Typically, a DPF is an extruded, cylindrical ceramic wall-flow monolith, with adjacent channels, plugged at opposite ends in a way that each open channel is always surrounded by plugged ones (Figure 6). This way exhaust gas is forced to flow through the porous walls (micro-porous scale), enabling a highly efficient particles filtration, which may exceed 90%.

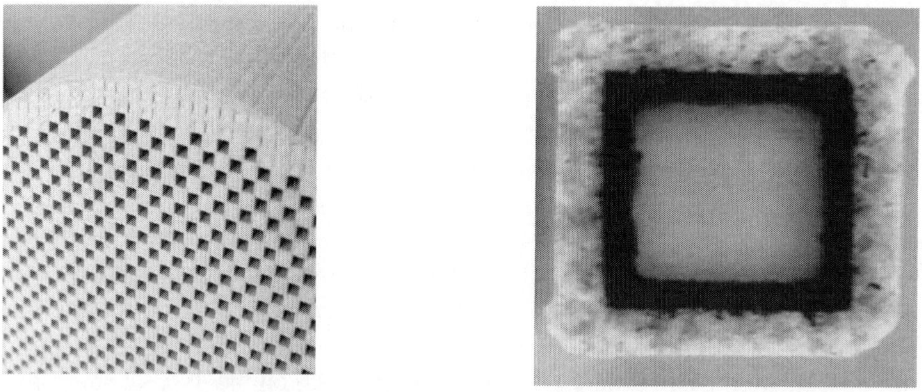

Figure 6. (Left) Close-up view of the inlet face of a cordierite DPF with alternatively plugged channels. (Right) Image of cross-section of a DPF channel with accumulated soot layer.

PM filtration can occur according to two mechanisms. Starting from clean state, particles passing through the walls are initially collected in the micropores inside the filter wall. This type of filtration is based on interception and/or diffusional deposition of particles on the porous medium and is called “deep bed filtration”. The filtration efficiency of a completely clean filter is in the order of 60 to 80%. Due to their low density, these particles quickly modify the apparent morphology of the initial wall microstructure, thus acting as filtering medium themselves. Effectively, the open pores on the wall surface are bridged and a continuous soot layer is formed. This constitutes a perfect filter for incoming soot, enabling to achieve filtration efficiencies, as high as 100%. For typical commercial filters, all incoming soot could be deposited on the cake when the accumulated soot mass exceeds approximately 0.2 – 0.5 grams per liter of filter. These values however, may vary depending on the filter microstructure. The latter filtration mechanism is often annotated as “cake filtration” and results in a rapid increase of the filtration efficiency, which is unfortunately accompanied with a steep increase of backpressure.

Using as a basis for calculation a typical soot emission rate in the order of 20 mg/km (corresponding to a modern mid-size passenger car 2 liter diesel engine) the accumulation of soot upon 1000 km would be approximately 20 grams. Considering that the DPF volume is usually selected to be in the same order of magnitude of the engine displacement, the specific soot loading of a 2 liter DPF after 1000 km would be 10 grams/liter. Such soot loading levels induce significant backpressure leading to increased engine pumping losses, which reduce its power output and fuel efficiency. Even worse, filter durability issues may arise, since a spontaneous oxidation of the accumulation soot of this quantity could result in extremely high exotherms damaging the filter itself. It is therefore necessary to continuously control the amount of soot loading in the filter, by periodically oxidizing the accumulated particles. This process, known as filter regeneration, involves oxidation of carbonaceous soot with oxygen, present in the exhaust gas and is kinetically favored above 550 to 600 ºC.

Under typical diesel engine operating conditions, exhaust gas temperature is rarely or never sufficiently high to oxidize the accumulated soot. Therefore, the task of regeneration can be faced following two different approaches: either the exhaust gas and/or the filter are heated up to reach the particulate oxidation temperature, or the ignition temperature is humbled through substances, which catalyze soot oxidation. The former regeneration strategy, known as “active regeneration” was possible to be applied upon the introduction of electronic fuel injection management combined with the flexibility offered by the common-rail technology in 2000 [41]. Using multiple injections in one engine cycle, part of the fuel may be actually used to produce heat instead of power, thus increasing the exhaust gas temperature to the desired level. This heat is partially generated in the combustion chamber, whereas the majority of it is usually released in an oxidation catalyst, located upstream of the DPF, which oxidizes the large amounts of hydrocarbons deliberately emitted during such a ''forced'' regeneration mode. This way, exhaust gas temperature can increase at values as high as 500 – 600 °C enabling the oxidation of accumulated soot. Soot oxidation may proceed through two main reaction mechanism involving O_2 (reac. 11 and reac. 12) [42,43] and/or NO_2 (reac. 13 and reac. 14) [44]. NO_2 formation can be achieved with a Pt-based diesel oxidation catalyst (DOC), placed upstream of the DPF. The regeneration concept, combining a DOC with a DPF is known since the late eighties and has been trademarked as ''Continuously Regenerating Trap'' or CRT® (commercialized in 1995 for heavy duty retrofit market) [45]. At this stage it has to be highlighted that forcing DPF regeneration too frequently via fuel post-injection is undesired not only due to fuel economy considerations but also because part of the post-injected fuel adsorbs in the lube oil and reduces the oil change intervals.

$$2C + O_2 \rightarrow 2CO \qquad \text{reac. 11}$$
$$C + O_2 \rightarrow CO_2 \qquad \text{reac. 12}$$
$$2NO_2 + C \rightarrow 2NO + CO_2 \qquad \text{reac. 13}$$
$$NO_2 + C \rightarrow NO + CO \qquad \text{reac. 14}$$

To facilitate DPF regeneration and minimize post injection requirements, it is possible to use catalytic promoters. Metal oxides can substantially promote soot oxidation, provided that they are in contact with soot. Very intimate oxide-soot contact can be achieved by introducing oxides in the fuel, which are emitted in contact with soot particles [46]. This fuel-borne catalyst (FBC) technology was first introduced by PSA Peugeot-Citroen in the first successful commercial DPF application in 2000 [47]. Cerium- and iron-based fuel dopants (30 ppm Ce and 10 ppm Ce/Fe) can decrease the temperature required for soot oxidation by 100 to 150 °C. Nevertheless, one of the drawbacks of the FBC technology is that the active oxides eventually stay in the DPF channels as inactive residual, blocking active filtration area. Thus a periodic cleaning of the DPF by removal at the workshop is required. In the first commercial DPF generation such maintenance was necessary after a mileage of approximately 80,000 km. Recently, PSA Peugeot-Citroen presented a new generation iron based FBC, which enables to further improve DPF regeneration characteristics, whereas it was estimated that for a car with a fuel consumption of 7 liters/100 km, the DPF ash cleaning interval is 300,000 to 400,000 km, depending on filter design [48].

Most manufacturers have followed the 'maintenance-free' option of the catalyzed DPF (CDPF) technology. This concept enables to promote soot regeneration by coating the DPF with a catalyst mixture, containing a platinum group metal (PGM), such as Pt and/or Rh and an alkaline earth metal oxide (eg magnesium oxide) [49]. Interestingly, Pt-based coatings indeed promote soot oxidation at low temperatures via an indirect mechanism involving NO to NO_2 oxidation on Pt [50]. NO_2 is an excellent soot oxidizer at temperatures as low as 300 °C. However, in order to exploit this regeneration mechanism two requirements have to be met: (a) the temperature should be higher than 300 °C and (b) enough NO_2 availability.

Warner et al. [51] have pointed out that passive regeneration is higher in the temperature window of 370 ºC, compared to that of 480 ºC, since at high temperature regime NO_2 may dissociate towards NO, hence limiting the rate of soot oxidation. As a rule of thumb the critical NO_x/soot ratio should be higher than 16 on a mass basis to sustain NO_x based regeneration. This condition can be met in the case of heavy-duty applications in which the engine combustion is calibrated for high NO_x and low PM. Many heavy-duty applications like long-haul trucks operate at high loads and exhaust gas temperatures which favor the NO_x based regeneration mode.

NO_2 used for soot oxidation is formed inside the CDPF catalytic wall, through oxidation of NO, existing in the exhaust stream [52]. Nevertheless, this mechanism is under debate due to the fact that most of the soot lies on the channel surface, which is upstream of the catalytic wall. Counter intuitively, soot is indeed oxidized by NO_2 produced in the wall, since nitrogen dioxide can diffuse opposite to the bulk flow due to the developed concentration gradient between the catalytic wall and the soot layer [53]. Finally, it is noteworthy that applying Pt-coatings in the DPF channels additionally serves as an oxidation catalyst for undesirable CO produced during regeneration which is converted to CO_2. Due to the high exothermal heat of this reaction, a catalyzed DPF will produce higher exotherms which in turn accelerates the regeneration process.

Irrespectively the regeneration strategy, a DPF is subjected to significant thermal loading and consequently thermal stresses during soot oxidation, which can lead to overheating and failure of the material. It can be easily shown by energy balance considerations that the peak temperature expected during soot oxidation increases with initial soot loading. The second critical parameter affecting the thermal loading of a filter is its thermal mass, which depends on the filter geometry and material. In this respect, higher density materials like SiC are thermally more stable and can withstand regenerations with initial soot loadings higher than 10 g/l. Filters have to be designed to sustain the so-called 'worst-case' regenerations, which occur when the oxidation is initiated at high temperature and the exhaust flow rate is too low to cool the filter as the reaction exothermy increases its temperature. In practice, this can happen when the engine drops to idle during the course of a regeneration event.

Chilumukuru et al.[54] proposed partial regenerations at high soot loadings, so as to minimize fuel penalty accompanying active DPF regeneration. This technique can lead to the formation of a residual soot layer on the filter, which may allow limit backpressure by preventing soot from penetrating into walls. Nevertheless, prolonged storage of residual soot on the DPF can cause not only filter accelerated ageing (graphitization and/or poisoning), but uncontrolled regeneration events as well. To eliminate the chances of DPF overheating and failure, the system has to be designed to regenerate at safe soot loadings. Commercial passenger car DPFs are allowed to operate until the soot loading reaches 6 to 8 g/l for the case of SiC materials. The safe soot limit in the case of cordierite filters is roughly 30 to 50% lower due to its lower heat capacity. To achieve DPF regeneration with high soot loadings, aluminum-titanate filters have been earlier proposed by Corning (with the commercial names DuraTrap AT and DEV AT filters) [55,56], which are characterized by increased thermal mass and limited backpressure.

Continuous monitoring of a DPF performance is a very useful technique in order to effectively control filter appropriate operation and to detect unlike material failures. In this context, on-board diagnostics (OBD) could offer several possibilities. Generally speaking, OBD provides a continuous examination of whether an aftertreatment device is properly operating, whereas it simultaneously evaluates whether exhaust emissions exceed a certain value, determined by regulations. So far, differential pressure sensors have been used for DPF applications. These sensors provide a voltage output to an engine management unit proportional to the differential pressure between two measurement sources. In the case of DPFs a pair of sensors is placed upstream and downstream of the filter in order to measure the actual filter load (and/or the fraction of accumulated soot mass and ash), as well as to detect material failures. The estimation of accumulated soot mass allows to preventing filter clogging by appropriate controlling the regeneration strategy [57]. Moreover, the measured pressure can be correlated with a pressure model at specific volumetric flow rates, allowing thus to detect material failures [58].

Nevertheless, stringent forthcoming legislation concerning OBD, calls for detection limits lower than the effective ones, achieved by pressure sensors. Therefore, PM sensors were recently introduced. Five different types of particles sensors have been reported, based on their operation principles. The reader is referred to [59,60] for more information. The simplest and less costly PM sensor is composed of two parallel electrodes between which PM collection can be achieved. The actual amount of soot is estimated by measuring changes in the sensor electrical conductivity. This type of sensor can be installed downstream of the DPF, allowing to detect filter malfunctions, whereas it can enable to check PM emissions to threshold compliance, upon sensor calibration for the particular engine/vehicle type [61].

Lean NO_x Trap (LNT)

Besides the control of PM emissions, the reduction of nitrogen oxides under lean conditions is one of the major challenges for diesel engines aftertreatment systems. NO_x Storage and Reduction Catalysts (NSRC) or Lean NO_x Traps (LNT) [62,63] is a well established technique, which allows to efficiently remove NO and NO_2 from diesel engines exhaust stream. Recently, lean burn gasoline engines (eg. FSI engines) have been proposed, in an attempt to limit CO_2 emissions from passenger vehicles. In this case NSR catalysts can be an attractive NO_x aftertreatment solution, since in excess of oxygen the abatement of NO and NO_2 cannot be efficiently performed through the conventional three-way catalytic converter. Commercial LNTs consist in flow through monoliths, which can achieve NO_x conversion efficiencies higher than 90%, within a temperature window of 200 – 450°C [64]. Therefore, they are the leading deNO_x technology for light duty vehicles, where additionally limited space for the engine aftertreatment system is provided.

The NSRC concept is based on a periodic exchange between NO_x storage and reduction processes. Under typical engine operating condition, oxygen rich exhaust gas, containing NO and NO_2 enters the catalyst. During this lean phase, which may last 60 – 90 s, the LNT is able to capture and adsorb NO_x on the catalytic surface. By running the engine periodically in rich mode (3 – 5 s), unburned hydrocarbons, CO and H_2 are emitted in the exhaust manifold, which can selectively reduce stored NO_x, towards N_2. Hence, the catalytic surface is regenerated, enabling NO_x readsorption during the subsequent lean operating cycle. At this stage it has to be pointed out that the inevitable fuel penalty related to catalyst regeneration is one of the most important limitations of the LNT technology. Indicatively, considering a 2.44 l NSR catalyst, used for a 2 l diesel engine, the measured fuel consumption for a complete cycle, including a 60 sec lean and a 3 sec rich operation period can exceed 35 g [65].

In order to achieve a cyclic lean/rich operation, catalytic formulations applied for LNTs have to exhibit a series of particular properties, including: appropriate surface sites, combined with high surface area for NO_x storage, as well as catalytic positions for nitrogen oxides reduction. Alkali (eg., Na, K), alkaline earth metals (eg. Ba, Sr) and rare earth (eg. La, Ce) are mainly used for NO_x storage [66], whereas γ-Al_2O_3 and CeO_2/ZrO_2 can play the role of the washcoat layer [67], providing high surface areas. Finally, as far as surface sites for NO_x reduction are concerned, PGM metals, such as Pt, Rh and Pd have been widely applied [68]. In order to achieve high NO_x conversion rates, the PGM loading should be quite significant, reaching around 10-12 g for a 2 l light duty vehicle diesel engine. To decrease the amount of used precious metal and hence the catalyst cost, rare earth oxides, known as perovskites could be applied. The latter exhibit similar to Pt based LNTs activity at temperatures higher than 300 °C, whereas at lower ones, their performance attenuates up to 10% [69].

The chemistry of NO_x reduction on a LNT is highly complicated. During the lean operating mode, NO and NO_2 are adsorbed on the NO_x storage component and further react with oxygen to form nitrite and nitrate adspecies. NO_2 can be adsorbed through different pathways as shown in reac. 15 and reac. 16 [70,71,72,73]. On the other hand, NO is suggested to adsorb on the surface either directly (reac. 17) [72,74] or via an initial oxidation on the PGM towards NO_2 (reac. 18) which is then stored on the catalyst as described above [70,71,72].

$$3NO_2 + BaO \rightarrow Ba(NO_3)_2 + NO \qquad \text{reac. 15}$$
$$2NO_2 + {}^1/_2\,O_2 + BaO \rightarrow Ba(NO_3)_2 \qquad \text{reac. 16}$$
$$2NO + {}^3/_2\,O_2 + BaO \rightarrow Ba(NO_3)_2 \qquad \text{reac. 17}$$
$$NO + {}^1/_2\,O_2 \rightarrow NO_2 \qquad \text{reac. 18}$$

When the engine is operated at rich conditions, CO, unburned hydrocarbons as well as H_2 are contained in the exhaust stream, through which NO_x conversion occurs. In the LNT concept, nitrogen oxides reduction can be performed in both gas and solid phase, whereas based on the reducing agent different reaction pathways are involved. Reaction steps through which surface NO_x species are removed, are summarized through reac. 19 to reac. 21 [72,75]. On the other hand, in case NO_x are initially released in the gas phase (reac. 22 and reac. 23), indicative reduction mechanisms shown through reac. 24 to reac. 26 can occur. Release of gaseous NO and NO_2 may be accomplished either due to thermal effects (instability of nitrates at high temperatures) and/or because of the fact CO, hydrocarbons and H_2 create a "net reducing environment", which acts as a driving force for NO_x species decomposition towards gaseous NO and NO_2.

$$Ba(NO_3)_2 + 5CO \rightarrow N_2 + BaO + 5CO_2 \qquad \text{reac. 19}$$
$$Ba(NO_3)_2 + {}^5/_9\,C_3H_6 \rightarrow N_2 + {}^5/_2\,CO_2 + BaO + {}^5/_3\,H_2O \qquad \text{reac. 20}$$
$$Ba(NO_3)_2 + 3H_2 \rightarrow 2NO + {}^5/_2\,CO_2 + BaO + 3H_2O \qquad \text{reac. 21}$$
$$Ba(NO_3)_2 \rightarrow BaO + 2NO + {}^3/_2\,O_2 \qquad \text{reac. 22}$$
$$Ba(NO_3)_2 \rightarrow BaO + 2NO_2 + {}^1/_2\,O_2 \qquad \text{reac. 23}$$
$$NO + CO \rightarrow CO_2 + {}^1/_2\,N_2 \qquad \text{reac. 24}$$

$$9NO + C_3H_6 \rightarrow 3CO_2 + {}^{9}/_{2}\,N_2 + 3H_2O \qquad \text{reac. 25}$$

$$NO + H_2 \rightarrow H_2O + {}^{1}/_{2}\,N_2 \qquad \text{reac. 26}$$

A schematic representation of the abovementioned reaction scheme on NSR catalysts is illustrated in Figure 7.

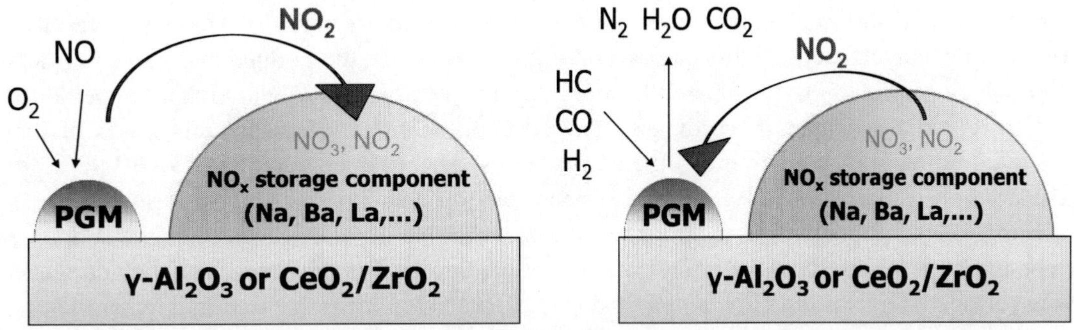

Figure 7. Schematic representation of the lean (left) and rich (right) operation of a lean NO_x trap.

As is the case for all catalytic converters, NSR catalyst selectivity is a very important issue. Based on the type of reducing agent, the targeted reduction of NO_x towards N_2 may not be the only reaction pathway. Reaction of surface NO_x species with H_2 lead to the formation of ammonia (reac. 27) [71,75]. As it is going to be discussed in the following section, this compound constitutes an alternative NO_x reducing agent and thus it can further reduce nitrogen oxides (reac. 28). Moreover, N_2O formation has been observed during NO, NO_2 reduction, which is a highly undesirable greenhouse gas. Nitrous oxide evolution may involve several reaction pathways which include interactions of NO_x with H_2 (reac. 29) [71,72,75], as well as reactions of gaseous NO with ammonia (reac. 30) [76].

$$Ba(NO_3)_2 + 8H_2 \rightarrow 2NH_3 + BaO + 5H_2O \qquad \text{reac. 27}$$

$$^{10}/_{3}\,NH_3 + Ba(NO_3)_2 \rightarrow {}^{8}/_{3}\,N_2 + BaO + 3H_2O \qquad \text{reac. 28}$$

$$Ba(NO_3)_2 + 4H_2 \rightarrow N_2O + BaO + 4H_2O \qquad \text{reac. 29}$$

$$NO + NH_3 \rightarrow N_2O + {}^{3}/_{2}\,H_2O \qquad \text{reac. 30}$$

Exhaust gas enrichment for NSR catalyst regeneration can be performed through two main strategies: i) direct fuel injection in the exhaust pipe and ii) exhaust gas enrichment in the engine cylinder. Concerning the former, fuel is periodically dosed through a dedicated nozzle upstream of the NSR converter and is subsequently evaporated due to the exhaust gas high temperature. Finally, injected hydrocarbons can be oxidized on the precious metal component and further react with H_2O to form reducing agents, (eg. H_2, CO, short chain hydrocarbons). Nevertheless, exothermicity produced due to HC oxidation can lead to excessive thermal loadings which may not be sustained by the LNT substrate. Alternatively, in-cylinder gas enrichment can be accomplished through different techniques, including inlet air throttling, proper calibration of the EGR and fuel post-injection. Such techniques however, may induce significant challenges. During the periodic lean/rich phase cycles, engine has to be precisely managed in order to achieve low fuel consumption, with a simultaneous hindrance of undesirable emissions (eg. particulates) [77].

As previously mentioned, NSR catalysts can achieve NO_x conversion efficiencies in the order of 80 – 90 %, within a temperature range of 200 – 450 °C [64]. At low temperatures, catalyst activity is limited by the difficulty to regenerate, whereas at higher ones, conversion efficiency is limited due to the decomposition of thermally unstable nitrites/nitrates. Moreover, high space velocities may negatively affect catalyst performance, due to low NO_x storage rate and short residence times for efficient reduction. Recently, Toyota has introduced a highly deNO_x efficient fuel dosage technology, know with the commercial name "Di-air" [78], which enable to further expand lean NO_x traps operating window to higher temperatures and increased flow rates. According to this concept, short period, oscillating doses of fuel upstream of the NSR converter can maintain catalyst performance in the order of 80-90%, for temperatures as high as 650 °C, which is impossible for conventional LNT systems. The main principle of this regeneration technique is the fact that unburned hydrocarbons are supposed to be partially oxidized on the PGM component of the lean NO_x trap and thus they can react with surface NO_x forming C-N containing species, which are characterized by greater thermal stability compared to that of nitrites/nitrates [79].

One of the key concerns in NSR technology is its vulnerability to Sulfur. Diesel fuel and lubrification oil are main sources of sulphur, existing in diesel engines exhaust stream, mainly in the SO_2 form. Storage components of the NSR catalyst act as an excellent sulphur oxides (sulfates) trap, which leads to a significant attenuation of catalyst NO_x storage ability (sulfation). Different types of sulfate surface species can be formed, which exhibit thermal stability in the range of 500 – 800 °C [80,81]. Sulfur removal from the catalyst can be achieved at net reducing conditions and at temperatures higher than 550-600 °C. (desulphation) [80]. Nevertheless, at such high temperatures PGM components can be sintered and consequently become inactive for NO_x conversion (thermal degradation) [82]. A complete desulfation event in a real passenger car application may last 10-20 minutes and is apparently associated with fuel penalty in order to produce rich gas mixtures. Depending on the fuel sulphur level, the intervals between desulfation events could be in the order of 1,000 to 5,000 km of mileage.

Selective Catalytic Reduction (SCR)

Given the fact that the operation of the lean NO_x trap is related to a significant fuel penalty and limited activity at high temperature regime, its commercial use is significantly questioned, particularly due to forthcoming stringent emissions standards. An alternative to the LNT technology is the Selective Catalytic Reduction (SCR) of NO_x through ammonia, which enables to achieve high NO_x conversion rates over a broad range of diesel engines operating conditions. Since liquid NH_3 is difficult to be stored and handled, the SCR technology employs a standardized 32.5% aqueous urea solution ($(NH_2)_2CO + H_2O$), widely known with the commercial name AdBlue, which is used as an ammonia precursor. At typical exhaust temperatures, urea is initially decomposed to NH_3, which can selectively and continuously reduce NO_x on the catalytic surface. New materials ensure NO_x conversion as high as 100%, over a wide range of operating conditions. On the other hand, system complexity, as well as the urea storage and management are important limitations to be overcome.

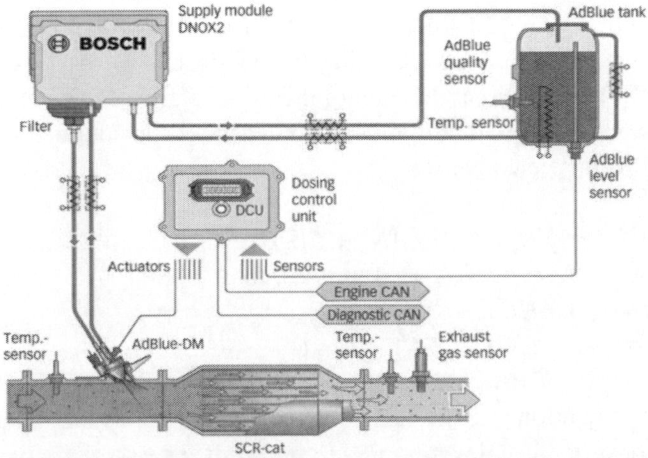

Figure 8. Urea-SCR system configuration [83].

A typical Urea-SCR system configuration is illustrated in Figure 8. The urea aqueous solution, stored in an auxiliary on board tank, is directly injected inside the exhaust pipe, upstream of the SCR catalyst. First, urea has to be homogeneously mixed with the exhaust gas. For this purpose static mixers may be used. Subsequently, thanks to exhaust gas high temperature, urea undergoes hydrolysis and thermal decomposition leading to the formation of gaseous ammonia. Urea decomposition towards NH_3 occurs at temperatures higher than 150 °C, which can be easily achieved under typical driving conditions. However, during engine cold-start operation, exhaust gas temperature may not be sufficiently high to allow high decomposition rates. In such cases undesirable urea byproducts deposits can be formed, including solid biuret and cyanuric acid [84]. These compounds can accumulate on several surfaces of the Urea-SCR system (including tube walls, injector, mixer surfaces and even the inlet cross-section of the actual SCR catalyst) and lead to an increase of exhaust line backpressure and materials erosion. To avoid the abovementioned phenomena, optimization of the injector positioning can be performed, hence ensuring the SCR catalyst sufficient dosing and the SCR system durability [85,86]. Alternatively, the recent Amminex technology has been proposed, which involves the direct injection of gaseous NH_3 [87]. In this concept, ammonia contained in a metal based amine complex ($M_a(NH_3)_nX_z$:M=Li, Mg, Ce etc and X=one or more ions) is safely stored in an auxiliary storage tank. Carefully dosing H_2O in this tank, NH_3 is replaced by water and thus the former can be released in the gas phase. This way, gaseous NH_3 is directly available for NO_x reduction, avoiding the formation of undesirable byproducts related to urea decomposition.

Generated gaseous ammonia, accesses then the SCR catalytic converter, which is a flow through monolith. There, NH_3 can selectively react with NO and NO_2 forming nitrogen and water. The SCR of NO_x through ammonia is a quite complicated process, involving several reaction pathways and intermediate species. A global reaction scheme is summarized through reactions 22 to 26. In case NO is in excess in the diesel exhaust stream, (reac. 31) predominantly occurs, which is called the "Standard SCR" reaction. When equimolar amounts of NO and NO_2 are contained in the feed, the (reac. 32) takes place. This reaction is characterized by a much higher reaction rate than the "Standard-SCR" and hence it is named as "Fast-SCR". For NO_2/NO_x ratios higher than 50%, NH_3 mainly reacts with NO_2 according to the (reac. 33), annotated as "NO_2-SCR". Finally, a very slow NO reduction by NH_3 may occur without oxygen consumption, according to the (reac. 34). At this stage it has to be pointed out that the NO/NO_2 ratio at the inlet of the SCR catalyst can be controlled by a diesel oxidation catalyst being systematically close coupled with the internal combustion engine, upstream of the SCR catalytic converter.

$$4NH_3 + 4NO + O_2 \rightarrow 4N_2 + 6H_2O \qquad \text{reac. 31}$$
$$4NH_3 + 2NO + 2NO_2 \rightarrow 4N_2 + 6H_2O \qquad \text{reac. 32}$$
$$8NH_3 + 6NO_2 \rightarrow 7N_2 + 12H_2O \qquad \text{reac. 33}$$
$$4NH_3 + 6NO \rightarrow 5N_2 + 6H_2O \qquad \text{reac. 34}$$

Besides the abovementioned SCR reactions, undesirable chemical phenomena may take place in the SCR converter, leading to a deterioration of catalyst activity. NH_3 oxidation is considered as one of the most critical ones, since it leads to an over-consumption of the reducing agent, at temperatures higher than 420 °C [88]. Ammonia oxidation is reported to occur through an initial evolution of NO (reac. 35), which is subsequently reduced by NH_3, according to the Standard SCR reaction (reac. 31), yielding N_2 and H_2O [89].

$$4NH_3 + 5O_2 \rightarrow 4NO + 5H_2O \qquad \text{reac. 35}$$

Low temperature catalyst deactivation, known as "the ammonia blocking effect" is another significant SCR inhibiting mechanism. At temperature regime as low as 50 ºC, interactions between NH_3 and NO_x can lead to the formation of intermediate species, particularly ammonium nitrate (NH_4NO_3) [90]. The latter remains stored on the catalytic surface up to 170 ºC, which limits catalyst light-off ability at this temperature range. Moreover, at higher temperatures the latter intermediate species are decomposed, leading to the formation of undesirable gaseous, species such as NO_2, N_2O and HNO_3 [91]. The ammonia blocking effect attenuates with increasing temperature and it totally disappears at 400 °C [92].

So far, several catalytic formulations have been tested for automotive SCR catalysts, including PGM, vanadium/tungsten and transition metal exchanged zeolites. Pt-based catalysts (Pt/Al_2O_3) have been reported to exhibit high $deNO_x$ performance within a temperature range of 150 – 230 °C, whereas further temperature increase leads to a deterioration of NO conversion due to the catalytic oxidation of ammonia [93]. In order to achieve high $deNO_x$ efficiency at increased temperatures, extruded vanadium-based catalysts were introduced [94]. These catalysts constitute homogeneous mixtures of titanium dioxide (TiO_2), tungsten trioxide (WO_3) (or molybdenum trioxide - MO_3) and divanadium pentoxide (V_2O_5), formulated in honeycomb shapes. SCR activity of such catalysts becomes limited however above 450 °C because of deactivation, related to modifications of the catalyst structure (phase transformation of TiO_2 anatase into TiO_2 rutile), combined with vanadia species release in the atmosphere [95]. Recently, transition-metal exchanged zeolites have been proposed as a promising candidate for automotive SCR applications, due to their enhanced deNO_x performance over a broad range of temperatures, as well as significant resistance to hydrothermal ageing. Cu- Fe- and Mn exchanged zeolites based catalysts have already been applied in the automotive sector, showing very satisfactory performance [96,97,98]. At low temperatures their activity is based on their high NO_x selectivity, whereas at high ones it is limited due to durability.

Concerning the latter, exposure of metal exchanged zeolites to temperatures higher than 750 °C and water vapor causes irreversible modifications of catalysts properties, leading to permanent deactivation [99]. This phenomenon, known as hydrothermal ageing may have two major effects on the catalyst: i) collapse of the zeolite structure due to dealumination (migration of Al^{3+} ions out of the zeolitic frame), which leads to a significant deterioration of catalytic acitivity and loss of active surface area and ii) aggregation of metallic species towards SCR inactive metallic particles. Thus catalyst selectivity is negatively affected, whereas undesirable side reactions (such as NH_3 oxidation) are promoted.

Irrespectively the type of catalytic material of an SCR catalytic converter, the complexity of the Urea-SCR system, particularly the urea storage and dosing management call for on board electronic management. Regarding the former, AdBlue freezes at temperatures lower than -11 °C, which can lead to a complete blocking of urea supply during low temperature winter conditions. To minimize the risk of urea injection problems, urea supply lines are heated, whereas the dosing system is emptied through by-pass ducts, when the engine is turned off [100]. Finally, as far as the urea injection management is concerned, AdBlue injection rate is typically managed through a closed control system, based on a pre-programmed map of engine NO_x emissions [101]. This control strategy ensures not only NO_x conversion efficiency as high as 98%, but enables to protect the catalytic converter against ageing or poisoning as well.

Ammonia Slip Catalyst (ASC)

The main challenge for the NH_3-SCR systems is to maximize the de-NO_x efficiency and simultaneously minimize the ammonia emissions to the environment. However, under certain operating conditions, an increase in NO_x conversion efficiency can only be achieved by increasing the reductant quantity and consequently allowing for higher ammonia slip [102]. One solution to reduce the amount of the released NH_3 is to add an ammonia slip catalyst (ASC) downstream of the SCR device. The NH_3 oxidation catalyst can be either a separate monolith or a small zone integrated in the rear part of the SCR catalyst.

The desired reaction in the ammonia slip catalyst is the oxidation of NH_3 towards N_2:

$$4NH_3 + 3O_2 \rightarrow 2N_2 + 6H_2O \qquad \text{reac. 36}$$

Ammonia slip catalysts for automotive applications are PGM (Platinum Group Metal) based catalysts, supported on inorganic oxides [103]. Unfortunately, ammonia oxidation on platinum is not selective for nitrogen and at high temperatures other non-desired reactions forming NO and N_2O are also observed [104]:

$$4NH_3 + 5O_2 \rightarrow 4NO + 6H_2O \qquad \text{reac. 37}$$

$$4NH_3 + 4O_2 \rightarrow 2N_2O + 6H_2O \qquad \text{reac. 38}$$

Due to the requirements for low packaging volume and compact system design, the NH_3 oxidation functionality is usually directly implemented as a short zone in the rear part of the SCR. Two different ASC configurations are illustrated in Figure 9. In the case of the single-layer ASC (see Figure 9a), the NH_3 slipping from the SCR is oxidized on the ASC layer. The oxidation products however may include NO and N_2O, besides N_2, which results in a deterioration of the overall NO_x conversion efficiency. Recently, ammonia slip catalysts with a dual-layer configuration have been proposed (see Figure 9b). In these systems, the lower layer is a platinum-based catalyst to oxidize the NH_3 while the top layer is an SCR catalyst. The unselective NO_x products, formed in the lower ASC layer, must diffuse back to the upper SCR layer where they react with NH_3 to form N_2, according to the NH_3-SCR reactions. In this way, the addition of the top SCR layer improves the selectivity of the ASC towards N_2 but it may lead to a decreased NH_3 conversion efficiency. This is because the SCR layer acts as a diffusive barrier for the NH_3 to reach the ASC layer, resulting in an optimization problem of the SCR layer thickness and morphology [105].

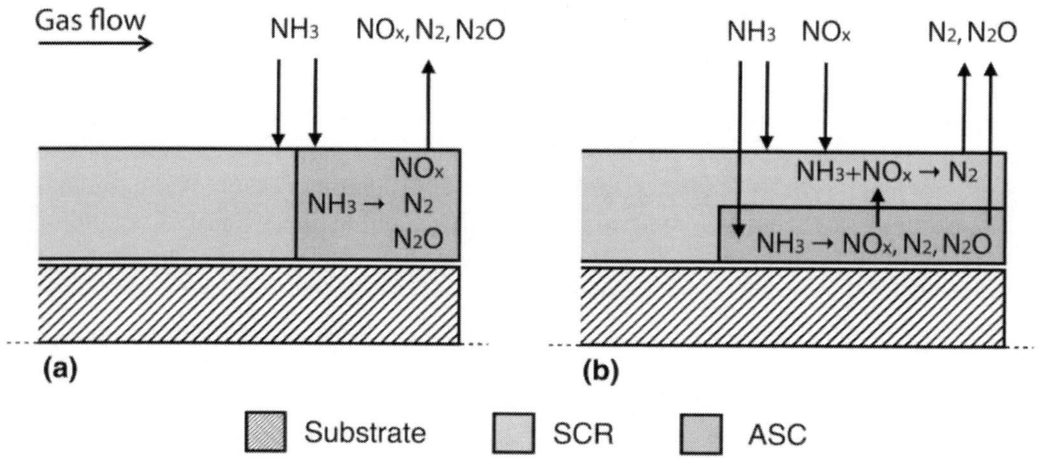

Figure 9. Schematic representation of two ASC configurations, single layer (a) and dual layer (b), integrated in the rear part of the SCR monolith.

A possible solution to limit the decrease in NH_3 conversion, due to diffusional limitations, is to create a single washcoat layer in which the PGM and SCR catalyst particles are mixed together [102]. The slipping NH_3 reacts on the SCR sites with the unreacted NO_x of the feed stream or with the NO produced by the NH_3 oxidation, increasing the overall selectivity towards N_2. The amount of Pt that is located in the combined washcoat layer is low and does not significantly affect the NO_x conversion efficiency. In this way, the positive interaction between PGM and SCR chemistries is still exploited while the decrease in NH_3 conversion is much less pronounced.

Combined SCR/DPF Systems (SCRF®)

The efficient removal of NO_x and particulate matter from the diesel exhaust is typically performed by separate devices [106]. In order to decrease the number of the components and the associated packaging volume and costs, there is great interest in combining various functionalities into one aftertreatment device. Recently, catalyst manufacturers focused on the development of combined DPF+SCR systems (referred to as SCRF® or SDPF) in which the porous walls of the DPF are impregnated with SCR catalytic material. Cu- and Fe-zeolite based catalysts are more appropriate and commonly used as SCR catalysts on DPF [107] due to their higher thermal stability. For light-duty applications, Cu-zeolite catalysts are more suitable due to the higher activity at low temperatures and the lower dependency on NO_2. Apart from the technological challenge of merging the de-NO_x and PM control functionalities, the performance of the SCRF®, compared to that of the individual SCR and DPF systems, is a key aspect [108]. The cross-interactions between NO_x conversion and soot oxidation for the general case of a soot loaded SCRF® must be considered.

The integration of SCR coating on DPF results in a competitive NO_2 consumption in both soot passive regeneration and SCR reactions. Watling et al. [109] performed a simulation study of the de-NOx activity effect on filter regeneration. The authors examined the rate of soot oxidation for a Cu-SCRF®, both in the presence and absence of NH_3 (see Figure 10). For exhaust temperatures up to 400°C the rate of soot removal was significantly decreased as SCR reactions greatly reduce the available NO_2 for soot oxidation. A different picture was observed for higher temperatures where the soot burn rate is actually enhanced by the exothermicity generated by the de-NO_x reactions.

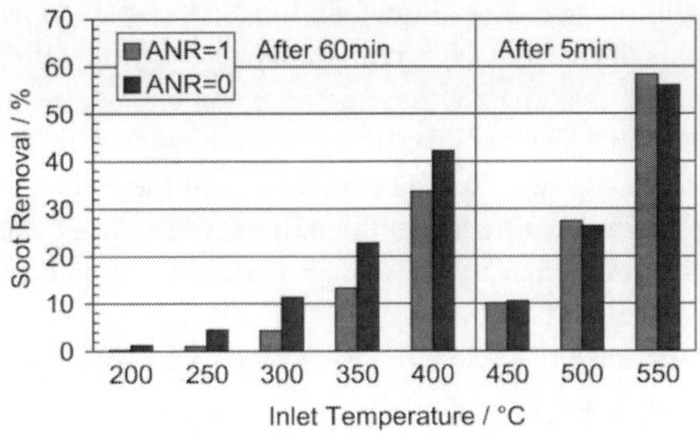

Figure 10. Predicted soot removal in a 60 (200-400°C) min period or a 5 (450-550°C) min period with (ANR=1) and without (ANR=0) SCR activity. (Feed: 500$\mathbf{ppm\ NO_x}$, 15% $\mathbf{O_2}$, 55000$\mathbf{h^{-1}}$ space velocity). Reprinted from Watling et al. [109].

Another aspect of the competitive interactions taking place in a wall-flow SCR is the effect of soot on NO_x conversion. According to the study of Colombo et al. [108], soot has a negative impact on NO_x conversion for typical exhaust conditions ($NO_2/NO_x<0.5$). The passive regeneration of soot by NO_2 leads to an alteration of the local NO_2/NO_x ratio towards smaller values while the NO_x conversion rate is found to be maximum for $NO_2/NO_x=0.5$. Furthermore, the gaseous NO_x and NH_3 species face a diffusional barrier when flowing through the soot layer, with the same effect also caused by the accumulated ash in long-term use. The negative effect of mass-transfer limitations on NO_x conversion efficiency is much more pronounced at high temperatures and space velocities.

GASOLINE TECHNOLOGIES

Three-Way Catalyst (TWC)

The three main pollutants found in the exhaust gas of gasoline engines are nitrogen oxides (NO_x), carbon monoxide (CO) and unburned hydrocarbons (HCs). For the efficient control of these emissions, three-way catalytic converters (TWC) have been widely used for more than 30 years. When catalysts were first introduced in the late 1970s they were basically 2-way, oxidizing CO and HC towards CO_2. The additional functionality for the simultaneous reduction of NO_x, incorporated a few years later, necessitated the operation at a stoichiometric air/fuel ratio which resulted in more complex systems utilizing closed-loop control of the combustion stoichiometry. Today, three-way catalysts are a standard and mature technology, being the most successful aftertreatment system for spark-ignited engines.

The main substrates used in TWC applications are honeycomb ceramic monoliths, typically made of cordierite. For the washcoat material, alumina (Al_2O_3) is the common choice due to its high surface area and good resistance to hydrothermal ageing. Components such as cerium (Ce) and zirconium (Zr) are added to the washcoat in order to promote the oxygen storage capability of the catalyst. These oxygen storage components (OSC) act as buffers, storing oxygen in lean conditions and releasing it in rich conditions [110]. The noble metals, impregnated into the washcoat, are a combination of platinum (Pt), palladium (Pd) and rhodium (Rh). Platinum and palladium are the active components for CO and HC oxidation while rhodium is primarily a NO_x reduction catalyst. Since Pt and Pd serve different roles than Rh, it is of interest to separate these active components and minimize the negative interactions between them. This is achieved by physical separation of two or more layers. Usually Rh is located in the top layer and is exposed to all the reductant species that reduce NO_x before the exhaust gases are diffused to the bottom Pt-Pd layer where they are oxidized [4]. The multiple layer design enhances the overall performance and increases the durability by preventing sintering and alloy formation.

The main reactions that take place in a three-way catalytic converter and contribute to the conversion of the main SI engine pollutants are [4,5]:

Oxidation reactions

$$CO + {}^{1}/_{2}\,O_2 \rightarrow CO_2 \qquad \text{reac. 39}$$

$$H_2 + {}^{1}/_{2}\,O_2 \rightarrow H_2O \qquad \text{reac. 40}$$

$$C_xH_y + \left(x + \frac{y}{4}\right) O_2 \rightarrow xCO_2 + \frac{y}{2} H_2O \qquad \text{reac. 41}$$

NOx reduction

$$CO + NO \rightarrow CO_2 + \frac{1}{2} N_2 \qquad \text{reac. 42}$$

$$C_xH_y + \left(2x + \frac{y}{2}\right) NO \rightarrow xCO_2 + \frac{y}{2} H_2O + \left(x + \frac{y}{4}\right) N_2 \qquad \text{reac. 43}$$

$$H_2 + NO \rightarrow H_2O + \frac{1}{2} N_2 \qquad \text{reac. 44}$$

Steam reforming

$$C_xH_y + xH_2O \rightarrow xCO + (x + \frac{y}{2})H_2 \quad \text{reac. 45}$$

Water-gas shift

$$CO + H_2O \rightarrow CO_2 + H_2 \quad \text{reac. 46}$$

In order for the NO_x, CO and HC emissions to be treated at the same time in the TWC, the engine has to be operated at a stoichiometric air/fuel ratio. In this case, enough reducing species to reduce the NO (reac. 42 to reac. 44) and enough O_2 to oxidize the CO and HC (reac. 39 and reac. 41) can be found simultaneously in the exhaust gas. As an exact stoichiometric air/fuel ratio cannot be achieved at all times, the catalyst should be able to reduce NO under slightly lean conditions and remove CO and HC when a slight deficiency of oxygen is observed. The water-gas shift and the steam-reforming reactions can additionally consume CO and HC respectively, providing extra reaction pathways.

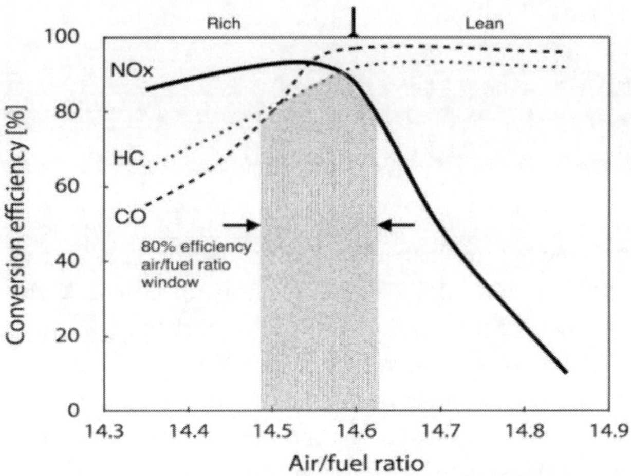

Figure 11. Conversion efficiency of NO, CO and HC as a function of air/fuel ratio, in a three-way catalytic converter.

Figure 11 shows the conversion efficiency of CO, NO and HC as a function of the air-fuel ratio. It is evident that there is a narrow window of air-fuel ratios, between 14.5 and 14.6, where high conversion efficiencies for all the 3 pollutants can be achieved. In order for the engine to operate in such a restricted regime, a closed loop fuel injection control is required. An oxygen (λ) sensor is placed upstream of the TWC which indicates if the engine is operating in lean or rich conditions. The signal from this oxygen sensor is used as a feedback from the electronic control unit (ECU) in order to adjust the fuel injection system and achieve the desired air/fuel ratio. However, the system's response is not instantaneous mainly due to the gas travel time and the response delay of the λ sensor. This lag of the control system causes the air/fuel ratio to oscillate around the stoichiometric value and has high practical interest. The conversions of NO, CO and HC in a TWC operated with cyclic variations of the air/fuel ratio are larger than the estimates based on the summation of steady-state values during the cycle [5]. The improved performance is mainly attributed to the presence of the oxygen storage components, which have the ability to undergo reduction-oxidation reactions [111]. In its oxidized state the storage component can provide oxygen for CO and HC oxidation when exhaust-gas conditions are rich. When the exhaust cycles to lean conditions the storage component, in its reduced state now, is able to store NO or O_2 and release them in the next rich cycle. The washcoat component that plays the most important role in such dynamic oxidation-reduction phenomena is cerium. Under oxidizing conditions the following Ce oxide reactions take place:

$$Ce_2O_3 + \frac{1}{2}O_2 \rightarrow 2CeO_2 \qquad \text{reac. 47}$$

$$Ce_2O_3 + NO \rightarrow 2CeO_2 + \frac{1}{2}N_2 \qquad \text{reac. 48}$$

$$Ce_2O_3 + H_2O \rightarrow 2CeO_2 + H_2 \qquad \text{reac. 49}$$

$$Ce_2O_3 + CO_2 \rightarrow 2CeO_2 + CO \qquad \text{reac. 50}$$

On the other hand, under reducing conditions, CeO_2 functions as an oxidizing agent according to the following reactions:

$$2Ce_2O_2 + CO \rightarrow Ce_2O_3 + CO_2 \qquad \text{reac. 51}$$

$$2Ce_2O_2 + H_2 \rightarrow Ce_2O_3 + H_2O \qquad \text{reac. 52}$$

$$C_xH_y + 2\left(x + \frac{y}{2}\right)CeO_2 \rightarrow \left(x + \frac{y}{2}\right)Ce_2O_3 + xCO + \frac{y}{2}H_2O \qquad \text{reac. 53}$$

Legislation dictates the periodical monitoring of the TWC functionality and the indication of a malfunction, in case poor performance is observed. A good indication of the catalytic performance is the ability of the catalyst to store oxygen. In order to assess the oxygen storage capacity, a second oxygen sensor is added downstream of the catalyst. The rear sensor is also used for a finer control of the stored oxygen amount in the catalyst.

With ever tightening emissions regulations there is an increasing pressure to improve the catalyst performance and at the same time keep the costs to a minimum. The development of new high-performance technologies, with better thermal stability and more efficient use of precious metals, is required. In order to create a catalyst with improved performance it is essential to suppress the degradation of precious metals as much as possible, while to lower the system cost the main focus is to reduce the amount of rhodium used. A zone-coated TWC with ~20% higher NO_x conversion efficiency and ~40% less rhodium content was proposed by Aoki et al.[112]. The zone-coating technology applies different coatings to different zones of the catalyst layer in order to convert the NO, CO and HC emissions more efficiently. Moreover, the introduction of an alumina diffusion barrier suppresses the Rh grain growth resulting in a decreased catalyst degradation. In another study, a two brick TWC system with a 50% decrease in rhodium usage has been proposed by Matsuzono et al. [113]. This system makes extensive use of palladium whose price is significantly lower than Pt and Rh. It comprises of a Palladium-only closed couple catalyst and a Palladium-Rhodium underfloor catalyst. The latter was developed through the use of ZrO_2, an oxygen storage component which has the ability to limit the deactivation of rhodium. In another approach, a new catalyst formulation with advanced washcoat materials and low PGM loading was developed [114]. This formulation showed that it is possible to achieve high performances and meet the legislative limits even with low or ultra low rhodium loadings.

Gasoline Particulate Filter (GPF)

Gasoline direct injection (GDI) is a widely applied engine technology allowing to effectively reduce CO_2 emissions from light duty vehicles. Nevertheless, this fuel dosage technique leads to significant particles emissions, greater than those related to conventional port-fuel injection engines. Interestingly, even though the soot mass emitted by gasoline engines is orders of magnitude lower compared to diesel PM, the number of particles is significantly high [115]. Indicatively, for a 1.6 l turbocharged engine, the number of tailpipe particles could be in the order of $2 \cdot 10^{12}$ particles/km [116], more than 3 times higher the Particle Number limit of the European legislative standards. Similarly to the diesel particulate filter concept, gasoline particulate filters, known as GPFs are currently considered as a sustainable and cost effective aftertreatment technology, allowing to efficiently control both mass and number of particles emitted by gasoline internal combustion engines [117]. Filtration efficiencies of optimized filters may be as high as 90% during NEDC cycles [118]. The most challenging requirement of the GPF is the capability to achieve filtration of a large number of particles with a simultaneous limited pressure drop, in order to maintain vehicle low consumption and hence low CO_2 emissions.

The simplest configuration of a GPF consists of a single component wall flow monolith, typically placed downstream of a three-way catalyst (add-on GPF). Kattouah et al. have recently presented filtration efficiency measurements over different GPFs (5-6mil/200-360 cpsi), used on 1.4 and 1.8 GTDI engines [119]. Filters performance was tested over a broad range of driving cycles, showing particles emissions reduction within the EURO 6 standards and a simultaneous very slight increase in CO_2 emissions. Alternatively, catalyzed gasoline particulate filters can be applied. This concept includes a cordierite, wall flow monolith, washcoated with catalytic formulations similar to those used for TWCs. The reader is referred to [120] for more details on the functional aspects of the wall-flow versus the flow-through reactor design. Moreover, non-conventional metal foams with nominal pore sizes of 1200, 800, 580 and 450 μm from the inside out have been reported elsewhere [121]. Catalyzed gasoline particulate filters can be washcoated with active PGM materials, acting hence as three-way catalytic converters themselves. Therefore, CO, HC and NO_x conversion could be performed in parallel with particles filtration in a single aftertreatment device [122]. The PGM component enhances filter regeneration through oxygen. Similarly, other researchers [119,116] have reported very encouraging results concerning catalyzed GPFs filtration efficiency and backpressure during homologated driving cycles.

In order to achieve enhanced GPF filtration and regeneration, filter geometry and material properties optimization, as well as OBD systems are required. Regarding the former, increased GPF diameters, combined with thin walls and low cells density are preferable in order to minimize backpressure. Moreover, filter length can be properly adjusted so as to meet ash storage requirements. Finally, as far as filter regeneration is concerned, challenges similar to those earlier reported for DPFs have to be tackled. Monitoring the air/fuel ratio oscillations downstream of the GPF can enable to identify suitable conditions for the filter regeneration, as described in [123].

CONCLUSION

Since the 1975 Clean Air Act and the 1980s early catalytic systems, exhaust gas aftertreatment technologies have evolved impressively. Advanced catalytic converters ensure highly efficient emissions control over a broad range of operating conditions, also exhibiting resistance against ageing and chemical deactivation.

Given the importance of emission control in the design of current and future vehicles, the powertrain engineer should have a global understanding on catalytic systems operating principles, applications and challenges. In this respect, the fundamentals of flow through and wall flow monoliths were discussed and the governing phenomena were mathematically described. Automotive catalytic converters are multi-functional units, in which a series of phenomena simultaneously occur, including: flow phenomena, mass and heat transfer phenomena as well as chemical reactions. From an engineering point of view, the deep understanding of the abovementioned aspects is essential in order to optimize catalysts operation and ensure durability. Mathematical modeling and the use of CAE is an essential tool for design optimization as the complexity of the exhaust system increases.

We briefly reviewed the state of the art of aftertreatment technologies. Concerning diesel vehicles, DOC, DPF and LNT or Urea-SCR catalytic converters are the commercially available catalytic systems for pollutants abatement under lean conditions. DPFs are the leading technology for removing PM from diesel engines exhaust stream for more than a decade, whereas Urea-SCR catalysts are dominating the field of NO_x emissions control in heavy-duty applications. Lean NO_x traps continue to play an important role in passenger car applications for diesel and lean-burn gasoline engines. Currently, significant work is ongoing in order to combine CDPF and SCR functionalities in the same catalytic device, in an attempt to enhance $deNO_x$ performance and efficient particulate matter filtration in limited space. On the other hand, catalyzed GPFs tend to replace conventional three-way catalysts in gasoline lean burn engines. This emerging technology could be essential in the future to allow automotive manufacturers to comply with GDI engines PM emissions.

Electronic engine management is substantial in order to achieve optimal operation of modern highly sophisticated catalytic systems, as well as to detect system failures. Finally, on-board diagnostics for each aftertreatment technology were briefly discussed. Forthcoming legislation imposes certain requirements for OBD systems, leading to the development of more reliable and catalytic systems.

Integration of different catalytic converters into the same exhaust layout or even into the same device is a very challenging work, which calls for extensive testing. Nowadays, the task of optimizing the performance of modern exhaust aftertreatment systems can be effectively faced through computer aided engineering. Combination of mechanical and chemical engineering with advanced material sciences can provide OEMs and automotive manufacturers with solutions for enhancing catalytic activity, ensuring system reliability and reducing costs. Predictive mathematical models enable the optimization of catalysts geometry, whereas phenomenological kinetic models can accurately estimate pollutants conversion, under realistic driving conditions. The development of innovative and cost-effective exhaust after-treatment technologies is a key enabler for the long-term sustainability of the internal combustion engine.

NOMENCLATURE

A. Latin letters

a_1	constant in channel pressure drop correlation	$-$
A_{ze}	pore size distribution constant	$(mol/J)^2$
c	concentration	mol/m^3
C_P	specific heat capacity	$J/(kg \cdot K)$
d_h	hydraulic diameter of a channel	m
D_{knud}	Knudsen diffusivity	m^2/s
D_{mol}	molecular diffusivity	m^2/s
d_{pore}	mean pore size	m
D_w	effective diffusivity	m^2/s
E	activation energy of reaction	J/mol
f_w	geometric parameter	$-$
h	heat transfer coefficient	$W/(m^2 \cdot K)$
H	heat source component	W/m^3
k	kinetic constant of reaction	units depend on reaction
k_j	mass transfer coefficient	m/s
M	molecular weight	kg/mol
n	stoichiometric coefficient	$-$
p_j	partial pressure	Pa
$p_{sat,j}$	saturation pressure	Pa
R	universal gas constant	$J/(mol \cdot K)$
r	radial coordinate	$-$
R_j	reaction rate	$mol/(m^3 \cdot s)$
S	heat source term	W/m^3
S_F	monolith specific surface area	m^2/m^3
T	temperature	K
t	time	s
v	velocity	m/s
w	dimension perpendicular to wall surface	$-$
W_0	total volume of micropores per reactor volume	m^3/m^3
w_c	washcoat layer thickness	m
w_p	soot layer thickness	m
w_w	wall thickness	m
x_{eq}	adsorbed mass at equilibrium	kg
y_j	molar fraction	$-$
z	axial coordinate along monolith	m

B. Greek letters

β	reaction order with regard to O_2	–
β_j	affinity coefficient	–
γ	surface coverage dependence factor	–
ΔH	reaction heat	J/mol
ε	macroscopic void fraction	–
ε_{pore}	porosity of the washcoat	–
θ_{NH_3}	surface coverage of NH_3	–
λ	thermal conductivity	$W/(m \cdot K)$
μ	dynamic viscosity	$Pa \cdot s$
ρ	density	kg/m^3
τ	tortuosity	–

C. Subscripts and Superscripts

g	exhaust gas
j	species index
k	reaction index
s	solid
w	substrate wall

REFERENCES

[1] Sher, E. (1998). Handbook of air pollution from internal combustion engines: pollutant formation and control. USA, Access Online by Elsevier, pp. 5.

[2] Environmental Protection Agency: EPA History: The Clean Air Act of 1970. http://www2.epa.gov/aboutepa/epa-history-clean-air-act-1970. accessed 23-10-2013.

[3] Environmental Protection Agency: Summary of the Clean Air Act: 42 U.S.C. §7401 et seq. (1970). http://www2.epa.gov/laws-regulations/summary-clean-air-act. accessed 23-10-2013.

[4] Twigg, M. V. (2007). Progress and future challenges in controlling automotive exhaust gas emissions. *Applied Catalysis B: Environmental*, 70(1), 2-15.

[5] Koltsakis, G. C., Stamatelos, A. M. (1997). Catalytic automotive exhaust aftertreatment. *Progress in Energy and Combustion Science*, 23(1), 1-39.

[6] Schlatter, J. C., Mitchell, P. J. (1980). Three-way catalyst response to transients. *Industrial & Engineering Chemistry Product Research and Development*, 19(3), 288-293.

[7] European Environment Agency: http://www.eea.europa.eu/data-and-maps/indicators /eea-32-nitrogen- oxides-nox-emissions-1. accessed 13-1-2013.

[8] Takahashi, N., Shinjoh, H., Iijima, T., Suzuki, T., Yamazaki, K., Yokota, K., Suzuki, H., Miyoshi, N., Matsumoto, S. i., Tanizawa, T., Tanaka, T., Tateishi, S. S. and Kasahara, K. (1996). The new concept 3-way catalyst for automotive lean-burn engine: NO_x storage and reduction catalyst. *Catalysis Today*, 27, 63-69.

[9] Matsumoto, S. (1996). $DeNO_x$ catalyst for automotive lean-burn engine. *Catalysis Today*, 29, 43-45.

[10] Morita, I., Hirano, M. and Bielawski G. T. (1998). Development and Commercial Operating Experience of SCR $DeNO_x$ Catalysts for Wet-Bottom Coal-Fired Boilers. Conf.: Power-Gen International '98, 12/1998, Orlando, Florida, U.S.A..

[11] Gibson, J. and Groene, O. (1991). Selective catalytic reduction on marine diesel engines. *Automotive Engineering*, 99, 18-22.

[12] Schommers J, Enderle C, Breitbach H, Lindemann B, Stotz M and Paule M. (2006) 15. Aachener Kolloquium. Aachen, Germany.

[13] Clemmens, W. B., Sabourin, M. A., Rao, T. (1990). Detection of catalyst performance loss using on-board diagnostics. *SAE transactions*, 99(4), 43-60.

[14] Heywood, J. B. (1988). Internal combustion engine fundamentals. New York: McGraw-Hill, Inc.

[15] Majewski, W. A., Khair, M. K. (2006). Diesel emissions and their control: SAE International.

[16] Heck, R. M., & Farrauto, R. J. (2002). Catalytic air pollution control: John Wiley & Sons, Inc.

[17] Murzin, D., Salmi, T. (2005). Catalytic kinetics: Elsevier Science & Technology Books.

[18] Santos, H., Costa, M. (2008). The relative importance of external and internal transport phenomena in three way catalysts. *International Journal of Heat and Mass Transfer*, 51, 1409-1422.

[19] Koltsakis, G. C., Konstantinidis, P. A., Stamatelos, A. M. (1997). Development and application range of mathematica models for 3-way catalytic converters. *Applied Catalysis B: Environmental*, 12, 161-191.

[20] Hawthorn, R. D. (1974). Afterburner catalysts: effects of heat and mass transfer between gas and catalyst surface. *A.I.Ch.E. Symposium Series*, 70, 428-438.

[21] Wheeler, A. (1955). Catalysis, p. 105.

[22] Wakao, N., Smith, J. M. (1962). Diffusion in catalyst pellets. *Chemical Engineering Science*, 17, 825.

[23] Konstandopoulos, A. G., Kostoglou, M., Skaperdas, E., Papaioannou, E., Zarvalis, D., Kladopoulou, E. (2000). Fundamental studies of diesel particulate filters: transient loading, regeneration and aging. *SAE technical paper*, 01-1016.

[24] Bischof, C., Boger, T., Gunasekaran, N., Bhargava, R., (2012). Advanced Particulate Filter Technologies for Direct Injection Gasoline Engine Applications. Presentation at US Department of Energy Directions in Engine Efficiency and Emissions Research (DEER) Conference, Dearborn, Michigan, USA

[25] Kattouah, P., Kato, K., Ohara, E., Vogt, C. D., (2013). Ceramic Wall Flow Filter for Particulate Emission Reduction of Petrol Engines. Presentation at Cambridge Particle Meeting, Cambridge, UK.

[26] Choi, K., Kim, J., Ko, A., Myung, C. L., Park, S., Lee, J. (2012). Size-resolved engine exhaust aerosol characteristics in a metal foam particulate filter for GDI light-duty vehicle. *Journal of Aerosol Science*, 57, 1–13.

[27] Bissett, E. J. (1984). Mathematical Model of the Thermal Regeneration of a Wall-Flow Monolith Diesel Paticulate Filter. *Chemical Engineering Science*, 39, 1233-1244.

[28] Koltsakis, G., Haralampous, O., Depcik, C., & Ragone, J. C. (2013). Catalyzed diesel particulate filter modeling. *Reviews in Chemical Engineering*, 29(1), 1-61.

[29] Haralampous, O., Koltsakis, G. (2002). Intra-layer temperature gradients during regeneration of diesel particulate filters. *Chemical Engineering Science*, 57(13), 2345-2355.

[30] Lafossas, F., Matsuda, Y., Mohammadi, A., Morishima, A., Inoue, M., Kalogirou, M., Koltsakis, G., Samaras, Z. (2011). Calibration and Validation of a Diesel Oxidation Catalyst Model: from Synthetic Gas Testing to Driving Cycle Applications. *SAE technical paper* 2011-01-1244.

[31] Güthenke, A., Chatterjee, D., Weibel, M., Krutzsch, B., Kočí, P., Marek, M., . . . Tronconi, E. (2008). Current status of modeling lean exhaust gas aftertreatment catalysts. *Advances in Chemical Engineering, 33*, 103-211.

[32] Watling, T., Ahmadinejad, M., Tutuianu, M., Johansson, A., Paterson, M. (2012). Development and Validation of a Pt-Pd Diesel Oxidation Catalyst Model. *SAE technical paper* 2012-01-1286.

[33] Güthenke, A., Chatterjee, D., Weibel, M., Krutzsch, B., Kočí, P., Marek, M., . . . Tronconi, E. (2008). Current status of modeling lean exhaust gas aftertreatment catalysts. *Advances in Chemical Engineering*, 33, 103-211.

[34] Kim, C., Schmid, M., Schmieg, S., Tan, J., Li, W. (2011). The Effect of Pt-Pd Ratio on Oxidation Catalysts Under Simulated Diesel Exhaust. *SAE technical paper* 2011-01-1134.

[35] Després, J., Elsener, M., Koebel, M., Kröcher, O., Schnyder, B., Wokaun, A. (2004). Catalytic oxidation of nitrogen monoxide over Pt/SiO2. *Applied Catalysis B: Environmental, 50*, 73-82.

[36] Kandylas, I., Haralampous, O., Koltsakis, G. (2002). Diesel Soot Oxidation with NO2: Engine Experiments and Simulations. *Industrial & Engineering Chemistry Research*, 41, 5372-5384.

[37] Kamasamudram, K., Currier, N. W., Chen, X., Yezerets, A. (2010). Overview of the practically important behaviors of zeolite-based urea-SCR catalysts, using compact experimental protocol. *Catalysis Today*, 151, 212-222.

[38] Grossale, A., Nova, I., Tronconi, E., Chatterjee, D., Weibel, M. (2008). The chemistry of NO/NO_2-NH_3 "fast" SCR reaction over Fe-ZSM5 investigated by transient reaction analysis. *Journal of Catalysis*, 256, 312-322.

[39] Verdier, S., Rohart, E., Larcher, O., Harle, V., Allain, M. (2005). Innovative Materials for Diesel Oxidation Catalysts, with High Durability and Early-Light-Off. *SAE technical paper* 2005-01-0476.

[40] Paulson, T., Moss, B., Todd, B., Eckstein, C., Wise, B., Singleton, D., Zemskova S., Silver, R. (2008). New Developments in Diesel Oxidation Catalysts. *SAE technical paper* 2008-01-2638.

[41] Salvat, O., Marez, P., Belot, G. (2000). Passenger car serial application of a particulate filter system on a common-rail, direct-injection diesel engine. *SAE transactions*, 109(4), 227-239.

[42] Neeft, J., Nijhuis, T. X., Smakman, E., Makkee, M., Moulijn, J. A. (1997). Kinetics of the oxidation of diesel soot. *Fuel*, 76(12), 1129-1136.

[43] Yezerets, A., Currier, N. W., Eadler, H. A., Suresh, A., Madden, P. F., Branigin, M. A. (2003). Investigation of the oxidation behavior of diesel particulate matter. *Catalysis today*, 88(1), 17-25.

[44] Stanmore, B. R., Tschamber, V., Brilhac, J. F. (2008). Oxidation of carbon by NO_x, with particular reference to NO_2 and N_2O. *Fuel*, 87(2), 131-146.

[45] Johnson and Matthey, The Continuously Regenerating Trap. http://www.matthey.com (accessed 9/10/2013)

[46] Burtscher, H., Matter, U. (2002). Particle Formation Due to Fuel Additives (2000-01-1883). *Progress in Technology*, 86, 517-522.

[47] Seguelong, T., Rigaudeau, C., Blanchard, G., Salvat, O., Colignon, C. Y. (2002). Passenger Car Series Application of a New Diesel Particulate Filter System Using a New Ceria-Based, Fuel- Borne Catalyst: From the Engine Test Bench to European Vehicle Certification.

[48] Rocher, L., Seguelong, T., Harle, V., Lallemand, M., Pudlarz, M., Macduff, M., (2011). New Generation Fuel Borne Catalyst for Reliable DPF Operation in Globally Diverse Fuels. *SAE technical Paper*, (2011-01), 0297.

[49] Dettling, J. C., Skomoroski, R. (1992). U.S. Patent No. 5,100,632. Washington, DC: U.S. Patent and Trademark Office.

[50] Kandylas, I. P., Haralampous, O. A., Koltsakis, G. C. (2002). Diesel soot oxidation with NO2: engine experiments and simulations. *Industrial & engineering chemistry research*, 41(22), 5372-5384.

[51] Warner, J. R., Dobson, D., Cavataio, G. (2010). A Study of Active and Passive Regeneration Using Laboratory Generated Soot on a Variety of SiC Diesel Particulate Filter Formulations. *SAE International Journal of Fuels and Lubricants*, 3(1), 149-164.

[52] Jelles, S. J., Krul, R. R., Makkee, M., Moulijn, J. A. (1999). The influence of NO_x on the oxidation of metal activated diesel soot. *Catalysis Today*, 53(4), 623-630.

[53] Haralampous, O. A., Koltsakis, G. C. (2004). Back-diffusion modeling of NO2 in catalyzed diesel particulate filters. *Industrial & engineering chemistry research*, 43(4), 875-883.

[54] Chilumukuru, K. P., (2009). An Experimental Study of Particulate Thermal Oxidation in a Catalyzed Filter During Active Regeneration. Presentation at US Department of Energy Directions in Engine Efficiency and Emissions Research (DEER) Conference, Dearborn, Michigan, USA.

[55] Boger, T., Jamison, J., Warkins, J., Golomb, N., Warren, C., Heibel, A. (2011). Next Generation Aluminum Titanate Filter for Light Duty Diesel Applications. *SAE technical paper*, 2011-01-0816.

[56] Boger, T., Heibel, A. K., Rose, D., Cutler, W. A., & Tennent, D. L. (2005). Evaluation of new diesel particulate filters based on stabilized aluminium titanate. *MTZ worldwide*, 66(9), 14-17.

[57] Tashiro, Y., Imai, T., Suzuki, T., Ochi, N., Gabe, M. (2003). U.S. Patent No. 6,622,480. Washington, DC: U.S. Patent and Trademark Office.

[58] Alkemade, U. G., Schumann, B. (2006). Engines and exhaust after treatment systems for future automotive applications. *Solid State Ionics*, 177(26), 2291-2296.

[59] Lavy, J., (2013) PM Sensor Development and Simulation for Diesel Particulate Filter On-Board Diagnostic. Presentation at *2013 DOE Crosscut Workshop on Lean Emissions Reduction Simulation*, Dearborn, Michigan, USA.

[60] Johnson, T. V. (2011). Diesel Emissions in Review (2011-01-0304). *SAE International Journal of Engines*, 4(1), 143.

[61] Ntziachristos, L., Fragkiadoulakis, P., Samaras, Z., Janka, K., Tikkanen, J. (2011). Exhaust particle sensor for OBD application. *SAE technical paper*, 2011-01-0626.
[62] Kim, K., Cho, S. (2010). The Emission Development for Lean NO_x Trap System to Meet Tier2Bin5. *SAE technical paper*, 2010-01-0566.
[63] Shinichi, M. (1996). DeNO_x catalyst for automotive lean-burn engine. *Catalysis Today*, 29, 43-45.
[64] Lambert, C., Hammerle, R., McGill , R., Khair, M., and Sharp, C. (2004). Technical Advantages of Urea SCR for Light-Duty and Heavy-Duty Diesel Vehicle Applications. *SAE technical paper*, 2004-01-1292.
[65] Alimin, A.J., Benjamin S.F., Roberts, C.A. (2007). Fuel Penalty from Periodic Rich Combustion of a Diesel Fuelled Engine during Lean NO_X Trap Purging. Proceedings of EnCon2007: 1st Engineering Conference on Energy & Environment, December 27-28, 2007, Kuching, Sarawak, Malaysia
[66] Takeuchi, M., Matsumoto, S. I. (2004). NO_x storage-reduction catalysts for gasoline engines. *Topics in catalysis*, 28(1-4), 151-156.
[67] Alkemade, U. G., Schumann, B. (2006). Engines and exhaust after treatment systems for future automotive applications. *Solid State Ionics*, 177(26), 2291-2296.
[68] Epling, W. S., Campbell, L. E., Yezerets, A., Currier, N. W., Parks, J. E. (2004). Overview of the fundamental reactions and degradation mechanisms of NOx storage/reduction catalysts. *Catalysis Reviews*, 46(2), 163-245.
[69] Qi, G., Kim, C., Li, W. (2010) Pt-Free, Perovskite-Based LNT Catalyst. US Department of Energy "Directions in Engine Efficiency and Emissions Research (DEER) Conference", September 27-30, 2010, Detroit.
[70] Kabin, K. S., Khanna, P., Muncrief, R. L., Medhekar, V., Harold, M. P. (2006). Monolith and TAP reactor studies of NO_x storage on $Pt/BaO/Al_2O_3$: Elucidating the mechanistic pathways and roles of Pt. *Catalysis today*, 114(1), 72-85.
[71] Bhatia, D., Clayton, R. D., Harold, M. P., Balakotaiah, V. (2009). A global kinetic model for NO_x storage and reduction on $Pt/BaO/Al_2O_3$ monolithic catalysts. *Catalysis Today*, 147, S250-S256.
[72] Lietti, L., Forzatti, P., Nova, I., Tronconi, E. (2001). NO_x Storage Reduction over Pt-Ba/γ-Al_2O_3 Catalyst. *Journal of Catalysis*, 204(1), 175-191.
[73] Larson, R. S., Chakravarthy, V. K., Pihl, J. A., Daw, C. S. (2012). Microkinetic modeling of lean NO_x trap chemistry. *Chemical Engineering Journal*, 189, 134-147.
[74] Nova, I., Castoldi, L., Lietti, L., Tronconi, E., Forzatti, P., Prinetto, F., Ghiotti, G. (2004). NOx adsorption study over Pt–Ba/alumina catalysts: FT-IR and pulse experiments. *Journal of catalysis*, 222(2), 377-388.
[75] Koci, P., Plat, F., Stepanek, J., Bartova, S., Marek, M., Kubicek, M., Smeißer, V., Chatterjee, D., Weibel, M. (2009). Global kinetic model for the regeneration of NO_x storage catalyst with CO, H_2 and C_3H_6 in the presence of CO_2 and H_2O. *Catalysis today*, 147, S257-S264.
[76] Lietti, L., Artioli, N., Righini, L., Castoldi, L., Forzatti, P. (2012). Pathways for N_2 and N_2O Formation during the Reduction of NO_x over Pt–Ba/Al_2O_3 LNT Catalysts Investigated by Labeling Isotopic Experiments. *Industrial & Engineering Chemistry Research*, 51(22), 7597-7605.
[77] Gill, L. J., Blakeman, P. G., Twigg, M. V., Walker, A. P. (2004). The use of NO_x adsorber catalysts on diesel engines. *Topics in catalysis*, 28(1-4), 157-164.

[78] Bisaiji, Y., Yoshida, K., Inoue, M., Umemoto, K., Fukuma, T. (2012). Development of Di-Air-A New Diesel deNO$_x$ System by Adsorbed Intermediate Reductants. *SAE International Journal of Fuels and Lubricants*, 5(1), 380-388.

[79] Inoue, M., Bisaiji, Y., Yoshida, K., Takagi, N., Fukuma, T. (2013). deNO$_x$ Performance and Reaction Mechanism of the Di-Air System. *Topics in Catalysis*, 1-4.

[80] Elbouazzaoui, S., Corbos, E. C., Courtois, X., Marecot, P., Duprez, D. (2005). A study of the deactivation by sulfur and regeneration of a model NSR Pt/Ba/Al_2O_3 catalyst. *Applied Catalysis B: Environmental*, 61(3), 236-243.

[81] Choi, J. S., Partridge, W. P., Lance, M. J., Walker, L. R., Pihl, J. A., Toops, T. J., Finney, C. E. A., Daw, C. S. (2010). Nature and spatial distribution of sulfur species in a sulfated barium-based commercial lean NO$_x$ trap catalyst. *Catalysis Today*, 151(3), 354-361.

[82] Ji, Y., Choi, J. S., Toops, T. J., Crocker, M., Naseri, M. (2008). Influence of ceria on the NO$_x$ storage/reduction behavior of lean NO$_x$ trap catalysts. *Catalysis Today*, 136(1), 146-155.

[83] Bosch Gmbh, Diesel Systems: Denoxtronic 2.2 – Urea Dosing System for SCR Systems. www.bosch-diesel.de (accessed 19/11/2011)

[84] Zheng, G., Fila, A., Kotrba, A., Floyd, R. (2010). Investigation of Urea Deposits in Urea SCR Systems for Medium and Heavy Duty Trucks. *SAE technical paper*, 2010-01-1941.

[85] Gullett, B. K., Bruce, K. R., Hansen, W. F., Hofmann, J. E. (1992). Sorbent/urea slurry injection for simultaneous SO_2/NO$_x$ removal. *Environmental progress*, 11 (2), 155-162.

[86] Nishioka, A., Sukegawa, Y., Mukai, T., Katogi, K., Yokota, H., & Mamada, H. (2006). A Study of a New Aftertreatment System (2): Control of Urea Solution Spray for Urea-SCR. *SAE technical paper*, 2006-01-0644.

[87] Johannessen, T., Schmidt, H., Norskov, J. K., Christensen, C. H. (2012). European Patent No. EP 2051798. Munich, Germany: European Patent Office.

[88] Long, R. Q., Yang, R. T. (2001). Selective catalytic oxidation (SCO) of ammonia to nitrogen over Fe-exchanged zeolites. *Journal of Catalysis*, 201(1), 145-152.

[89] Akah, A., Cundy, C., Garforth, A. (2005). The selective catalytic oxidation of NH_3 over Fe-ZSM-5. *Applied Catalysis B: Environmental*, 59(3), 221-226.

[90] Grossale, A., Nova, I., Tronconi, E. (2009). Ammonia blocking of the "Fast SCR" reactivity over a commercial Fe-zeolite catalyst for Diesel exhaust aftertreatment. *Journal of Catalysis*, 265(2), 141-147.

[91] Brandenberger, S., Kröcher, O., Tissler, A., & Althoff, R. (2008). The State of the Art in Selective Catalytic Reduction of NO$_x$ by Ammonia Using Metal-Exchanged Zeolite Catalysts. *Catalysis Reviews*, 50(4), 492-531.

[92] Malmberg, S., Votsmeier, M., Gieshoff, J., Söger, N., Mußmann, L., Schuler, A., Drochner, A. (2007). Dynamic phenomena of SCR-catalysts containing Fe-exchanged zeolites–experiments and computer simulations. *Topics in Catalysis*, 42(1), 33-36.

[93] Baik, J. H., Yim, S. D., Nam, I. S., Mok, Y. S., Lee, J. H., Cho, B. K., Oh, S. H. (2004). Control of NO$_x$ emissions from diesel engine by selective catalytic reduction (SCR) with urea. *Topics in catalysis*, 30(1-4), 37-41.

[94] Tomašić, V. (2007). Application of the monoliths in DeNO$_x$. *Catalysis Today*, 119(1), 106-113.

[95] Kröcher, O. (2007) Challenges for urea-SCR systems: From vanadia-based to zeolite based SCR catalysts. 1st $MinNO_x$ conference, Berlin, Germany.
[96] Colombo, M., Nova, I. Tronconi, E. (2010). A comparative study of the NH3-SCR reactions over a Cu-zeolite and a Fe-zeolite catalyst. *Catalysis Today*, 151, 223-230.
[97] Greenhalgh, B., Fee, M., Dobri, A., Moir, J., Burich, R., Charland, J. P., Stanciulescu, M. (2010). $DeNO_x$ activity–TPD correlations of NH_3-SCR catalysts. *Journal of Molecular Catalysis A: Chemical*, 333(1), 121-127.
[98] In Sik Nam. (2011). Urea/SCR Technology for removing NOx from Diesel Engine. CLEERS Workshop 2011, Detroit, USA.
[99] Cavataio, G., Jen, H. W., Warner, J. R., Girard, J. W., Kim, J. Y., Lambert, C. K. (2009). Enhanced Durability of a Cu/Zeolite Based SCR Catalyst. *SAE International Journal of Fuels and Lubricants*, 1(1), 477-487.
[100] Thompson, J., De Beeck, J. O., Joubert, E., Wilhelm, T. (2008). Case Studies of Urea SCR Integration on Passenger Cars Monitoring of Urea Inside the Tank During Hot and Cold Environment Test Missions. *SAE technical paper*, 2008-01-1181.
[101] Van Helden, R., Verbeek, R., Willems, F., van der Welle, R. (2004). Optimization of urea SCR deNOx systems for HD diesel engines. *SAE technical paper*, 2004-01-0154.
[102] Colombo, M., Nova, I., Tronconi, E., Koltsakis, G. (2013). A Modeling Study of NH3 Slip Catalysts: Analysis of the SCR/PGM Interactions. *Topics in Catalysis*. doi: 10.1007/s11244-013-9948-x.
[103] Votsmeier, M., Scheuer, A., Drochner, A., Vogel, H., Gieshoff, J. (2010). Simulation of automotive NH3 oxidation catalysts based on pre-computed rate data from mechanistic surface kinetics. *Catalysis Today*, 151, 271-277.
[104] Scheuer, A., Hautpmann, W., Drochner, A., Gieshoff, J., Vogel, H., Votsmeier, M. (2012). Dual layer automotive ammonia oxidation catalysts: Experiments and computer simulation. *Applied Catalysis B: Environmental*, 111-112, 445-455.
[105] Colombo, M., Nova, I., Tronconi, E. (2012). A simplified approach to modeling of dual-layer ammonia slip catalysts. *Chemical Engineering Science*, 75, 75-83.
[106] Tan, J., Solbrig, C., & Schmieg, S. (2011). The Development of Advanced 2-Way SCR/DPF Systems to Meet Future Heavy-Duty Diesel Emissions. *SAE technical paper*, 2011-01-1140.
[107] Maunula, T. (2013). Intensification of Catalytic Aftertreatments Systems for Mobile Applications. *SAE technical paper*, 2013-01-0530.
[108] Colombo, M., Koltsakis, G., Koutoufaris, I. (2011). A modeling study of soot and de-NOx reaction phenomena in SCRF systems. *SAE technical paper*, 2011-37-0031.
[109] Watling, T. C., Ravenscroft, M. R., Avery, G. (2012). Development, validation and application of a model for an SCR catalyst coated diesel particulate filter. *Catalysis Today*, 188, 32-41.
[110] Shelef, M., Graham, G. W., McCabe, R. W. (2002). Catalysis by Ceria and Related Materials. *Imperial College Press*, 343-389.
[111] Koltsakis, G. C., Stamatelos, A. M. (1999). Modeling dynamic phenomena in 3-way catalytic converters. *Chemical Engineering Science*, 54, 4567-4578.
[112] Aoki, Y., Sakagami, S., Kawai, M., Takahashi, N., Tanabe, T., Sunada, T. (2011). Development of Advanced Zone-Coated Three-Way Catalysts. *SAE technical paper*, 2011-01-0926.

[113] Matsuzono, Y., Kuroki, K., Nishi, T., Suzuki, N., Yamda, T., Hirota, T., Zhang, G. (2012). Development of Advanced and Low PGM TWC System for LEV2 PZ EV and LEV3 SULEV30. *SAE technical paper,* 2012-01-1242.

[114] Arnold, L., Collins, N., Cooper, J., Eyre, D., Morris, D., Ravenscroft, A., Twigg, M. (2007-01-0046). Development and Application of New Low Rhodium Three-Way Catalyst Technology. *SAE technical paper,* 2007-01-0046.

[115] Myung, C. L., Park, S. (2012). Exhaust nanoparticle emissions from internal combustion engines: A review. *International Journal of Automotive Technology*, 13(1), 9-22.

[116] Bischof, C., Boger, T., Gunasekaran N., Bhargava R. (2012). Advanced Particulate Filter Technologies for Direct Injection Gasoline Engine Applications. Presentation at US Department of Energy Directions in Engine Efficiency and Emissions Research (DEER) Conference, Dearborn, Michigan, USA.

[117] Spiess, S., Wong, K. F., Richter, J. M., Klingmann, R. (2013). Investigations of Emission Control Systems for Gasoline Direct Injection Engines with a Focus on Removal of Particulate Emissions. *Topics in Catalysis*, 1-6.

[118] Mikulic, I., Koelman, H., Majkowski, S. Vosejpka, P. (2010). A study about particle filter application on a state-of-the-art homogeneous turbocharged 2L DI gasoline engine. 19th Aachen Colloquium, Aachen, Germany.

[119] Kattouah P., Kato, K., Ohara, E., Vogt C.D. (2013). Ceramic Wall Flow Filter for Particulate Emission Reduction of Petrol Engines. Presentation at Cambridge Particulate Meeting, Cambridge, U.K.

[120] Opitz, B., Drochner, A., Vogel, H., Votsmeier, M. (2013). An experimental and simulation study on the cold start behaviour of particulate filters with wall integrated three way catalyst. *Applied Catalysis B: Environmental*. 144, 203-15.

[121] Choi, K., Kim, J., Ko, A., Myung, C. L., Park, S., Lee, J. (2012). Size-resolved engine exhaust aerosol characteristics in a metal foam particulate filter for GDI light-duty vehicle. *Journal of Aerosol Science*. 75, 1-13.

[122] Blakemann, P. (2013) Catalyzed Exhaust Filters: Future directions. Presentation at 2013 DOE Crosscut Workshop on Lean Emissions Reduction Simulation, Dearborn, Michigan, USA.

[123] Ruona, W. C., Van Nieuwstadt, M. J. (2010). U.S. Patent Application 12/689,930.

In: Automotive Exhaust Emissions and Energy Recovery ISBN: 978-1-63321-493-4
Editor: Apostolos Pesiridis

Chapter 6

DIESEL PARTICULATE FILTER OVERVIEW: MATERIAL, GEOMETRY AND APPLICATION

Martin J. Murtagh* and Timothy V. Johnson
Corning Incorporated – Corning, NY, US

ABSTRACT

Prevailing global environmental concerns over mobile and stationary diesel emissions prompted strict legislation incrementally implemented over the first decade of the 21st century. The standards established by the new emissions legislation posed a significant materials science and engineering challenge. New requirements demanded greater efficiency from existing emissions after-treatment (catalyst support systems, i.e. catalytic converters), as well as engine modifications. The challenge of acquiring means to meet the stricter standards generated the emergence of some innovative approaches. As a result, heavy and light duty diesel engine manufacturers (in North America, Europe and Japan) were faced with the challenge of implementing rapidly developing technologies to meet the environmental legislation for particulate matter (PM) and nitrogen mono and dioxide (NOx) emissions. Diesel particulate filters (DPF) are a critical component enabling modern diesel engines to meet these challenges.

The chapter will review the state of the DPF past, present and future, covering wall flow filter material choices, i.e. cordierite, silicon carbide, and aluminium titanate, and the wall flow DPF design considerations, such as component filtration (efficiency & pressure drop), thermomechanical, and thermochemical properties derived from the fixed integration of both the macro and microstructural attributes.

Keywords: Diesel particulate filter; DPF; aluminium titanate; silicon carbide; cordierite; particulate matter; oxide of nitrogen

* Corning Incorporated – Corning, NY, USA; Email: mjm43@cornell.edu.

INTRODUCTION

Rudolf Diesel, inventor of diesel engine in 1892, was a well-respected thermal engineer and a social theorist. Diesel conceived the diesel engine to enable independent craftsmen and artisans to compete with large industry, and by a fascination with the second law of thermodynamics, i.e. maximum efficiency of the Carnot cycle [1].

The motivation for the construct on the through-the-wall (wall-flow) Diesel Particulate Filter (DPF) is simply that diesel engines emit particulate matter (carbon, hydrocarbon, inorganic ash). Some 86 years after Diesel's invention, GM patented the concept of the modern DPF in 1981 [2]. The DPF design is based on plugging an existing "honeycomb" flow-through automotive catalytic converter in a "checkerboard" manifold design. That is, every other honeycomb cell was plugged on each end of the substrate with alternating patterns to create the wall-flow DPF. Corning Incorporated then developed the DPF in the early 1980's under a royalty-free license from GM.

Diesel exhaust particulate matter (PM) is a regulatory concern [3] and the global diesel mobile emissions legislation (1994 to present) that was drafted secured the launch of the DPF as the critical element of the PM emission solution by 2006.

The global diesel mobile emissions legislation provided a significant material science and mechanical engineering challenge. It is common practice when discussing DPF's to identify the filter by the material description. In the early development of the DPF, the micro and macro geometry of the filter as well as the material, from which it is composed, played a substantial role in the filter survivability and filtration. While it was not disputed that the material properties play a major role in filter survivability, the filter geometry has an important influence upon temperatures reached during uncontrolled regeneration and thus upon survivability. If the filter was to provide the safety margin for uncontrolled regeneration with high soot loads, a high mass filter was required. If the system is to provide the safety margin, i.e. electronic control, by avoiding uncontrolled regenerations with moderate soot loads, a low mass filter became possible with and added through-the-wall pressure drop advantage. In either case, the filter requires good thermochemomechanical integrity.

This Chapter is a review of diesel particulate filters (DPF's) covering the legislation and regulatory history, filter material choice and geometry, and DPF with system design integration for application.

REGULATIONS

Figure 1 represents the diesel engine combustion tradeoff as a non-linear inverse relationship between PM and NOx. The emission legislation and regulations follow a vehicle year step down format of the PM and NOx criterion pollutants in gram per kilometer for the light duty diesel engines application and gram per kilowatt hour for the heavy duty diesel engine application. Shown in Table 1 is the United States Environmental Protection Agency (USEPA) overview of the regulated gross vehicle weight rating (GVWR) for light duty vehicles, light duty trucks, and heavy duty vehicles.

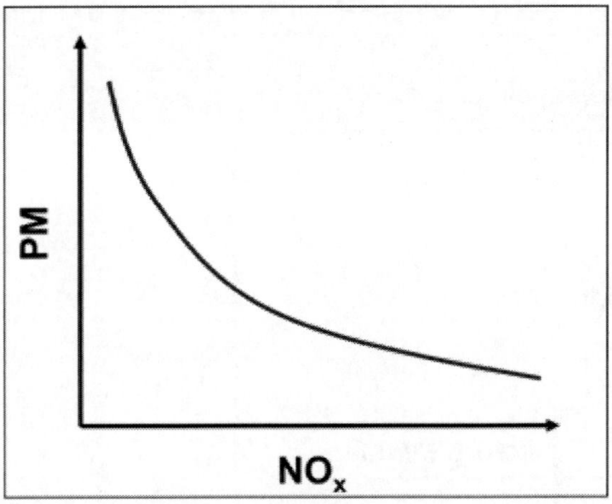

Figure 1. Schematic of Particulate Matter (PM) as a function of NOx diesel combustion tradeoff.

Table 1. Overview of diesel vehicle classifications [4]

Vehicle	Abbreviation	Definition
		Passenger Vehicles
Light-duty Vehicles (i.e. Passenger Cars)	LDV	maximum Gross Vehicle Weight Rating (GVWR) < 8,500 lbs.
Medium-duty Passenger Vehicles	MDPV	8,501 - 10,000 lbs GVWR
		Light-duty Trucks
Light-duty Trucks	LDT	Max 8500 lb GVWR, Max 6000 lb curb weight, and Max 45 ft^2 frontal area
light light-duty trucks	LDT1	Max 3750 lb LVW (loaded vehicle weight: curb weight + 300 lb)
light light-duty trucks	LDT2	Min 3750 lb LVW (loaded vehicle weight: curb weight + 300 lb)
heavy light-duty trucks	LDT3	Max 5750 lb ALVW (adjusted loaded vehicle weight: avg of GVWR and curb weight)
heavy light-duty trucks	LDT4	Min 5750 lb ALVW (adjusted loaded vehicle weight: avg of GVWR and curb weight)
		Heavy-duty Vehicles
Heavy Duty Vechicles	HDV 2b	8,501 - 10,000 lbs
Heavy Duty Vechicles	HDV 3	10,001 - 14,000 lbs
Heavy Duty Vechicles	HDV 4	14,001 - 16,000 lbs
Heavy Duty Vechicles	HDV 5	16,001 - 19,500 lbs
Heavy Duty Vechicles	HDV 6	19,501 - 26,000 lbs
Heavy Duty Vechicles	HDV 7	26,001 - 33,000 lbs
Heavy Duty Vechicles	HDV 8a	33,001 - 60,000 lbs
Heavy Duty Vechicles	HDV 8b	>60,001 lbs

Diesel engines used in heavy-duty vehicles are further divided into service classes by GVWR: *Light heavy-duty diesel engines (LHDDE): 8,501 - 19,500, for use in HDV classes 2b - 5 Medium heavy-duty diesel engines (MHDDE): 19,501 - 33,000, for use in HDV classes 6-7Heavy heavy-duty diesel engines (including urban bus) (HHDDE): > 33,000, for use in HDV class 8*

Figures 2 & 3 show the historical emission limits for light duty vehicles (LDV) in North America and Europe and heavy duty vehicles (HDV) in North America, Europe and Japan (ESC: European Steady State Cycle), respectively.

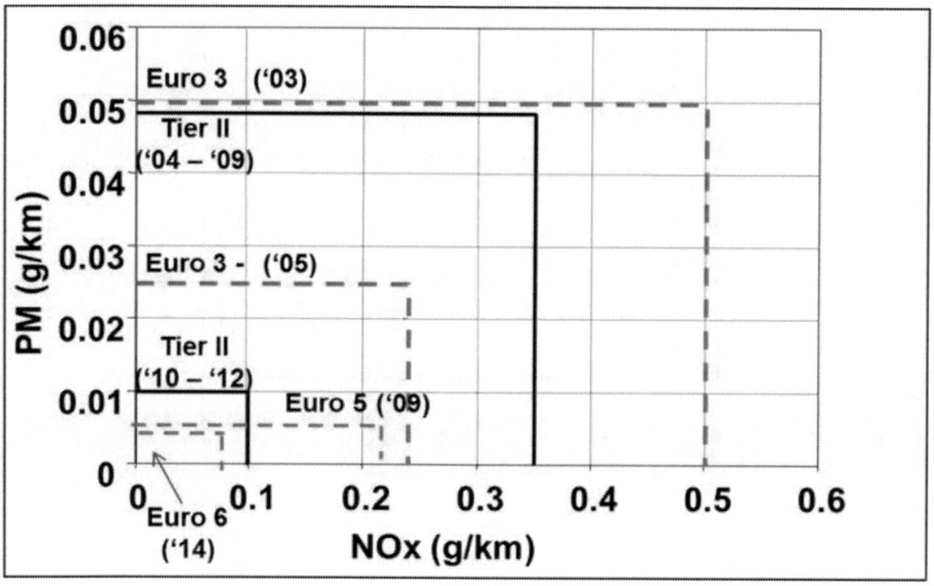

Figure 2. Historical limits of the light duty vehicle diesel regulations PM as a function of NO_X.

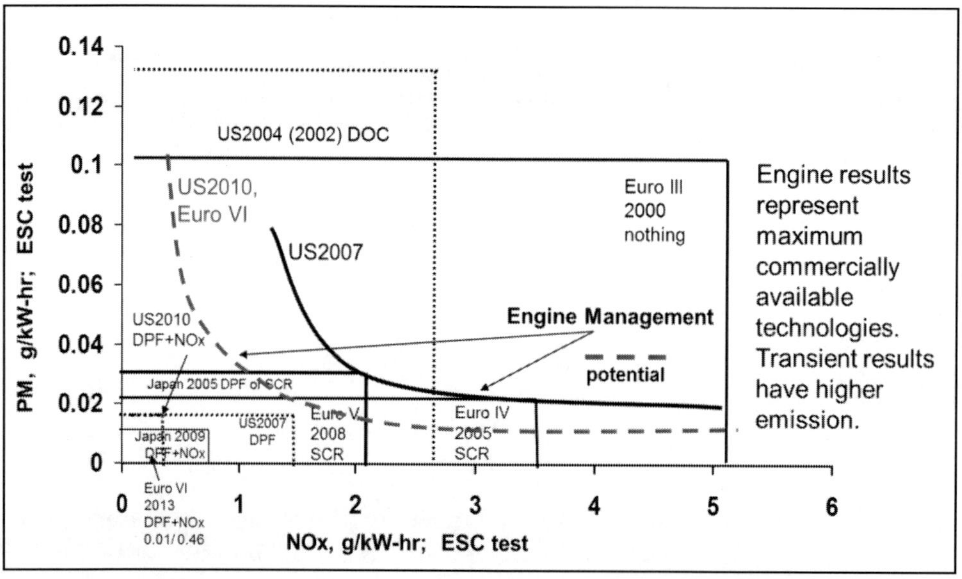

Figure 3. Historical limits of the heavy duty vehicle diesel regulations PM as a function of NO_X (ESC: European Steady State Cycle).

Light Duty

The emission reduction strategies for the US and Europe show similar change levels over time, however, the US accents PM, hydrocarbon and NOx, while Europe places a heavier emphasis on PM and particle number (PN). US EPA Tier I (up to the last 25% of the fleet in

2007) and Euro 4 legislation can be met with a diesel oxidation catalyst (DOC) and without a DPF.

At the time of this writing LDV regulatory outlook for Europe now includes petrol regulations PN that is finalized at the diesel level of $6x10^{11}$/km for 2017. Since the introduction of Euro 5 and USA EPA Tier 2 Bin 5 (fleet average) the DPF is standard on all diesel passenger cars.

Heavy Duty

The HDV emission limits in US, Europe, and Japan historically were more specifically managed by engine management and to some extent through selective catalytic reduction (SCR) up to US EPA 2007, Euro IV 2005, Japan 2005. The DPF is now standard on all HDV engines since the US EPA 2007 and Euro 6.

At the time of this writing, the HDV regulatory outlook for US EPA and elsewhere is shifting interest to real-world and off-cycle emissions such as urban or low-load NOx, including changes in the not to exceed (NTE) regulation. Japan is considering the next round of tightening for 2016 after their fuel consumption (FC) regulation are implemented in 2015. It is likely Japan will harmonize with Euro VI. China implemented Euro IV in July 2013, and is being attained using partial filters for urban applications and SCR for cargo. At the time of this writing, delay of full implementation is eminent to perhaps January 2015 when 50 ppm sulfur fuel will be widely available. India is implementing Euro IV in eleven major cities, and now has an expert senior panel deriving a 2025 roadmap for future regulations and fuel quality.

DIESEL SOOT REMEDIATION SYSTEM DESIGN REQUIREMENTS

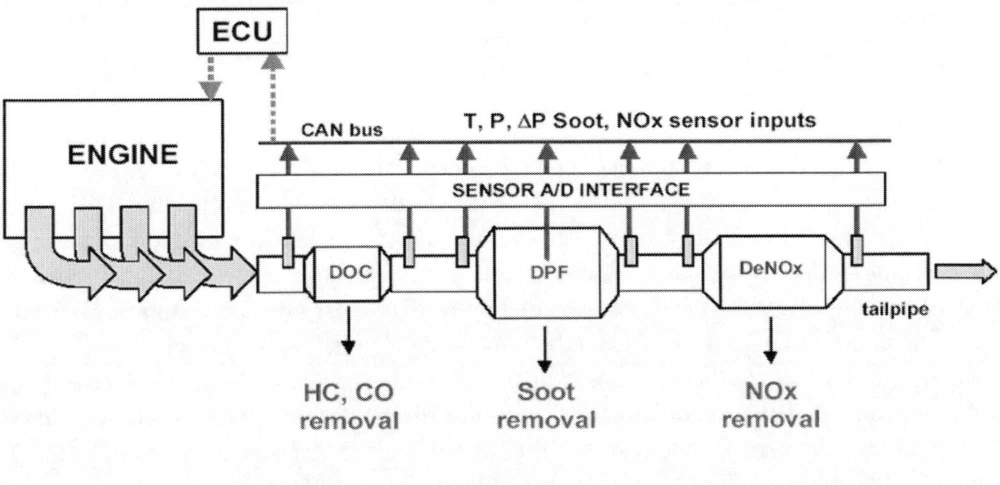

Figure 4. Schematic of a diesel exhaust remediation system [5].

The generic needs of the diesel exhaust remediation system, shown in Figure 4, is the integration of the specific components making up the diesel exhaust “after-treatment” (after the engine) such as; the diesel oxidation catalysis (DOC), DPF and a de-NOx solution with appropriate engine controls, managed by the engines electronic control unit (ECU). The component integration specifications were designed to meet the regulated criterion emission targets (with no secondary emissions) of hydrocarbons (HC), carbon monoxide (CO), NOx, (nitrogen monoxide (NO) and nitrogen dioxide (NO_2)), and PM (soot), as well as preserving maximum engine performance: power, torque, and minimizing fuel penalty. The vehicle integration: space limits, regulated durability and reliability limits: Pass Car: minimum 160,000 km (per Euro V), and HDV: 500,000 km - 1,200,000 km (minimum 435 kmiles in the US) with minimal maintenance defined expectations.

The DOC system placement in front of the DPF insures that the inlet auto ignition temperature of the core carbon (~ 550°C; lowered in practice due to the presence of hydrocarbons and NO_2) is maintained by taking advantage of the heat of reaction of NO to NO_2 (the enthalpy of the reaction $NO + O \rightarrow NO_2 = -143$ kJ) and the oxidation of diesel fuel (e.g. the enthalpy of the reaction) [Hydrocarbons] + $O_2 \rightarrow CO_2 + H_2O$ (specifically; $4\ C_{12}H_{23} + 71\ O_2 \rightarrow 48\ CO_2 + 46\ H_2O$); 9400 kJ/mol (42.76 MJ/kg) [6,7].

The placement of the DPF in front of the de-NOx solution provides for minimal PM contamination. NO_2 participates in the oxidation of the PM core carbon enhancing the oxidation of the core carbon by lowering the auto ignition temperature in the range of 250–400°C (hydrocarbon dependent), and for this reason most DPFs are catalysed to take advantage of the oxidation of the PM and to minimize NO_2 slip. The concept of an integrated 4-way component (criteria pollutants hydrocarbons (HC), carbon monoxide (CO), NO_x and PM) was rejected circa 2010 due to technical and economic reasons. It would not work with selective catalytic reduction (SCR) integration solution, since a 4-way component would oxidize the NH_3[8]. However, currently commercialized systems do combine 2-3 of the functions. In catalysed soot filters (CSF) the oxidation function (CO, HC) is combined with the filtration. In selective catalytic reduction filter (SCRF) the $DeNO_x$ function is integrated into the filter.

The following section covers only the DPF component design requirements. The DPF is the heart of the functioning "after-treatment" system, i.e. without the DPF there would be no system, per PM regulatory requirements.

DPF Design Requirements

As it was discussed in the introduction the current diesel particulate filter (DPF), conceived circa 1978, was an outgrowth of the successful launch (circa 1975), in the United States, of the spark ignition engine catalytic converter. However, the DPF design requirements were significantly more demanding due to the need of accommodating the periodic regeneration temperature (above steady state transient cycles) of the carbon core portion of the diesel exhaust PM. Shown in Figure 5 is the first manifestation of the DPF concept, the manifolding of a flow-through "honeycomb" patented by GM [2].

The unique DPF design requirements sufficient to abate a harsh thermochemomechanical environment included the need for "low" pressure drop (system specific), "high" filtration efficiency (> 90% by mass), chemical stability, thermal shock resistance, mechanical integrity for under body packaging, and minimal space utilization.

Since 1978 there have been many DPF variations on the original "honeycomb" manifold design theme. Shown in Table 2 is a collage of concepts and utilization statistics.

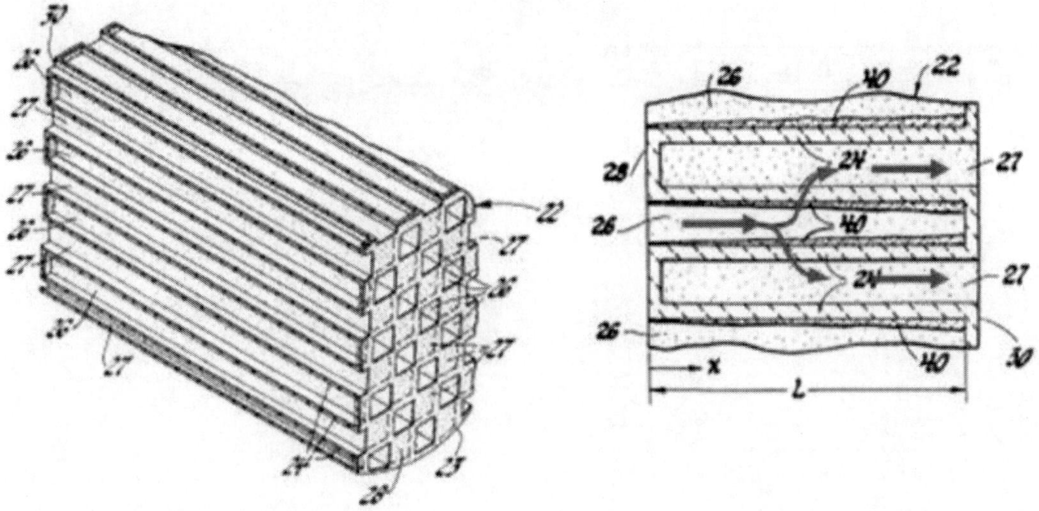

Figure 5. Schematic of a manifolded "honeycomb" DPF drawing showing through-the-wall inlet to outlet filtration [2].

Table 2. Collage of DPF concepts and utilization statistics [9-13]

	Utilization	
DPF Concept	**Mobile (%)**	**Stationary (%)**
Ceramic Monolith & Segmented Wall-Flow*	85	15
Ceramic Wound Fiber Monolith	<1	10
Ceramic Knitted Fiber Monolith	<1	10
Sintered Metal Monolith Wall-Flow	<1	<1
Electrostatic Metal Flow-Through Monolith	0	<1
Non-blocking Metal Flow-Through Monolith	< 5	<1
Woven Wire Mesh Filter	<1	<1
Ceramic Foam Monolith	<1	<1
Metal Wrap Wall-Flow Monolith	< 1	<1

* Current DFP technology per GM's "honeycomb" manifold concept

DPF Geometry and Material Choices

The DPF geometry starts with defining the local cell geometry as shown in Figure 6. The square cell geometry is preferred in order to maximize the orthogonal tensile and compressive strength of the monolithic DPF. The plug depth will typically range from 6 mm (~ 0.25 in) to 12 mm (~ 0.5 in) and is a function of the preferred over all channel length affecting the filtration pressure drop. The typical size (diameter x length) and volume (volume of a cylinder) of DPFs vary as a function of application from 12 cm (~ 5 in.) x 12 cm (~ 5 in.); 1.64L to 30 cm (~ 12 in.) x 35 cm (~ 14 in.); 49.5L.

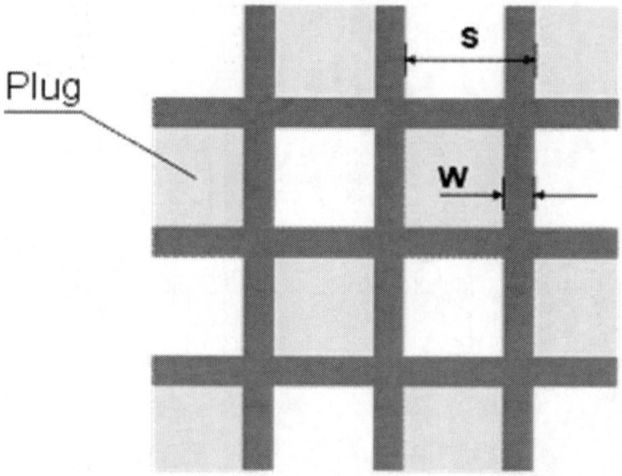

Figure 6. Schematic of the DPF plugging manifold pattern; s = cell repeat distance and w = wall thickness.

The cell density (cells/area) is calculated as shown in equation 1:

$$N = 1/S^2 \tag{1}$$

The hydraulic diameter is calculated as shown in equation 2:

$$d_h = (S - W) \tag{2}$$

The hydraulic diameter varies inversely with channel friction pressure drop loss to the 4th power and inversely with contraction and expansion pressure drop loss to the 1st power (see equation 3, later in the text). Table 3 shows the past, present, and future DPF cell density choices.

Table 3. Past, present, and future DPF cell density choices

Past 1980 – 2000	Present 2000 – 2013	Future > 2013
100/17	200/12	200/9
	200/16	
	200/20	
	300/16	

Geometric description is N/w; '/' is a punctuation mark, not a arithmetic symbol, N = cell density; cells per area, e.g. 100 cell/inch and w = wall thickness in thousands of an inch (mils; 1 mil = 25.4 μm), e.g. 0.0017 inch = 17 mils or 434 μm).

The DPF material choice extended over the years from Cordierite (2MgO $2Al_2O_3$ $5SiO_2$), circa 1981 to alpha silicon carbide (α-SiC), circa 1995 to aluminum titanate (Al_2TiO_5: AT), circa 2005. Figure 7 illustrates example geometries of the specific DPF material choice. Cordierite and aluminum titanate are produced as extruded monoliths while silicon carbide

DPFs are produced as an assembly of segmented extrudates. The materials selection is application specific and will be discussed next in the application section of this chapter.

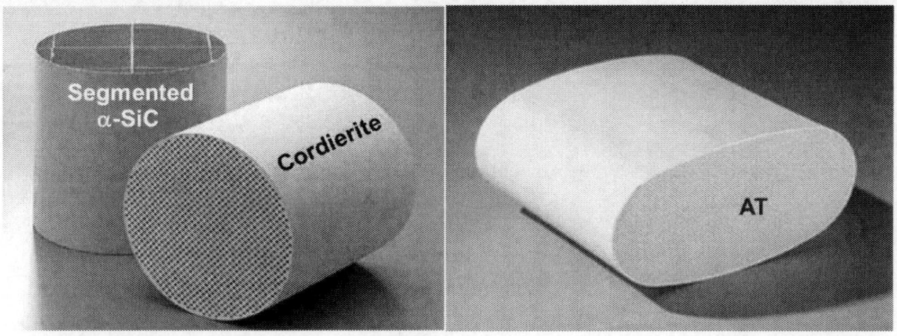

Figure 7. Selected DPF material choices and shapes.

APPLICATION

The DPF design function is to filter while surviving the thermochemomechanical environment of the application. PM filtering efficiency is primarily controlled by the wall pore size and porosity while cell geometry is secondary and both the through-the-wall microstructure and geometry can be tailored independent of material choice.

The DPF fits an application as a function of the engine size, as was discussed in the regulatory section referring to vehicle size, classified as light to heavy duty. In all sizes the DPF needs to filter while meeting the regulated lifetime in each vehicle size classification.

The next section will cover, in some detail, the attributes that defines the DPF geometry and material choices, which maximize filtration, while minimizing uncontrolled regenerations and maximizing survival.

Filtration (Filter Efficiency)

The factor influencing filtering efficiency is the through-the-wall pore profile (size and volume fraction) as the primary determinant (the filter geometry profile, cell and wall thickness, is secondary) of initial filtering efficiency, i.e. wall permeability, as a function of the pore size, volume, and connectivity. The PM membrane can raise efficiency above that of bare filter, once it begins to build [14]. Once again, it should be noted that filtering is mechanical and independent of material choice. Figures 8 and 9 are scanning electron microscopy (SEM) micrographs representing each DPF material class illustrating both the surface and through-the-wall pore profiles, respectively.

Shown in Figure 10 is a transmission electron microscopy (TEM) micrograph depicting the typical primary particle size and shape of PM. The primary particle size is on the order of 10s of nanometers and can agglomerate to submicron size per exhaust flow rate conditions. The DPF filtering mechanism is diffusional capture along the tortuous path of the micron sized porosity making up the DPF microstructure, which makes it possible to capture nano-size PM particles very efficiently through-the-wall [15].

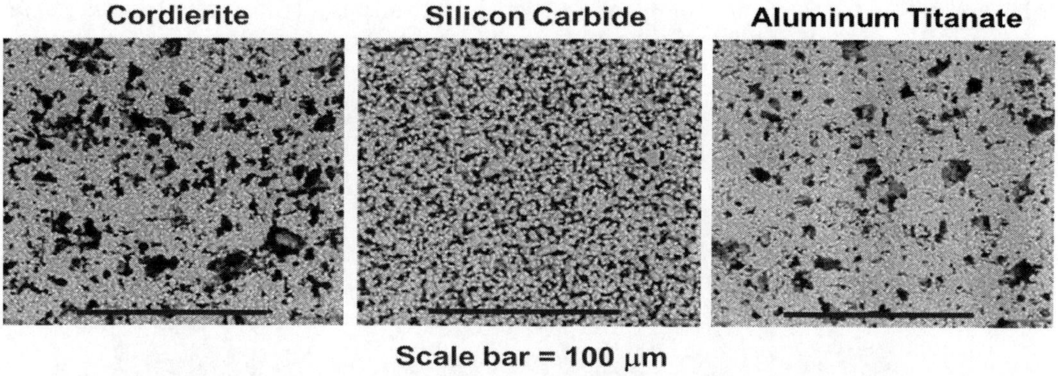

Figure 8. SEM micrographs representative of the as fired wall surface pore size and volume for the selected DPF materials.

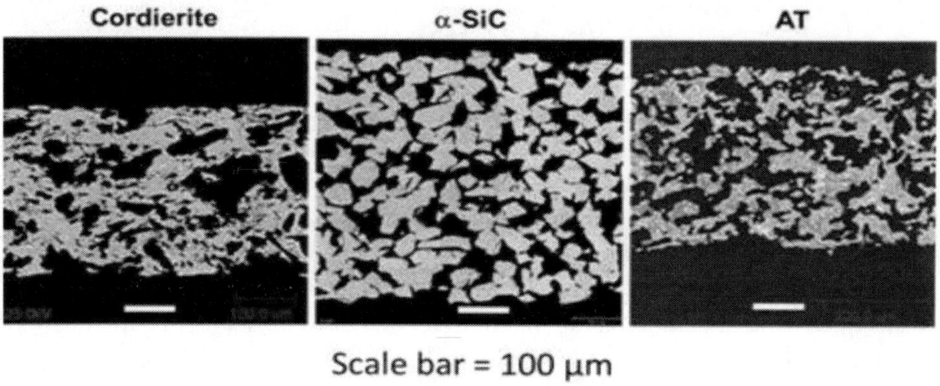

Figure 9. SEM micrographs representative of the through-the-wall pore size and volume for the selected DPF materials.

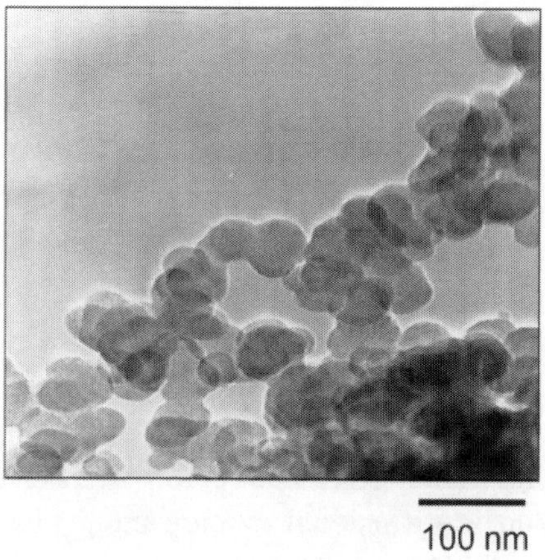

Figure 10. TEM micrograph of a representative of a typical PM particle size and morphology [15].

Shown in Figure 11 are optical micrographs of a cross sectional view of the through-the-wall pore size and volume with PM deposition illustrating a deep bed and surface filtration mechanism. It should now be clear form observing these optical micrographs in Figure 11 that once the PM membrane begins to build the membrane itself becomes the filtering medium.

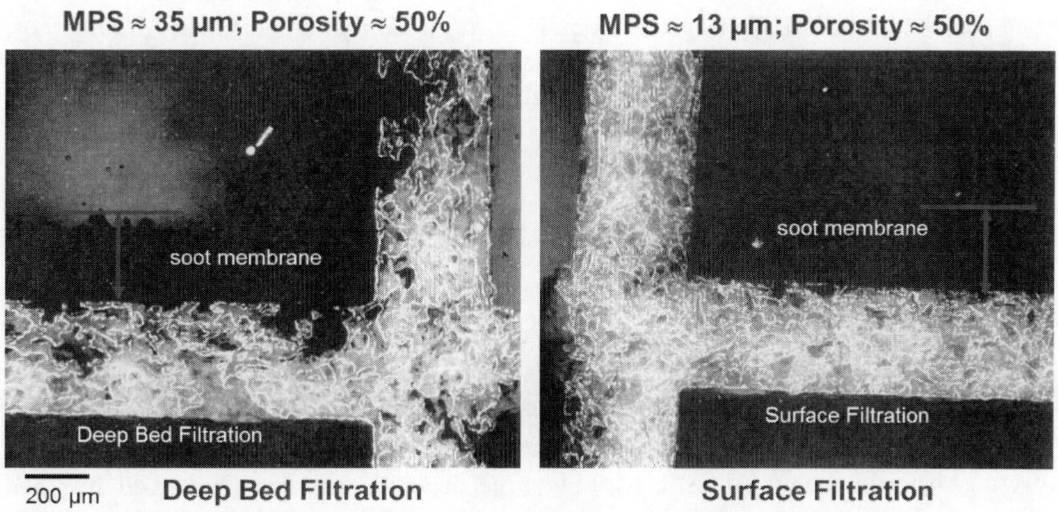

Figure 11. Optical micrographs of a cross sectional view of the through-the-wall median pore size (MPS) of 13 and 35 microns and 50% pore volume with PM deposition [14].

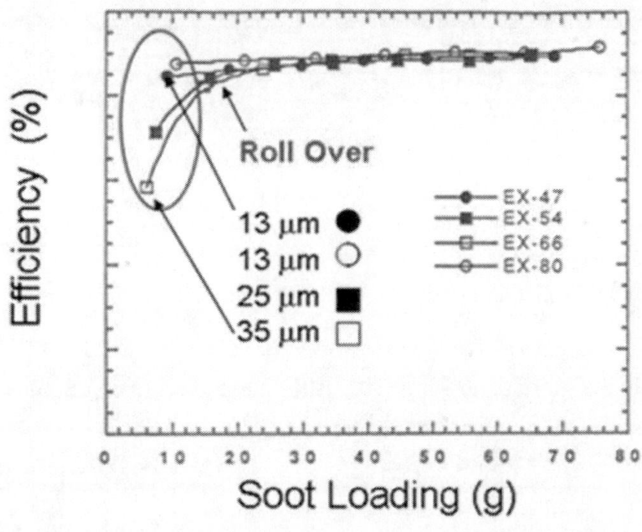

Figure 12. Filtration efficiency of selected cordierite wall flow filters as a function of PM (soot) loading and average pore size [14].

Shown in Figure 12 is the filtering efficiency of selected (Corning Incorporated) wall flow filters as a function of PM (soot) loading. DPF removes up to 99.9% of the elemental carbon and up to 90% of the organic fraction of the soot by mass [16]. All filters have a

100/17 cell geometry; 17 L filters at 50% porosity [14]. Corning Incorporated does not commercially offer the 100/17 cell geometry anymore as discussed in Table 3. The pre and post roll over represents the pore filling through-the-wall and PM membrane filtering dependency, respectively. All DPFs have the same filtration post roll over illustrating the significance of the PM membrane filtration towards improving the overall filtration efficiency. Efficiency is dominated by the through-the-wall pore size and porosity.

In summary, the PM filtering efficiency is lower initially at combinations of large pore volume (>50%) and mean pore size (>15μm), and is primarily deep bed filtration, high at small mean pore size (<14μm), and is primarily surface membrane filtration. The filtering efficiency has a secondary dependency on the choice of cell geometry, i.e. wall thickness, which can influence filtering efficiency due to the degree of through-the-wall pore bridging. The filtering efficiency is also influenced by the soot hydrocarbon make-up ("wet" and "dry") and flow rates. PM filtering is mechanical and independent of material choice.

Filtration (Pressure Drop)

Filtration results from the constriction of flow accompanied by an elevated pressure drop. The filter through-the-wall pore profile, as with filtering efficiency, is influenced by the PM membrane density and wall permeability and the material contribution to pore connectivity. The DPF cell geometry profile: cell density, wall thickness, channel length and plug length (affecting channel volume) is illustrated in Figure 13 depicting a schematic of a DPF inlet and outlet channel with corresponding exhaust gas flow pattern. The total change in pressure drop is described by equation 3 and demonstrates the mechanical nature of the pressure drop variance, which is independent of material choice.

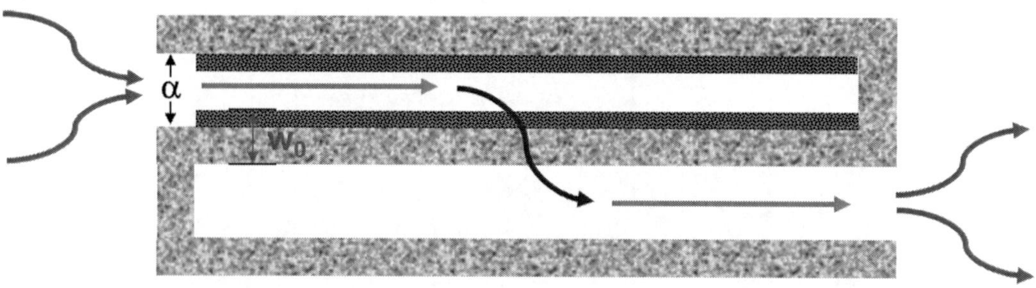

Figure 13. Schematic of a DPF inlet and outlet.

The total change in pressure drop for a DPF filter is derived from equation 3 [17]:

$$\Delta P = \underbrace{\boxed{(\Delta P_1 + \Delta P_2)}}_{\text{Wall Permeability}} + \underbrace{\boxed{\frac{\mu Q}{2V}(\alpha + w_o)^2 \left[\frac{8FL^2}{3}\left(\frac{1}{\alpha^4}\right)\right]}}_{\text{Channel Friction}} + \underbrace{\boxed{\frac{\rho Q^2(\alpha + w_o)^4}{V^2\alpha^2}\left(2\zeta\left(\frac{L}{\alpha}\right)^2\right)}}_{\text{Contraction and Expansion Losses}} \tag{3}$$

Where; μ = mass flow, Q = flow rate, V= channel volume, α = hydraulic diameter, W_o is wall thickness (without PM), F = wall friction coefficient, ρ = density of air, ζ= Forchheimer coefficient, L = channel length. The color code is related by the direction of flow arrows in Figure 13.

Shown in Figure 14 is a generalized depiction of a change in pressure drop as a function of PM loading. The general view is consistent with the understanding that filtering efficiency is enhanced with PM membrane filtration and with a moderate change in pressure drop if the PM membrane build up is surface specific per DPF through-the-wall pore size rather than deep bed filtration (PM penetrates pores) as illustrated. The rapid rise in pressure drop for deep bed filtration is caused by soot blocking the smaller pores, forcing more flow through the larger pores, and a higher pressure drop rise with loading as these pores also fill. Figure 15 shows the pressure drop as a function of soot (PM) loading for the same filter wall porosities and size as was illustrated in Figure 14. Figure 15 demonstrates what was presented as a generalized case in Figure 14. As a result, DPFs through-the-wall porosity is now design to maximize the surface PM cake filtration to minimize pressure drop over the range of PM (soot) loading.

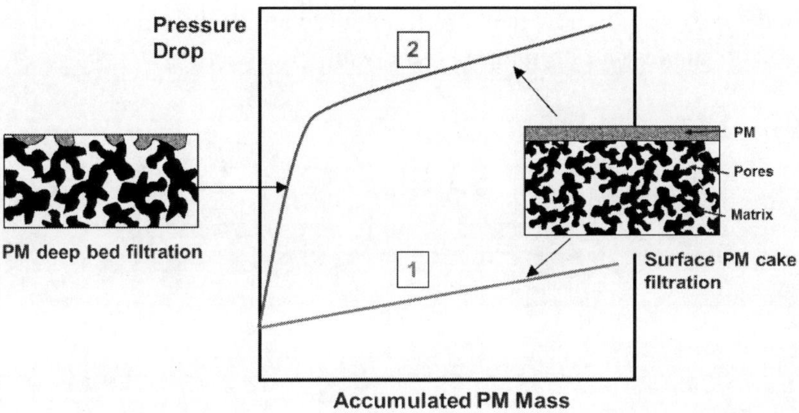

Figure 14. A generalized change in pressure drop (ΔP) as a function of accumulated PM.

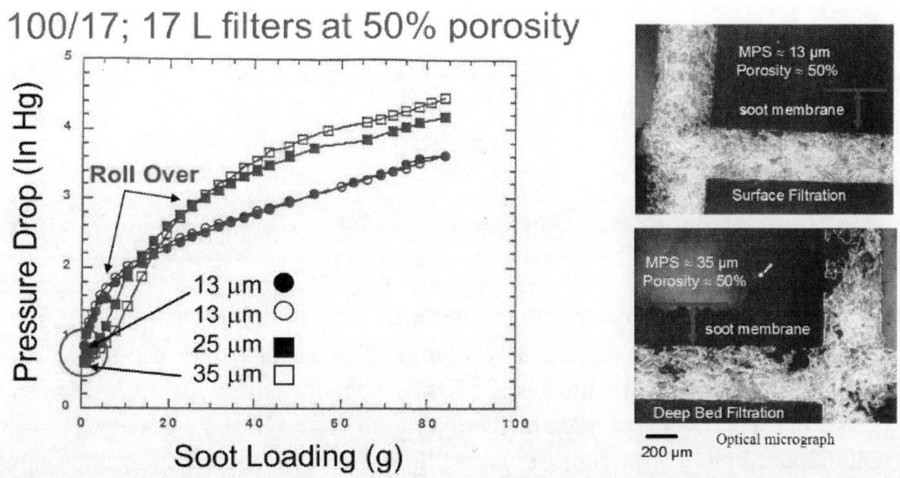

Figure 15. Pressure drop of selected wall flow filters as a function of soot (PM) loading [14].

Hickman et al. [18] published the effects a DPF cell and monolithic geometry contribution to DPF. The pressure drop as a function of cell density is shown in Figures 16 and 17. In Figure 16 the monolithic DPF geometry is 5.66" x 6" (14.38 cm x 15.2 cm) modelled with 25 grams of PM (soot) loading for a 12, 14, 16, 18 mil wall thickness. The lower pressure drop as a function of cell density for all wall thicknesses is dominated by the through-the-wall and soot pore volume.

In Figure 17 the monolithic DPF geometry is 5.66" x 12" (14.38 cm x 30.48 cm) modelled with 25 grams of PM (soot) loading for a 12, 14, 16, 18 mil wall thickness. The minimum in the pressure drop as a function of cell density is dominated by the down the channel, which proportionally makes a greater contribution to change in pressure drop than the through-the-wall and soot pore volume. It should be noted, that for both Figures 16 & 17 the increase in pressure drop is due to the increase in the through-the-wall tortuosity.

As a result of Hickman et al. [18] study, the DPF cell and global geometry is now design to minimize pressure drop by tailoring cell density, wall thickness, and aspect ratio (see table 3).

We will see in the next section how the design for surviving an uncontrolled regeneration can influence a trade off in pressure drop through-the-wall.

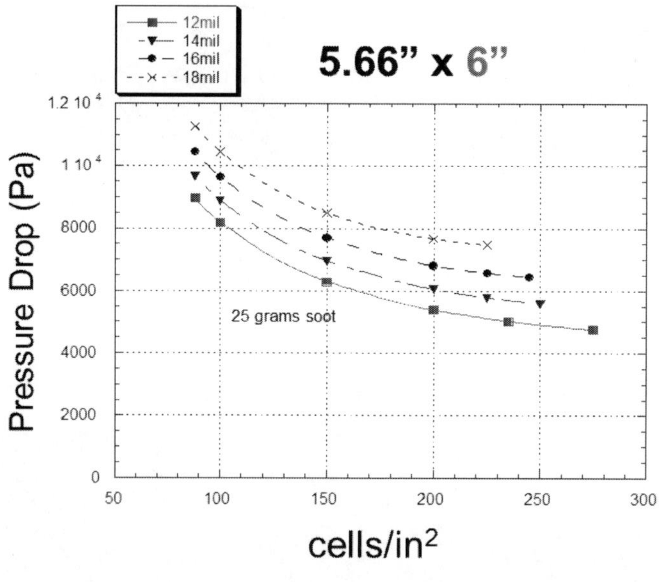

Figure 16. Pressure drop as a function of cell density modelled at 25 gram of PM (soot) for a 5.66" x 6" Filter [18].

Shown in Figure 18 is an optical micrograph of a DPF channel length and cell density cross section illustrating effect of nominal ash build up in use [19]. The ash distribution down the DPF inlet channel builds to the back of the channel; ash is residual after engine oil additives [20], and will, overtime, narrow the hydraulic diameter and reduce effective channel length (volume) adversely affecting in use DPF pressure drop. As a result, maintenance intervals were established to periodically clean and remove ash from the DPF [21].

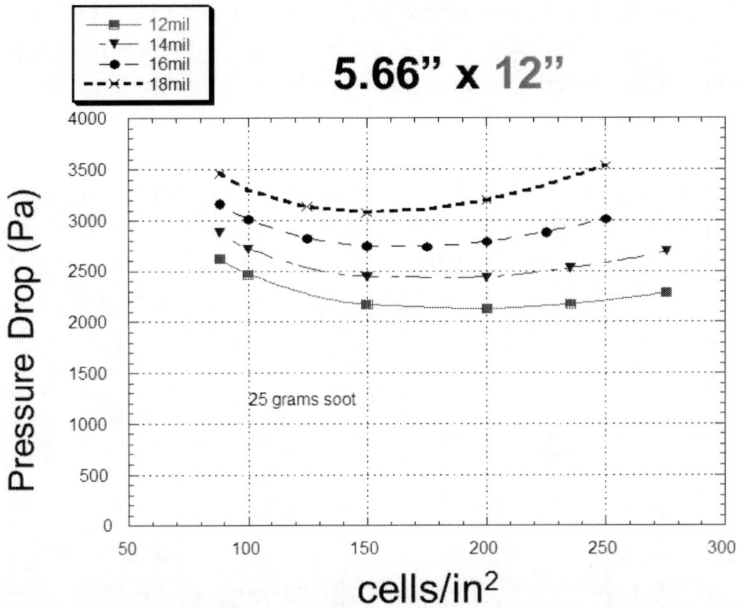

Figure 17. Pressure drop as a function of cell density modelled at 25 gram of PM (soot) for a 5.66" x 12" Filter [18].

Figure 18. Optical micrograph of a DPF channel length and cell density cross section illustrating effect of nominal ash build up in use [19].

Once maintenance intervals were established for ash removal for the DPF, alternative cell geometry designs were consider. The schematic of an asymmetric cell design, shown in Figure 19, allows for an increase in the inlet surface area by enlarging the inlet cell dimension for greater ash storage capacity in the inlet channels, reducing the impact contribution of the ash on the inlet hydraulic diameter and channel volume, which preserves pressure drop and minimizes ash cleaning cycles when compared to the standard cell design. The benefit of an asymmetric cell design to pressure drop is shown in Figure 20 as a function PM (soot) loading for a 5.66" x 10" (14.38 cm x 25.4 cm) with a 50 g/L ash loading at a space velocity of 78.700 $(hr)^{-1}$ for selection inlet to outlet ratios. The asymmetric inlet to outlet ratios, now known as asymmetric cell technology (ACT), are balanced to preserve strength as well as pressure drop in order to maximize ash storage [22 & 23].

Although adding length to the DPF would also add ash storage capacity, the down the channel friction (see equation 3) increases at the expense of increasing pressure drop. Therefore the down the channel trade off in pressure drop is avoided with the asymmetrical cell design for a more eloquent solution for enhancing longer intervals between ash removal maintenance. For further reading on the benefit of alternative asymmetrical cell designs see reference [24].

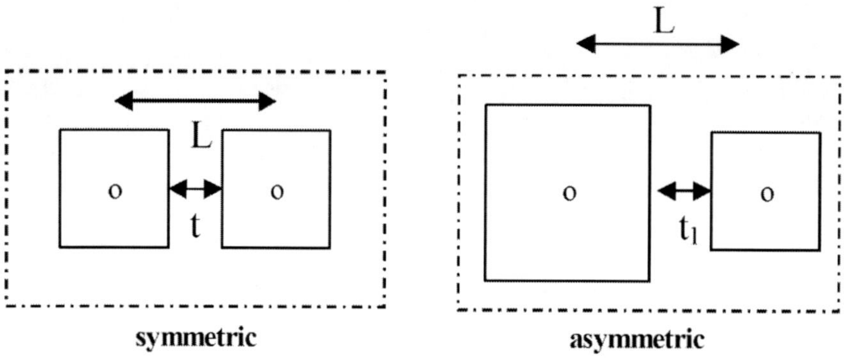

Figure 19. Schematic of the transformation from symmetric cell design (standard design) to an asymmetric cell design [22 & 23].

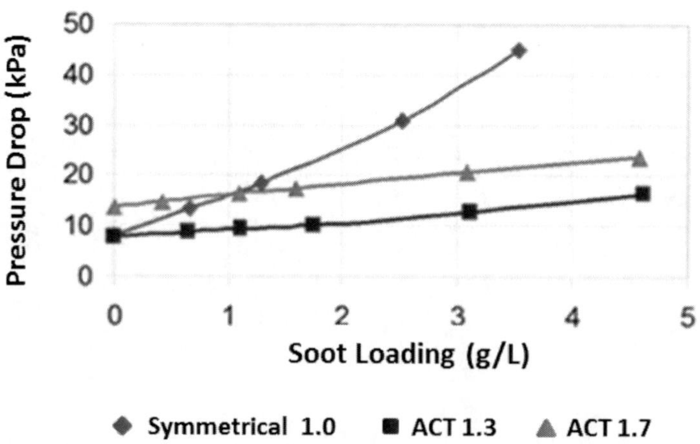

Figure 20. Pressure drop as a function PM (soot) loading for a 5.66" x 10" (14.38 cm x 25.4 cm) with 50 g/L ash loading with selected asymmetric inlet to outlet ratios.

In summary, pressure drop is mechanical and independent of material choice. The in use DPF pressure drop is low with narrow surface and through-the-wall pore size distribution (MPS > 14 microns) and well-connected porosity for a fixed cell and global geometry. The cell and global geometry contribution was shown to yield higher clean pressure drop at higher cell density and generally lower soot loaded pressure drop. For DPFs with an aspect ratio (diameter/length) of 1, the higher the cell density the lower the soot loaded pressure drop at a given wall thickness and pore profile. For DPFs with an aspect ratio of 0.5 there is an optimum cell density to provide lower soot loaded pressure drop for a given wall thickness and pore profile. Ash narrows hydraulic diameter and reduces effective channel length (volume) increasing in use DPF pressure drop. The effect of reducing pressure drop as a function ash has been overcome with innovative asymmetrical cell designs.

REGENERATION

The PM (soot) must be periodically burned to regenerate the filter (the forced removal of the core carbon portion of the PM) to recover the system backpressure. The auto-ignition of carbon with oxygen is normally 550°C however the auto-ignition will vary in practice due to adsorbed hydrocarbons and NO_2 content.

The DPF regeneration cycle considers two types of regenerations. A controlled regeneration is when the engine O_2 concentration is low at high flow rate with high convective heat transfer and inlet temperature. This type of regeneration can last for up to 10 to 20 minutes. An uncontrolled regeneration is when the engine drops into idle after regeneration is initiated. The O_2 concentration goes up and flow rate goes down resulting in low convective heat transfer. This type of regeneration can last for just 10s of seconds, which is essentially adiabatic. So, it is critical to properly manage the engine exhaust gas composition, NOx/PM ratio, time at temperature, O_2 concentration, flow rate, PM loading, and DPF inlet temperature to minimize uncontrolled regenerations.

Shown in Figure 21 is example of an uncontrolled regeneration showing the elevated temperature, per thermal couple placement, on the outlet center end of the DPF. The time at temperature is critical for stress and strain analysis and in this case the short duration supports the eariler discuss of an addiabatic assumption during a uncontrolled regeneration. The time at temperature is specific to the PM loading, souble organic fraction (SOF), and engine transient cycle. Large PM (> 6 g/L) loadings provide significant fuel to sustain an uncontrolled regeneration. Today, heavy duty PM loadings are controlled within the range of a 1 to 4 g/L before a controlled regeneration is induced.

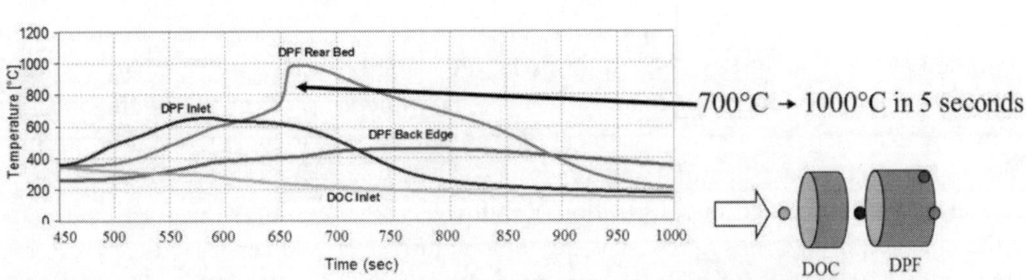

Figure 21. Example of an uncontrolled regeneration for 15 g/L loading of PM (soot).

The ease and efficiency of the regeneration is material and geometry dependent. The DPF thermal conductivity is critical and is influenced by the material choice and pore volume. The pore volume effect on thermal conductivity is shown by equation 4:

$$k_p = k_s[1-\upsilon] \quad (4)$$

Where k_p = thermal conductivity of a porous material, k_s = thermal conductivity of the solid material and υ = pore fraction (volume fraction). Shown in Figure 22 is the thermal conductivity as a function of temperature for cordierite, AT and α-SiC at selected pore volumes. α-SiC has an order of magnitude greater thermal conductivity over the application temperature range of 600°C to 900°C (controlled regeneration). A higher thermal conductivity dissipates heat away from the DPF during an uncontrolled regeneration to keep thermal gradients low.

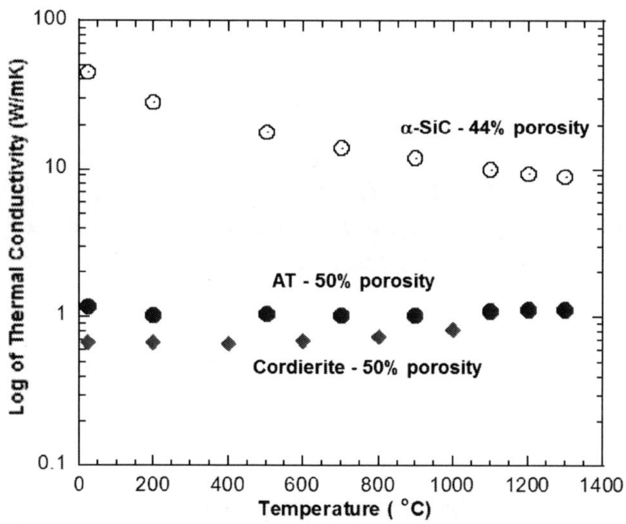

Figure 22. Thermal conductivity as a function of temperature for cordierite, AT and α-SiC at selected pore volumes.

Shown in Figure 23 is the representation of a nominal experimental thermocouple configuration per DPF geometry derived to model regeneration behavior. The thermocouples are placed from the exit (outlet) of the DPF at 1, 3, and 5 inches (2.54 cm, 7.62 cm, and 12.7 cm) down the channel. The radial placement, as shown, assumes radial symmetry with only the inlet position recording the full diameter effect.

The inlet, mid body, and outlet positions capture the regeneration behavior radially and down the channel. Shown in Figure 24 is the temperature after 500 seconds for a DPF inlet temperature of 650°C as a function of thermocouple location at a flow rate of 80L/min for Cordierite and α-SiC. The legend in Figure 24 shows the DPF material density, which is proportional to the DPF volumetric material heat capacity. The comparison, as shown in the legend, is intended to elucidate the effects of heat capacity and thermal conductivity on the heat up behavior of cordierite and α-SiC DPFs. For equivalent material density, the higher thermal conductivity makes it more difficult to heat up the α-SiC DPF at equivalent inlet temperatures and flow rate conditions compared to the cordierite DPF, as indicated by the radial thermal gradient differences between the two materials in Figure 24.

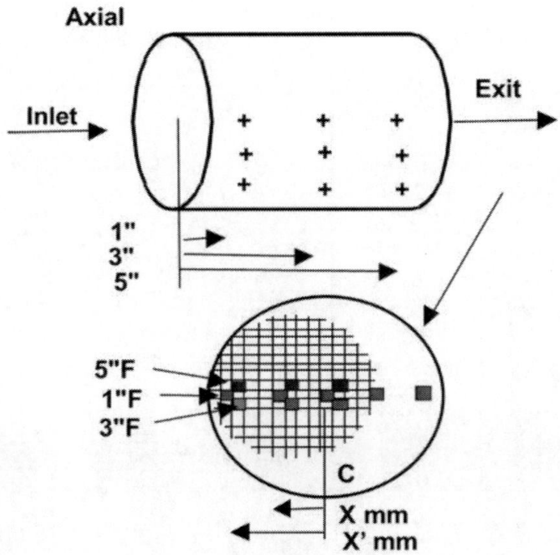

Figure 23. Schematic of example thermocouple configuration for regeneration modelling (thermocouple placed from exit end).

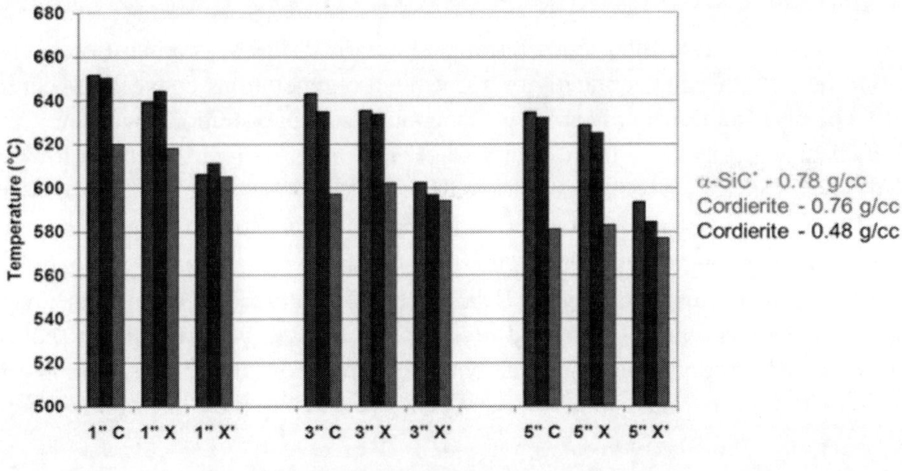

Figure 24. Temperature after 500 seconds for a DPF inlet temperature of 650°C as a function of thermocouple location (80L/min flow rate) for cordierite and α-SiC.

At low flow rates α-SiC is more effective than cordierite or AT to dissipate heat given the higher conductivity. Shown in Figure 25 is the time to reach 500ºC at a 650°C inlet temperature as a function of flow rate (80, 160 and 320 L/min) for α-SiC and cordierite. The benefit of the higher thermal conductivity diminishes as flow rate increases. At 80L/min it mimics an idle condition while the 320 L/min mimics highway driving. Controlled regeneration convective heat transfer (not thermal conductivity) keeps temperatures and gradients down. Uncontrolled regeneration thermal events are too rapid for thermal conductivity (W/mK; W = J/s) to have maximum benefit (see Figure 21).

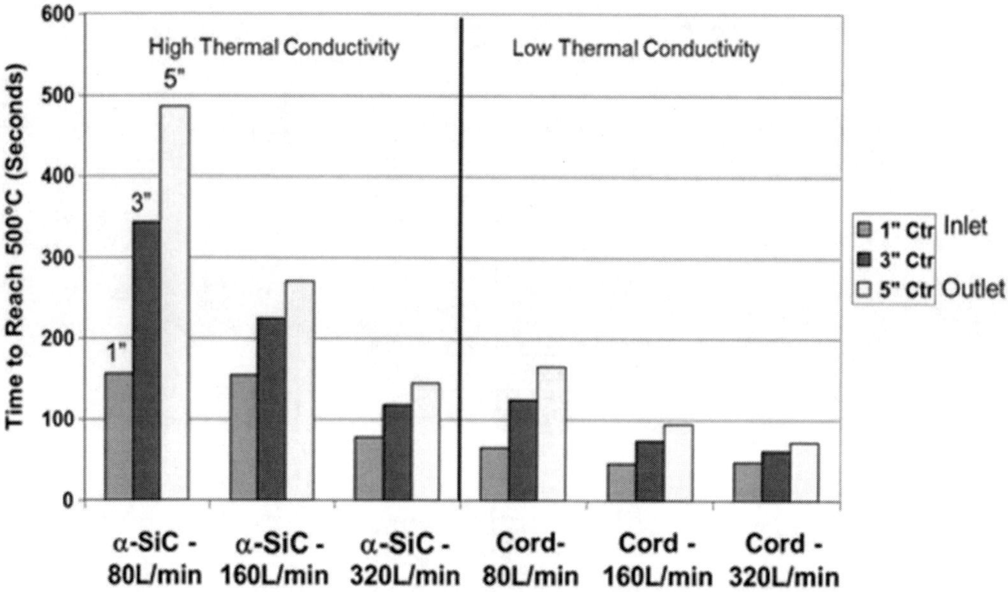

Figure 25. Time to reach 500 seconds at a 650°C inlet temperature as a function of flow rate for α-SiC and cordierite.

In summary, the internal filter temperature and rate of PM (soot) combustion is primarily a function of the DPF thermal conductivity. Controlled regenerations convective heat transfer (not heat capacity) minimizes heat rise. Uncontrolled regeneration, with an adiabatic assumption, heat capacity will limit heat rise. For Cordierite and AT their low thermal conductivity makes initiation of regeneration easier, i.e. requires lower inlet temperature. The low thermal conductivity results in increased temperatures and thermal gradients during an uncontrolled regeneration. α-SiC's high thermal conductivity makes it more difficult to initiate regeneration, i.e. requires higher DPF inlet temperature (the required segmentation diminishes radial conductivity). Controlled regeneration convective heat transfer (not thermal conductivity) keeps temperatures and gradients down. Uncontrolled regenerations thermal events are too rapid for thermal conductivity (W/mK; W = J/s) to have maximum benefit.

SURVIVABILITY

The DPF survival considerations are focused on surviving uncontrolled regenerations. Factors influencing survival are similar to those influencing ease and efficiency of regeneration. However, the DPF heat capacity plays a larger role in addition to the thermomechanical properties: coefficient of thermal expansion (CTE), modulus of rupture (MOR), and elastic modulus (Emod); and the chemi-mechanical properties, i.e. must be inert with respect to residual ash post regeneration at normal operating temperatures, ideally at or below 900°C. As was the case for ease and efficiency of regeneration, proper management of the engine exhaust gas composition, SOF fraction, NOx/PM ratio, time at temperature, O_2 concentration, flow rate, PM loading, and DPF inlet temperature is needed to minimize uncontrolled regenerations and the associated stress and strain profiles.

Cordierite and AT thermomechanical properties are influenced by pore structure, density of microcracks, and secondary phase while α-SiC thermomechanical properties are influence by pores and secondary phase(s) only.

Table 4 list the thermomechanical properties for cordierite, α-SiC, and AT. The low CTE of cordierite and AT combined with the low MOR and Emod provides excellent thermal shock resistance as a monolith. The higher thermal conductivity can partially compensate for α-SiC's higher CTE under moderate heat-up rates axially, but not radially, and not under high heat up rates. α-SiC needs to be segmented in order to manage thermal shock resistance. The small segments reduce the probability of failure and the segmented joints act as compliance seams to minimize the stress and strain associated with a thermal shock.

Table 4. List of thermomechanical for cordierite, α-SiC, and AT

	Cordierite	**α-SiC**	**AT**
Geometry* **(cells/in^2 /wall thickness (mil))**	200/12	200/16	300/13
CTE x 10^{-7}/ °C^{-1} **(23°C to 1000°C)**	5.00	45.0	9.00
Modulus of Rupture (MPa)^	2.50	18.0	1.50
Elastic Modulus (GPa)^	5.50	33.0	1.70

The DPF volumetric heat capacity is reduced by adding pore volume, similar to the thermal conductivity. The material pore volume effect on heat capacity is shown by equation 5:

$$Cp_p = Cp_s[1-\upsilon] \tag{5}$$

Where Cp_p = volumetric heat capacity of a porous material, Cp_s = volumetric heat capacity of the solid material and υ = pore fraction (fractional porosity).

Shown in Figure 26 is volumetric heat capacity as a function of temperature for cordierite, AT and α-SiC at selected pore volumes. The specific density of cordierite is lower than that of AT and α-SiC resulting in lower volumetric heat capacity. The volumetric heat capacity can be tailored by choice of pore volume in addition to varying the cell wall thickness at any given cell density.

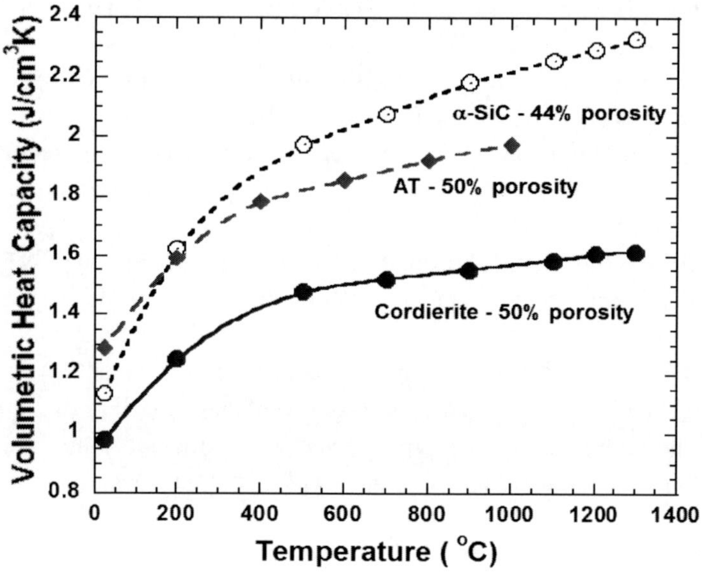

Figure 26. Volumetric heat capacity as a function of temperature for cordierite, AT and SiC at selected pore volumes.

Figure 27 shows the precent of closed frontal area (proportional to heat capacity) as a function of DPF wall thickness illustrating the approach to managing heat capacity via cell density for a given material choice. So, one must determine heat capacity for survivability per material and geometry choice, and then balance the pressure drop trade off.

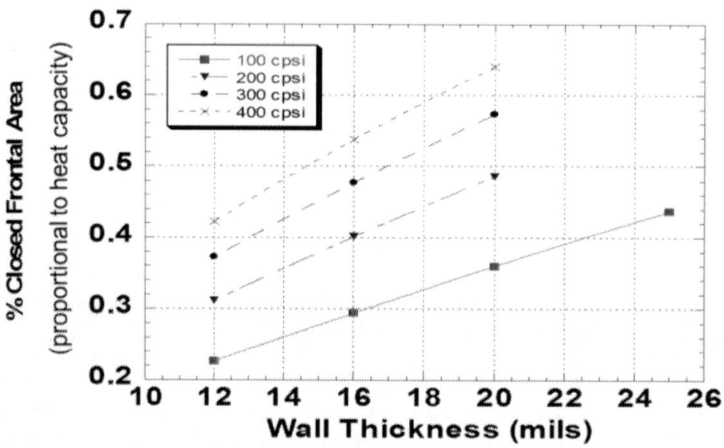

Figure 27. Precent of closed frontal area as a function of DPF wall thickness.

An approach to summarizing the effective heat capacity for the 2007 Clean Diesel launch in North America is shown in Figure 28. At the time of the launch high heat capacity was favoured due to the need to have the DPF provide the safety margin to insure the system survivability from an uncontrolled regeneration. As engine electronic controls improved, the system itself is designed to avoid uncontrolled regeneration. This allows for lower heat capacity DPFs, which can allow thin wall filters and reduced pressure drop.

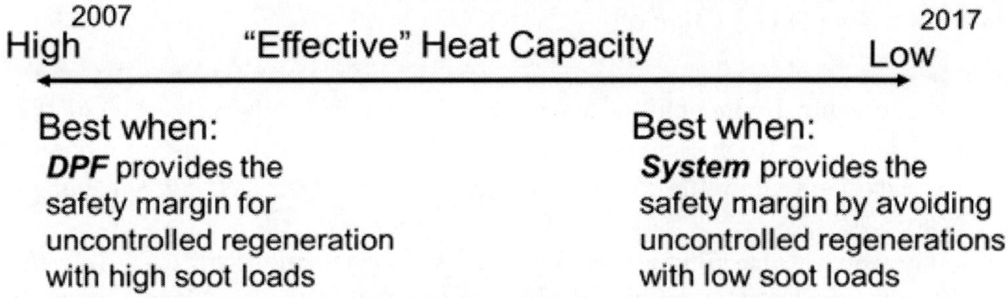

Figure 28. DPF "Effective" bulk heat capacity continuum.

Shown in Figure 29 is the regeneration exotherm temperature as a function of PM (soot) loading for cordierite, AT and α-SiC at selected wall thickness. The effect of tailoring DPF wall thickness to manage volumetric heat capacity is demonstrated, a 66% increase in the cordierite wall thickness (from 12 to 20 mils) overcame the difference in the material density of AT at 16 mil wall thickness. The α-SiC higher thermal conductivity in combination with the volumetric heat capacity, in this laboratory operating condition, keeps regeneration temperatures down but, as the previous section showed, it is more difficult to initiate regeneration for a given porosity and geometry due to the high thermal conductivity.

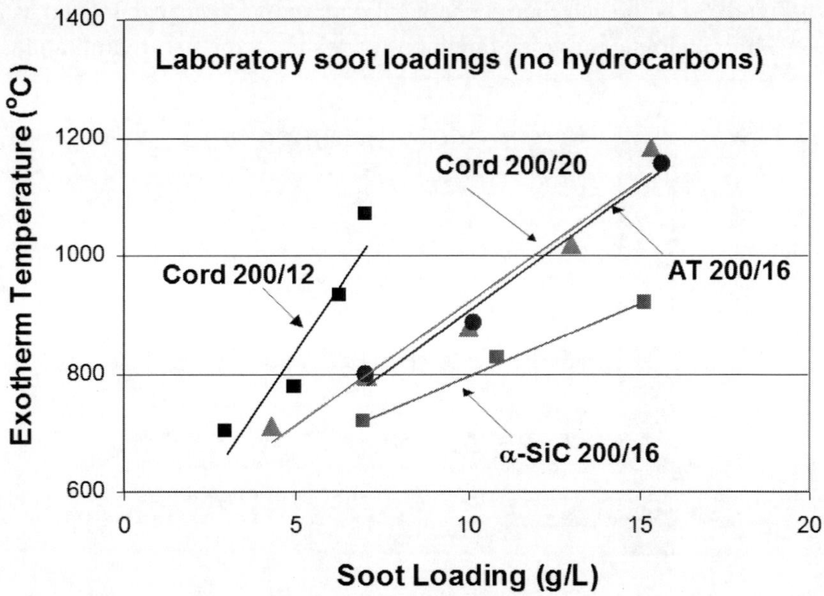

Figure 29. Regeneration exotherm temperature as a function of PM (soot) loading for cordierite, AT and SiC at selected wall thickness.

The thermomechanical survivability of the DPF (the resistance to a change in temperature associated with an uncontrolled regeneration) can be evaluated with a simple model by using the DPF material mechanical properties in an unstressed and stress state by the equation 6:

$$R = \Delta T(°C) = [\sigma k(1-\upsilon)][E\alpha]^{-1} \quad (6)$$

where T = temperature σ = modulus of rupture, k = thermal conductivity, υ = Poisson's ratio, E = elastic modulus, α = Coefficient of thermal expansion, and R = the resistance to initiating fracture; the maximum allowable ΔT in a steady state heating of a body containing non-conducting pores [25].

It was discussed in the Filtration section that ash accumulation will negatively impact pressure drop due to alterations to the cell and inlet channel geometry. When it comes to the influence of ash on survivability of the DPF, it is the chemical interactive nature of the ash composition with the filter material. The ash interaction is temperature and time dependent. The highest temperature events (from 1000°C to 1200°C) are brief lasting only 10s of seconds typically, as previously discussed.

As was discussed earlier, the residual ash source in the DPF is primarily from engine oil additives such as: anti-foaming agents, oxidation inhibitors, corrosion inhibitors, rust inhibitors, anti-wear and anti-scuff agents (ZDDP - zinc dithiodialkyphosphate), detergents to neutralize acids formed during combustion, sulfonate and/or phenates, (neutralizes acid combustion by products – contain calcium and/or magnesium), dispersants to hold contaminants in suspension and off the metal surfaces of the engine, and polymers to improve the ability of the oil performance in cold temperatures and over a wide range of temperatures [26].

Shown in Figure 30 is the ash thermochemical corrosion effects catalogued as a function of temperature. This catalogue provided direction for system temperature limitation.

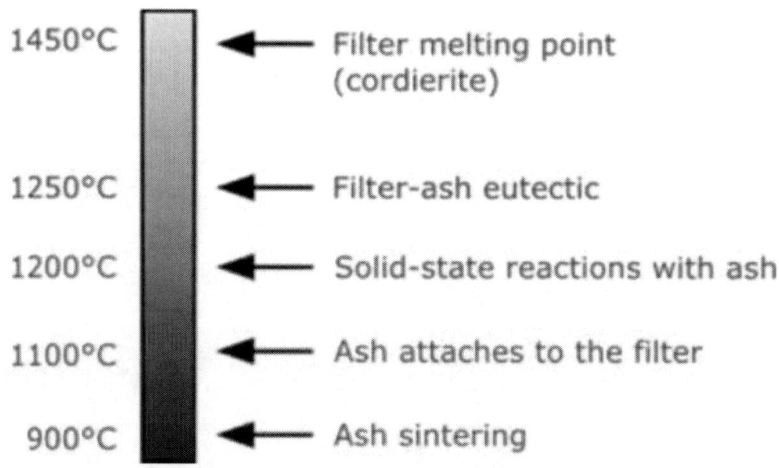

Figure 30. Ash thermochemical reactions catalogued as a function of temperature [18].

Listed in Table 5 is a list of selected ash compositions and corresponding sintering temperatures. The compositional makeup for A, C-D is specific to lubrication oil speciation and engine wear, while the compositional makeup of B includes the fuel additive cerium oxide (to promote lower auto-ignition temperature of the PM (soot)).

Table 5. List of selected ash compositions with sintering temperatures

	Oxide	A	B	C	D	E
●	MgO	5.16	0.08	16.0	4.26	3.34
●	CaO	27.6	1.33	30.0	23.3	34.4
X	Fe_2O_3	0.34	2.45	2.29	1.23	0.66
●	ZnO	9.20	0.65	8.01	8.76	11.3
■	CeO_2	-	87.8	-	-	-
●	SO_3	42.9	2.79	32.8	55.5	34.8
●	P_2O_5	14.8	4.90	10.9	6.95	15.5
	Totals	100.0	100.0	100.0	100.0	100.0
	T_s*(°C):	**1050**	**1200**	**1200**	**1100**	**1050**

● lube oil species, **X** engine wear, ■ fuel additive

T_s^* = approximate sintering temperature

The thermochemical reaction effect of the each ash, shown in Table 5 is dependent on the lubrication speciation concentration specific to the DPFs material composition. For cordierite the presence of iron induces local melting at ≥ 1190°C and for cerium ≥ 1210°C. The formation of gahnite ($ZnAl_2O_4$) in the ash on heat treatment as low as 900°C, is independent of the presence of iron or cerium [27]. For AT there are no deleterious interactions with the ashes for temperatures up to 1300°C [28], and for α-SiC, an amorphous phase was detected as a minor phase after heat treatment to 1100°C in the ash containing iron and after heat treatment to 1250°C in the ash containing cerium. Oxidation of α-SiC can be accelerated by reactions of ZnO, CaO, FeO, and Fe_2O_3, and CeO_2 with existing SiO_2 phase seen as early as 800°C [29].

It can be concluded that the three commercially available ceramic wall-flow filters are thermally inert to thermochemical corrosion by ash up to 1200°C. The significance of the inertness is shown in Figure 31, which is describing the suggested maximum use temperature of the DPF.

In summary, the system survival of an uncontrolled regeneration thermal response is a function of bulk heat capacity and thermal conductivity. Cordierite has low specific heat and conductivity. AT has a high specific heat and low thermal conductivity. α-SiC high specific heat and thermal conductivity. There are engineering tradeoffs between the bulk volumetric heat capacity and pressure drop, influenced by pore volume and geometry choices. Bulk heat capacity can be managed with pore volume and geometry.

There is a need to control amount of energy going into the regeneration sequence, i.e. exhaust gas composition and temperature, and the amount of soot on the filter before regeneration. The thermomechanical response is a function of thermal gradients and bulk material properties Cordierite and AT' strength (MOR), elastic modulus, and CTE minimize thermal stress and strain for a given thermal gradient. Thermal conductivity partially compensates for α-SiC's higher thermal expansion under moderate heat up rates axially, but due to segmentation not radially, and not under high heat up rates. Thermochemical response is a function of maximum temperature, time, and degree of ash deposits, influenced by heat capacity and thermal conductivity. Ash interactions are temperature and time dependent however, the highest temperature events (1000°C to 1200°C) are brief lasting only 10s of seconds, typically.

Shown in Figure 32 is the system functionality define by the interrelationship of the DPF material and geometry makeup. It is a pictorial summary depicting the complexity in the achievement to launch a very successful criterion emission abatement system.

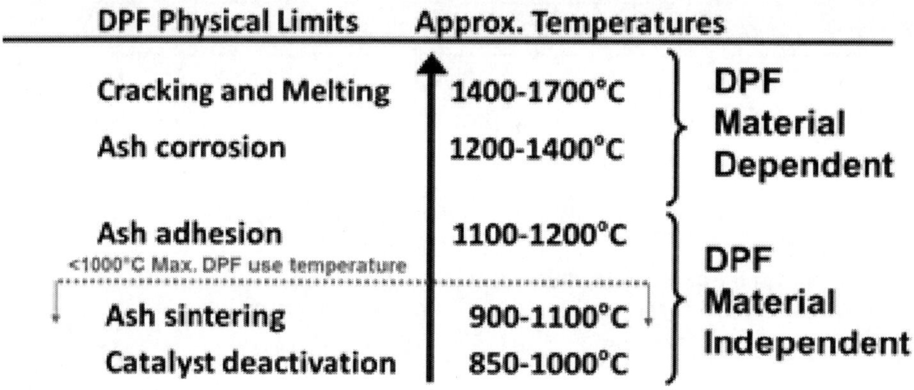

Figure 31. Projected DPF use temperature limits.

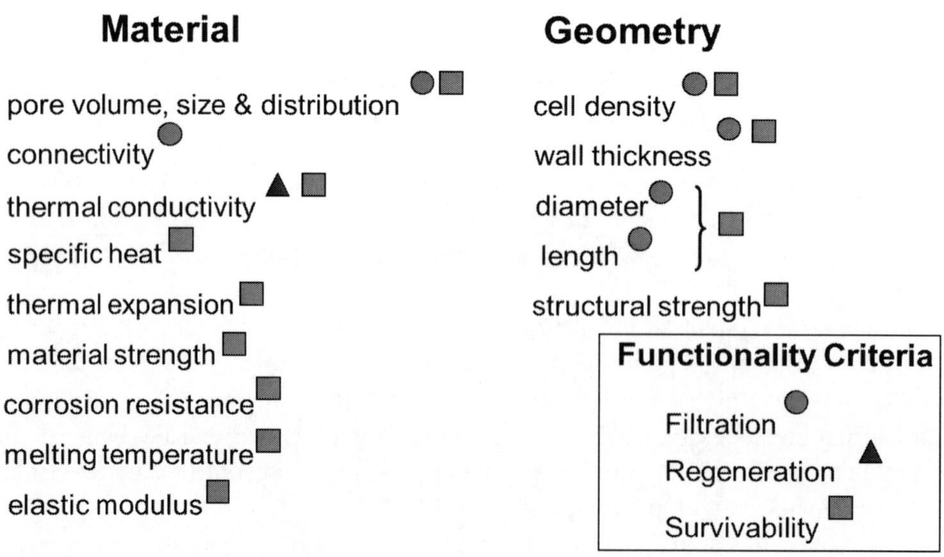

Figure 32. Interrelationship of the DPF material and geometry makeup with system functionality.

What is next? DPF regeneration strategies, catalysts, filter properties, material, and catalysts are advancing. At the time of this writing, the new developments in thin-wall DPFs are enabling low PM and NOx for improved fuel economy. The PN specification (at the time of this writing only in Europe) requires filter maximum pore size to be < 20 um for optimum capture. New developments are underway to tailor porosity profiles to improve catalyst loading. Integrating new understanding of ash permeability behavior to better define the effect on pressure drop impact and ash removal maintenance intervals are underway.

CONCLUSION

The wall-flow "checker board" manifolded DPF was developed by Corning Incorporated in the early 1980's. All new on-road diesel engines in the US, EU, and Japan utilize them. The DPF application material preferences are cordierite, alpha-silicon carbide, and aluminum titanate. Filtration (filtering and pressure drop) is mechanical and independent of material choice. The DPF cell geometry evolved from standard cell geometry to asymmetric cell geometry to enhance ash storage capacity while minimizing pressure drop impact. PM regeneration is dependent on DPF material choice, inlet exhaust chemistry and temperature as well as the product thermal conductivity: high thermal conductivity keeps regeneration temperatures down but makes it more difficult to initiate regeneration. Survivability is dependent on DPF material choice for thermal conductivity and bulk volumetric heat capacity (material specific heat) affects. There is an engineering tradeoff between bulk volumetric heat capacity and pressure drop, influenced by the through-the-wall pore profile and cell and monolithic geometry choices to manage the bulk volumetric heat capacity.

ACKNOWLEDGMENTS

The authors would like to thank all of our Corning Incorporated colleagues that contributed to the bulk of this work, mostly specifically; Dr. Thorsten Boger, Dr. Douglas Beall, Dr. David Hickman, Dr. Greg Merkel, and to Rodney Frost for the many fruitful discussion of the early years of the DPF development at Corning Incorporated, to Professor David Kittlesen, of the University of Minnesota, for the many insightful discussions on the nucleation of diesel PM, and finally to our colleagues in the Materials Science & Technology Department at the University of Limerick, Limerick Ireland for all the perceptive results regarding thermochemical ash interactions with the DPF materials; cordierite, AT, and α-SiO_2.

REFERENCES

[1] Kaiser W, "Rodolf Diesel", *German News Magazine,* Aug/Sept 1997.

[2] US Patent: 4,276,071; Assignee: General Motors Corporation (Detroit, MI) Inventor: Outland; Robert J. (Grosse Pointe Woods, MI) Appl. No.: 099935 Filed: December 3, 1979. Issued: June 30, 1981.

[3] Samet J.M., “Fine particulate air pollution and mortality in 20 U.S. cities,” New England Journal of Medicine 343, 1742-1749, 2000.

[4] http://www.epa.gov/oms/standards/weights.htm.

[5] Konstandopoulos A, G., “Growth Dynamics & Microstructure of Soot Deposits in Diesel Particulate Filters”, *Presentation at the 6th ETH Conference on Nanoparticles Measurements, Zurich,* Swisserland, August 19-21, 2001.

[6] Müller J. O., Frank B., Rolf E. Jentoft R. E., Schlögl R., and Su D. S., “The oxidation of soot particulate in the presence of NO2,” Catalysis Today, Volume 191, Issue 1, 15, Pages 106–111, 2012.

[7] Eugene S. Domalski, “Selected Values of Heats of Combustion and Formation of Organic Compounds Containing the Elements C, H, N, O, P and S,” Chemical Thermodynamics Data Center, National Bureau of Standards, Washington, D.C., *J. Phys. Chem. Ref. Data,* Vol. 1, No. 2, 1972

[8] Well-to-Wheels analysis of future automotive fuels and powertrains in the European context; TANK-to-WHEELS Report; Version 2c, March 2007; http://ies.jrc.ec.europa.eu/WTW

[9] T. Boger, personal communication, Corning Incorporated, May 2014.

[10] Brück R., Hirth P., Meike R., Treiber P., Breuer J.,“Metal Supported Flow-Through Particulate Trap; a Non-Blocking Solution”, *SAE Technical Paper* 011950, 2001.

[11] Zelenka, P., Reczek, W., Mustel, W., Rouveirolles, P., "Towards Securing the Particulate Trap Regeneration: A System Combining a Sintered Metal Filter and Cerium Fuel Additive", *SAE* 982598 1998.

[12] Mayer A., “Available Particulate Trap systems for Diesel Engines”, VERT, Report TTM W04/4/98, October 22, 1998.

[13] Mayer A., “Verified Particulate Trap Systems for Diesel Engines”, VERT, Distributed by DieselNet; http://www.dieselnet.com/tech/papers/filterliste_e_10_2000.pdf.

[14] Murtagh M. J., Socha L., Sherwood D., “Development of a Diesel Particulate Filter Composition and Its Effect on Thermal Durability and Filtration Performance”, SAE Technical Paper 940235, 1994. Republished in Diesel Particulate Filter Technology, editor T. V. Johnson (selection of the top 29 SAE papers covering the most significant research in this technology), International Society of Automotive Engineers, ISBN Number: 978-0-7680-1707-6, 2007.

[15] Courtesy of Prof. David Kittlesen (personal conversation), University of Minnesota, Twin Cities Minneapolis, MN, circa 2010.

[16] W. A. Majewski, “Diesel Particulate Filters,” Diesel Net Technology Guide, Ecopoint Inc. Revision 2011.03b.

[17] Konstandopoulos, A. G., Skaperdas, E., and Masoudi, M., "Microstructural Properties of Soot Deposits in Diesel Particulate Traps," *SAE Technical Paper* 2002-01-1015, 2002.

[18] Hickman, D.L., Ebener, S., Zink, U., "Reduktion der Partikelemissionen bei Diesel PKW und Nutzfahrzeugen", 22. Internationales Wiener Motorsymposium, Fortschritt-Berichte VDI (VDI Verlag, Duesseldorf), Reihe 12, Nr. 455, Band 2, pg. 267-285, 2001

[19] Aravelli K, Jamison J., Robbins K., Gunasekaran N., Heibel A., “Improved Lifetime Pressure Drop Management For DuraTrap® RC Filters with Asymmetric Cell Technology (ACT)” DEER conference presentation, Detroit, MI, August 24, 2006.

[20] Sappok A.,"Ash Accumulation in Diesel Particulate Filter," DieselNet Technology Guide, www.DieselNet.com, Copyright© Ecopoint Inc. Revision 2013.01.

[21] MECA, "Diesel Particulate Filter Maintenance: Current Practices and Experience", Manufacturers of Emission Controls Association, www.meca.org, Washington, D.C., June 2005,

[22] Marcher J. H., "Method of making a BITE Die" – US 6,570,119, Issued May 27, 2003.

[23] Beall D. M. and Marcher J. H.,"BITE honeycomb and Die" US 2003/0044572 A1, Filed August 30, 2001.

[24] Majewski A.," Wall-Flow Monoliths," DieselNet Technology Guide, www.DieselNet.com, Copyright© Ecopoint Inc. Revision 2005.09b.

[25] Kinergy W. D. "Factors Affecting Thermal Stress Resistance of Ceramic Materials," *J. Am. Cer. Soc.* Vol. 38. No. 1, January 1955.

[26] Jääskeläinen, H. and Majewski A.," Diesel Engine Lubricants," DieselNet Technology Guide, www.DieselNet.com, Copyright© Ecopoint Inc. Revision 2013.09.

[27] Pomeroy, M. J, O'Sullivan, D., Hampshire S., and Murtagh M. J.,"Degradation Resistance of Cordierite Diesel Particulate Filters to Diesel Fuel Ash Deposits" *J. Am. Ceram. Soc.,* 95 [2] 746–753 2012.

[28] Ogunwumi, S.B., Tepesch, P.D., Chapman, T., Warren, C. J.,"Aluminum Titanate Compositions for Diesel Particulate Filters," SAE *Technical Paper* 2005-01-0583, 2005.

[29] O'Sullivan D., Pomeroy, M. J., Hampshire S, and Murtagh M. J., "Degradation Resistance of Silicon Carbide Diesel Particulate Filters to Diesel Fuel Ash Deposits", *J. Mater. Res.,* Vol. 19, No. 10, Oct. 2004.

In: Automotive Exhaust Emissions and Energy Recovery ISBN: 978-1-63321-493-4
Editor: Apostolos Pesiridis

Chapter 7

TURBOCHARGING AND EXHAUST ENERGY RECOVERY

Hua Chen*
National Laboratory of Engine Turbocharging Technology, Tianjin, China

ABSTRACT

This chapter describes exhaust energy recovery using turbocharging and turbochargers. Firstly, the exhaust energy in reciprocating internal combustion engines is illustrated through an ideal engine cycle, and the reason the engine itself is unable to recover this energy is explained. The principle of turbocharging - the most successful way of exhaust energy recovery so far - is then introduced. The benefits of turbocharging the internal combustion engine are provided and discussed. A description of a turbocharger that performs the function of turbocharging is then provided, and the performance of the turbocharger turbine and compressor is described. The principles of matching the turbocharger to an engine, and the difficulty in doing so, are discussed together with a description of various turbochargers and turbocharging systems designed to overcome this difficulty. Other functions of turbochargers, as air management systems to the engine such as driving exhaust gas recirculation and engine braking are also presented. Finally, more recent applications of turbochargers in engine waste heat recovery are discussed.

Keywords: Internal combustion engine, turbocharging and turbocharger, waste heat recovery

* National Laboratory of Engine Turbocharging Technology, 96 Yongjin Road, Beichen District, Tianjin, China, 300400. Email: hchen@nlett.com

NOMENCLATURE

Abbreviations

AFR	Air fuel ratio
BDC	Bottom dead centre
BW	Borg Warner Turbo Systems
CTT	Commins Turbo Technologies
ER	Expansion ratio
ETC	Electrical turbo compound
EVC	Exhaust valve closed
EVO	Exhaust valve open
HTT	Honeywell Turbo Technologies
IVC	Inlet valve close
IVO	Inlet valve open
TDC	Top dead centre

Letters

C_{is}	Isentropic spout speed
C_p	Specific heat capacity under constant pressure
D	Diameter
h	Specific enthalpy
$\dot{m}$	Mass flow rate
N	Rotating speed
p	Pressure
T	Temperature
U	Blade speed
V	Absolute velocity, Engine piston displacement volume
$\dot{W}$	Power
γ	Ratio of specific heat capacities
ε	Effectiveness of intercooler
η	Efficiency
ρ	Density

Subscripts

0	Total or stagnation state
1	Inlet
2	Outlet
3	Intercooler air outlet
a	Ambient
c	Compressor, Corrected

f	Fuel
is	Isentropic
mech	Mechanical loss
t	Turbine
tm	Thermal mechanical
t-s	Total-to-static
t-t	Total-to-total
u	Tangential
w	Cooling medium at intercooler inlet

ENERGY IN ENGINE EXHAUST SYSTEM

The exhaust process in a reciprocating internal combustion engine starts when engine exhaust valves open. To enable a rapid clean up of combustion gas from the engine cylinders, the exhaust gas must leave the cylinders at pressures and temperatures higher than those at ambient conditions. The discharge of this high pressure and high temperature gas to ambient wastes energy, and constitutes approximately 25-35% of the total energy input to the engine. With ever increasing demand for fuel economy from the engine and vehicle due to environmental concerns, can one do something to recover some of this wasted energy without affecting engine performance or even enhance it?

Figure 1 shows an idealised engine thermodynamic cycle in P-V diagram form and the potential energy available in the exhaust system. The exhaust valve opens at BDC, point 5, where the cylinder pressure is much greater than the ambient pressure at the end of the exhaust pipe. If the contents of the cylinder at point 5 were somehow allowed to expand isentropically down to the ambient pressure (to point 6), then the work that could be done is represented by the cross-hatched area 5-6-1. This work could be recovered by allowing the piston to move further than normal from point 5 to point 6. This energy recovering method has been used in large, low speed marine engines with exceptionally long expansion strokes [**1**]. However, this method increases the size and weight of the engine, and is inconvenient to automotive and other on-highway applications. In these applications, engine speed is also much higher than that for marine applications so the additional piston frictional loss offsets the work gained by an ultra-long expansion stroke [**2**].

While this exhaust energy is difficult to be recovered by the reciprocating, internal combustion engine itself, it can be recovered by rotating turbomachinery in the form of a turbine. The turbine can be either a radial flow, axial flow or mixed flow type (inbetween the radial and axial flow directions). These turbines can work efficiently at relatively low pressure ratios provided by the engine exhaust gas, and they are compact and light weight and therefore can rotate at high speeds: the tip speed of a typical turbocharger turbine rotor can reach more than 500m/s. There are several ways to utilise the mechanical work produced by the turbine, some of these will be discussed towards the end of this chapter, but the most successful one is turbocharging of the engine which is the main topic of this chapter.

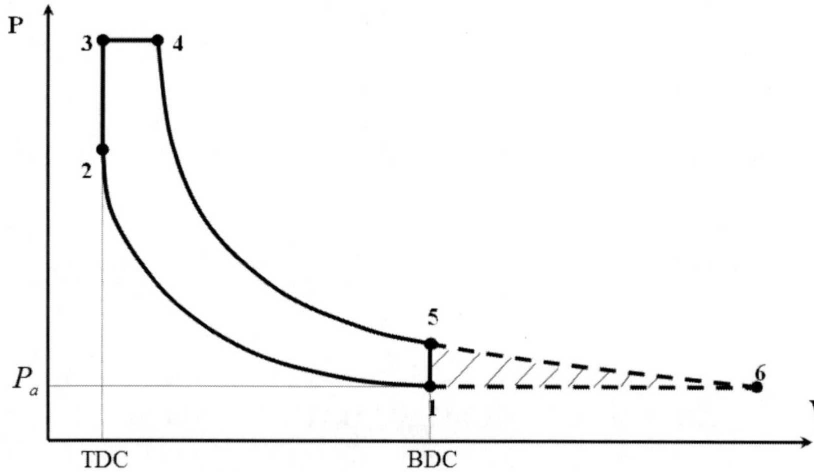

Figure 1. Ideal limited pressure cycle – naturally aspirated.

TURBOCHARGING THE INTERNAL COMBUSTION ENGINE

Figure 2 is a schematic showing how turbocharging works. The engine exhaust is directed to a turbocharger composed of a turbine and a compressor. A part of the exhaust energy is converted to turbine shaft work by the turbine. This work is then absorbed by the compressor on the same shaft of the turbine. The compressor, typically a centrifugal one, is also a turbomachine capable of rotating at the same high speed as the turbine. Using the turbine power, the compressor rotates at the same high speed to increase the air pressure going to the engine cylinders. An intercooler is often placed between the compressor and the engine to reduce charge air temperature to further increase charge air density, and to reduce the air temperature at the end of engine compression stoke. On a P-V diagram, the idealised thermodynamic cycle of a turbocharged engine is shown in Figure 3 reproduced from [2]. Turbocharging raises the inlet manifold pressure, hence the inlet process (12 – 1) is at pressure P_1, where P_1 is above ambient pressure P_a. The exhaust manifold pressure P_7 is also above the ambient pressure P_a. The exhaust process from the cylinder is represented by line 5-13-11, where 5-13 is the 'blown-down' period when the exhaust valve opens and high pressure gas expands out into the exhaust manifold. The 'blown-down' energy is represented by area 5-8-9. Process 13-11 represents the remainder of the exhaust process, when the piston moves from BDC to TDC displacing most of the gas from the cylinder to the exhaust manifold. This gas is above ambient pressure and therefore also has the potential to expand down to ambient pressure whilst doing useful work. The potential work that could be done is represented by the cross-hatched area 13-9-10-11. This work is done by the piston but could be recovered by a turbine in the exhaust. It will be called the piston pumping component of exhaust energy.

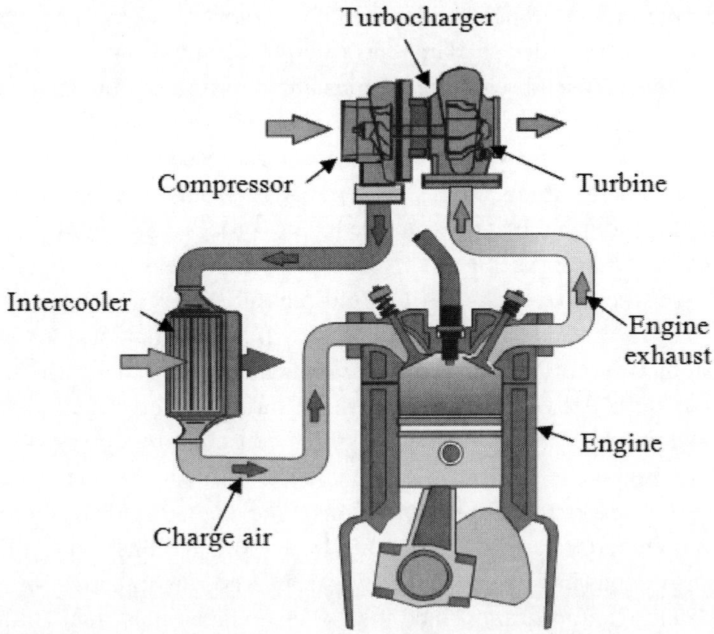

Figure 2. Turbocharged internal combustion engine – a schematic.

The maximum possible energy available to drive a turbine will be the sum of areas 5-8-9 and 13-9-10-11, but it is impossible to devise a practical system that will harness all this energy. To achieve this, the turbine inlet pressure must instantaneously rise to P_5 when the exhaust valve opens, followed by isentropic expansion 5-7-8 of the exhaust gas through P_7 to the ambient pressure P_8 (= P_a). During the displacement part of the exhaust process, the turbine inlet pressure would have to be held at P_7 (> P_a). Such a series of processes is impractical.

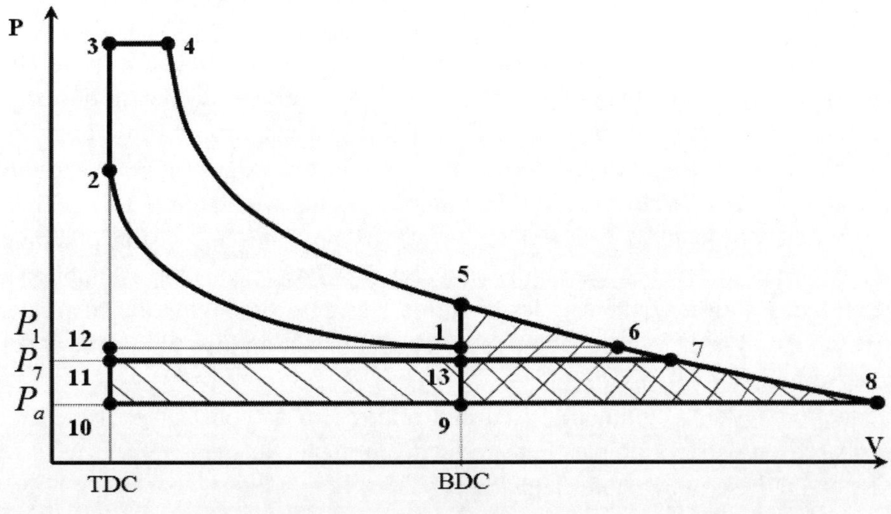

Figure 3. Ideal limited pressure cycle – turbocharged [2].

Consider a simple process that would occur if a large chamber were fitted between the engine and turbine inlet in order to damp down the pulsations in exhaust gas flow. The turbine acts as a flow restrictor creating a constant pressure P_7 in the exhaust manifold chamber. The available energy at the turbine is given by area 7-8-10-11. This is the ideal '*constant pressure turbocharging system*'. The energy represented by area 5-7-13 cannot be used by turbine. This energy is not lost, since energy loss only occurs by heat transfer, but since no work is done during the pressure reducing process 5-7, it represents a loss of potential turbine work. The proportion of this loss to the total possible energy depends on the exhaust manifold pressure P_7 or the inlet manifold pressure P_l as can be seen from Figure 3. The larger the ratio of P_l to P_a is the smaller the loss will be. In practice, the constant pressure turbocharging system is largely used by marine application and others where high charging pressure up to 5 to 6 bar is delivered by a single-stage turbocharger system or to higher values by a two-stage system [**3-4**]. A major advantage of the constant pressure system is that turbine inlet conditions are steady and known, hence the turbine can be matched to operate at optimum efficiency at specified engine conditions. The main disadvantage is that the full advantage of the pulse energy has not been capitalized upon, and this can significantly reduce the available energy entering the turbine when charge air pressure is low. Another disadvantage is system response time, since the engine exhaust must first fill a large exhaust chamber before reaching the turbine, serious delay in turbine acceleration can happen in response to engine demand of charge air pressure at low engine speeds when engine exhaust flow rate is low, and it takes a longer time to fill the chamber. This presents less of a problem to marine and most stationary power applications where auxiliary electric blowers can be employed to temporarily increase boost air pressure if required. To on- and off-highway applications, particularly passenger car applications, however, this is a serious issue as good system response is a key parameter for acceptable engine performance, and where there is less space and insufficient electricity resource to install the blower. For such applications, an alternative system called '*pulse turbocharging*' is used.

In the pulse turbocharging system, the turbine is placed as close to the engine exhaust valve as possible. The objective is to make maximum use of the high pressure and temperature which exists in the cylinder when the exhaust valve opens, even at the expense of creating highly unsteady flow through the turbine. In most cases the benefit from increasing the available energy will more than offset the loss in turbine efficiency due to unsteady flow. In an ideal situation, the gas would expand directly through the turbine along line 5-6-7-8 in Figure 3, assuming isentropic expansion and no losses in the exhaust port. If the turbine is sufficiently large, both cylinder and turbine inlet pressure would drop to equal ambient pressure before the piston moved significantly from BDC. Thus the piston pumping work would be zero during the ideal exhaust stoke and area 5-8-9 represents the available energy at the turbine. In such a system, there is less volume between the engine exhaust valves and turbine inlet, so the system response time is quicker. For automotive application, the pulse turbocharging system is solely used.

In practice the systems commonly used and referred to as constant pressure and pulse systems are based on these principles but are far from ideal. Readers are referred to [**2**] for more discussion on these two systems and others.

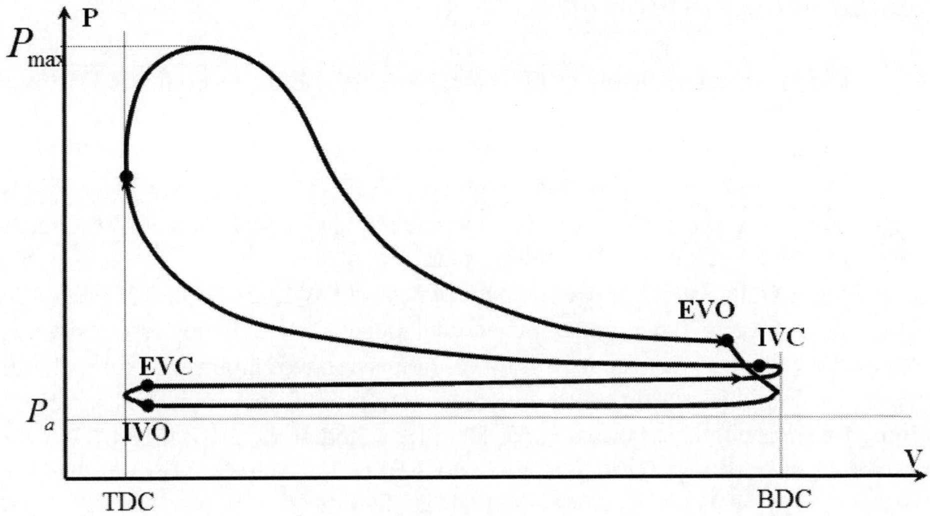

Figure 4. P-V diagram of a turbocharged diesel engine.

Although the diagrams illustrating the energy available at the turbine are based on the four-stroke engine cycle, similar diagrams can be constructed for two-stroke engines. Apart from the change in valve or port timing, the work done by the piston is replaced by energy transferred from the compressor to scavenge air. For clarity, scavenging in the overlap period in the case of a four-stroke engine has been ignored. The real P-V diagram of a turbocharged, four-stroke diesel engine is given in Figure 4.

BENEFITS OF TURBOCHARGING

Increase in Engine Power Density

By increasing the air pressure going into the engine cylinders, turbocharging increases the air density drawn into the cylinders. If the compression process of air inside the compressor is isentropic, then the ratio of air density after and before the compression ρ_2/ρ_1 is expressed as $\rho_2/\rho_1 = (P_2/P_1)^{1/\gamma}$, where P_2 and P_1 are the air pressure after and before the compression respectively and γ is the isentropic index at γ~1.4. So the air density increases with air pressure. The real compression process inside the compressor is not isentropic, so the air temperature will be higher, and the air density will be lower after the compression than those after an isentropic process, but the air density is still increased after the compressor. For the same flow velocity and the same area at the engine intake valves, the mass flow rate going through the valves is proportional to air density, so more air will enter the engine cylinders. With more air inside the cylinders, more fuel can be burned to produce more power. So one of the benefits of turbocharging is increased engine power output.

Improvement in Engine Efficiency

This power increase immediately brings other benefits. Because certain power losses the of internal combustion engine, such as piston ring friction loss and the power to drive auxiliary equipment are largely independent of engine power, so their negative influence on engine efficiency will reduce with increased engine power; alternatively for a given power, a smaller engine can be used. With less cylinder friction loss and heat transfer loss, higher engine efficiency can be achieved. A lighter engine also means better fuel economy. Modern turbocharged engines can reach a power density of 80-100kW/litre and higher power density engines are on the way to the market. The recent pursuit of high power density engines through engine downsizing has been driven by CO_2 emission regulations and turbocharging is key to this trend. For this reason more and more automotive manufacturers have adopted turbocharging for their engines. For example, Daimler stated at the 2012 IMechE Conference on Turbocharging that all their future gasoline cars will be turbocharged (in Western Europe, the majority of passenger diesel cars are already turbocharged).

Reduction of Engine Emissions

The widespread use of turbocharging on automotive engines in the past twenty years has mainly been driven by emission regulations. Without turbocharging or other means, the air intake of diesel engines depends only on engine speed and is independent of engine load. When the load is increased and more fuel is injected there is often insufficient air to burn them, and this results in poor emissions. Turbocharging provides additional air to enable a better and more controlled combustion, thus improving engine emissions. Emissions control of automotive engines is covered in other chapters of this book.

Improvement of Vehicle Drivability

Turbocharging can also improve drivability of vehicles by shifting the engine peak torque point to lower engine speeds. The torque of naturally aspirated gasoline engines typically peaks at about 4000 engine rpm, whilst a modern turbocharged engine can have its torque peak at 1500 engine rpm, Figure 5. So less gear down shift is necessary with turbocharged engines when the load of vehicles is increased.

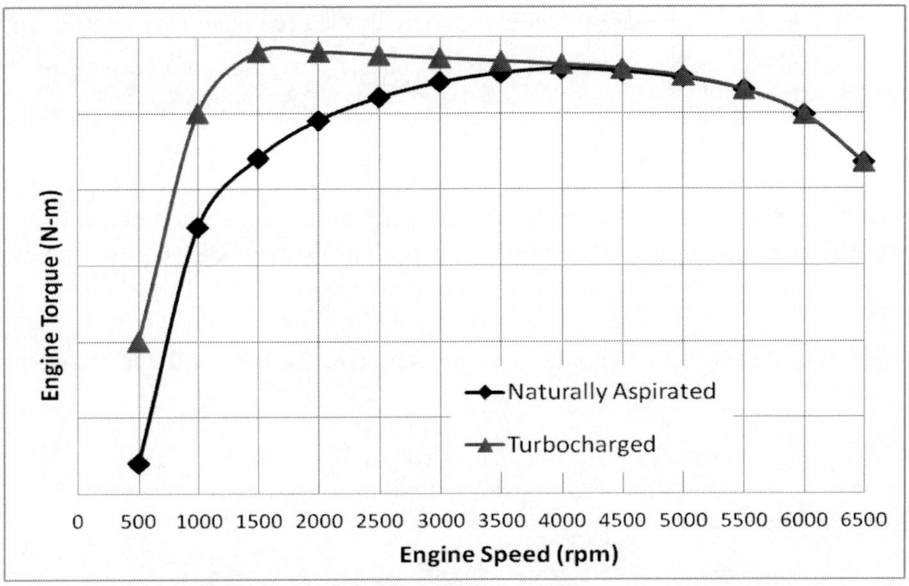

Figure 5. Torque of turbocharged and naturally aspirated engines.

Altitude Compensation

One additional benefit of turbocharging is altitude compensation. Air pressure and density reduces with altitude. At 5000m above sea level, the air pressure is only about 53%, and the air density only 59%, of those at sea level, respectively. For naturally aspirated engines, this reduction is critical: there is insufficient air in the cylinders to burn the necessary amount of fuel, so the engine power is significantly reduced, and emissions increased. For a turbocharged engine, however, the reduction of ambient pressure also means that the turbine outlet pressure is reduced and turbine expansion ratio is increased. The turbine therefore produces more power driving the compressor to deliver higher pressure ratios. So as long as the turbine and compressor can withstand the increase in speed, the engine inlet pressure and engine performance will be less affected by altitude change.

Charge Air Cooling

Air temperature rises after the compressor. This happens regardless of the losses incurred as a result of the compression process inside the compressor. A less efficient compressor results simply in a higher compressor exit temperature. Compressor outlet air temperature can be calculated by the following equation:

$$\frac{T_{02}}{T_{01}} = 1 + \frac{(p_{02}/P_{01})^{\frac{\gamma-1}{\gamma}} - 1}{\eta_{t-t}}, \tag{1}$$

where T_{01} and T_{02}, P_{01} and P_{02} are the inlet and outlet total temperatures, the inlet and outlet total pressures of the compressor respectively and η_{t-t} is the compressor total-to-total efficiency. So for a given compressor efficiency, compressor outlet air temperature will increase with compressor pressure ratio.

Increased air temperature means less dense charge air and less air inside engine cylinders. Higher inlet air temperature also increases the temperature of combustion hence thermal load of engine components as well as NOx production. For gasoline engines, higher temperature of air-fuel mixture increases the risk of knocking combustion and should be avoided.

An intercooler can be used to cool charge air. The effectiveness of the intercooler may be expressed as:

$$\varepsilon = \frac{T_2 - T_3}{T_2 - T_w}, \text{ i.e. actual heat transfer / maximum possible heat transfer} \tag{2}$$

where,

T_2 = intercooler hot air inlet temperature (from compressor);
T_3 = intercooler air outlet temperature;
T_w = temperature of cooling medium at inlet.

Since T_2 is close to T_{02} and T_1 is close to T_{01}, from equations (1) and (2), the following equation is obtained for the air temperature after the intercooler:

$$\frac{T_3}{T_1} = (1-\varepsilon)\frac{T_2}{T_1} + \varepsilon\frac{T_w}{T_1} \approx (1-\varepsilon)\left(1 + \frac{(p_{02}/P_{01})^{\frac{\gamma-1}{\gamma}} - 1}{\eta_{t-t}}\right) + \varepsilon\frac{T_w}{T_1} \tag{3}$$

This equation shows that as long as the temperature of the cooling medium T_w is less than the compressor outlet air temperature T_2, air temperature after the intercooler will reduce linearly with the intercooler effectiveness ε.

The air pressure drop across the intercooler should be minimised to maximise the turbocharging effect. The use of the intercooler can lead to better engine fuel economy, see [2]. More than one intercoolers may be needed when a two-stage turbocharging system is used, with one intercooler for each compressor stage.

A supplemental but important technology for charge air cooling is the *'Miller cycle'* through engine inlet valve timing. That is, the inlet valve closes before the piston reaches BDC. Between the closure of the valve and BDC, the charge air expands inside the cylinder and the air temperature is reduced. Figure 6 compares the P-V diagrams of the Miller and conventional cycles. Miller cycle can be used in both diesel and gasoline engines.

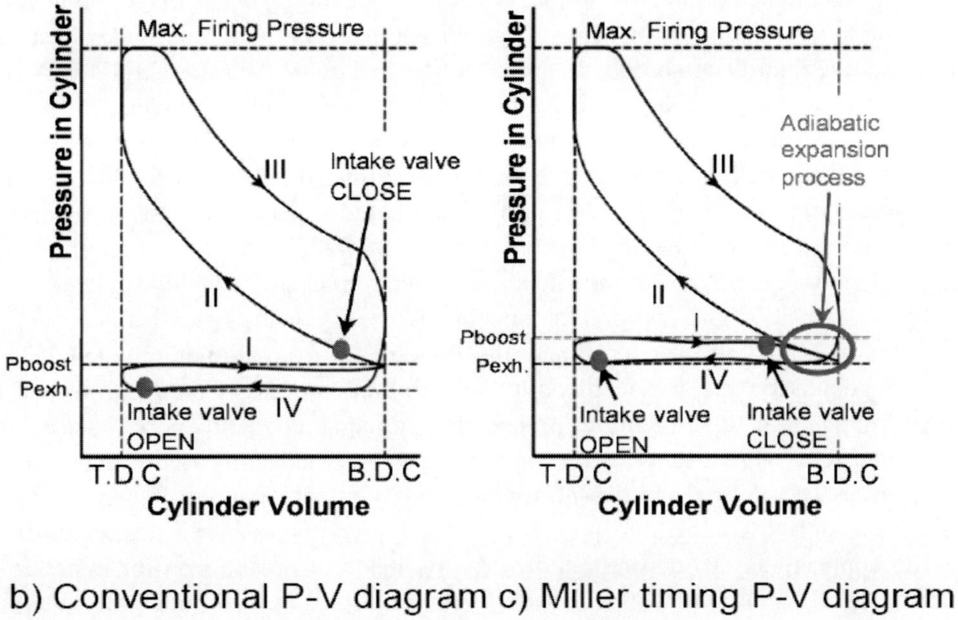

Figure 6. P-V diagrams of conventional and Miller cycles from [4].

TURBOCHARGERS AND PERFORMANCE MAPS

Turbochargers

In 1905, Dr Alfred Buchi from Switzerland patented the first exhaust driven 'supercharger' and in 1909 the first turbocharger. Since 1990, the use of turbocharging and turbochargers in passenger cars and trucks has exploded driven by emissions regulations. Today turbocharging is a multibillion US dollar business using state-of-the-art technologies. See [5] for a brief historic review of turbocharging and turbochargers. Although working under high inlet temperatures at 650~860 deg C for diesel engines and up to 1050deg C for gasoline engines, and at high rotating speeds of up to 550m/s turbine tip speed and near 600m/s compressor tip speeds, small turbochargers are often part of the modern automotive engine. This means that it must be cheaply made. The cost-reduction pressure on larger turbochargers for trucks and others applications is less but still intensive. The sensitivity to the cost has dictated the automotive turbocharger industry, and driven it to develop highly efficient technologies for product development, manufacturing and service, and to develop a simple and robust configuration for turbochargers.

A cutaway of a typical turbocharger for road vehicles is shown in Figure 7. In the centre of the turbocharger is the rotating assembly composed of a turbine rotor, a compressor impeller and a shaft connecting the two. The shaft is housed in the central housing where bearings and lubricant oil passages are also located. Water passages may also be present in the central housing to cool it, and this is often the case for gasoline turbochargers. Two single-piece, plain journal bearings are commonly used to support the shaft. The thrust bearing located near the compressor side is employed to absorb the unbalanced axial force generated by the impeller and the rotor. A turbine housing outside the rotor is used to distribute and to provide the correct flow conditions at the rotor inlet, to provide cover for the rotor passage (the rotor is un-shrouded), and to collect and guide the rotor exit flow. A similar housing is used for the impeller: it guides the flow into the impeller, provides the cover for impeller passages (the impeller is again un-shrouded), acts as a (vaneless) diffuser to diffuse the flow out of the impeller, and to collect and further diffuse the flow in its volute.

The turbine rotor is usually made of cast, nickel-based super alloys such as Inconel 713. For high temperature gasoline applications, cast Mar-M2468 can be used. The rotor is friction-welded or electronic beam-welded to a shaft made of low alloy steel such as AISI4140 or 42CrMo4. The compressor impeller is usually made of cast aluminium alloy such as C355 or C354, but fully machined impellers forged from aluminium alloy 2618-T61 are gaining popularity due to the modern flank milling process. Compared with casting process, the impeller made through this process has similar aerodynamic performance, similar cost if produced in large quantities, but is more durable and geometrically more accurate (hence produces less noise), and has no tooling process. Cast Titanium impellers are more expensive, and only used for high pressure ratio (high compressor exit temperature), long service life applications; it often requires a housing made of ductile iron for impeller burst containment, while aluminium impellers only require housings made of lighter aluminium alloys such as C356 or C319.

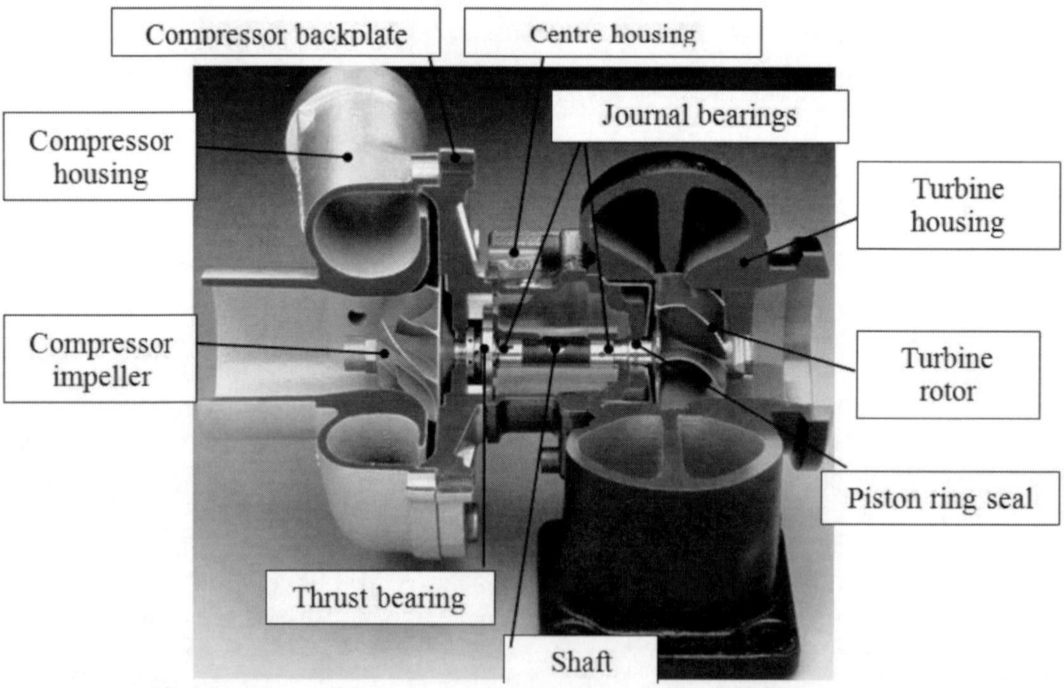

Figure 7. A cutaway turbocharger (courtesy of HTT).

Turbine Performance Map

The performance of a turbocharger is often given in the form of a turbine map and a compressor map. The turbine map shows turbine performance such as flow capacity and efficiency as functions of turbine expansion ratio and turbine corrected speed, and may be obtained from a hot gas-stand [6]. One such map is shown in Figure 8.

In the map, turbine speed is corrected to a 288° K SAE standard temperature. In a truly non-dimensional form, this corrected speed would be the ratio of rotor tip speed to the speed of sound at the 288° K temperature, but some of the parameters in the expression are considered constant and omitted from the expression. Turbine mass flow rate is also corrected to the same temperature and to 101,325Pa SAE standard pressure. In its truly non-dimensional form, this corrected mass flow rate would be the Mach number based on turbine inlet mass flow rate, the speed of sound at the standard temperature and a gas density evaluated at the standard temperature and pressure. The corrections make maps comparable and are necessary because turbine performance is greatly affected by inlet total temperature and total pressure. Turbine exit kinetic energy is assumed wasted so the turbine efficiency and turbine expansion ratio are both expressed as total-to-static values. Turbine corrected mass flow rate initially increases with turbine expansion ratio, then becomes constant when the expansion ratio is sufficiently large and the turbine is choked. Under the choking condition, turbine physical mass flow rate is increased linearly with turbine expansion ratio, but turbine corrected mass flow rate remains constant and is only slightly affected by turbine speed. Turbine efficiency generally increases at first then reduces with turbine expansion ratio.

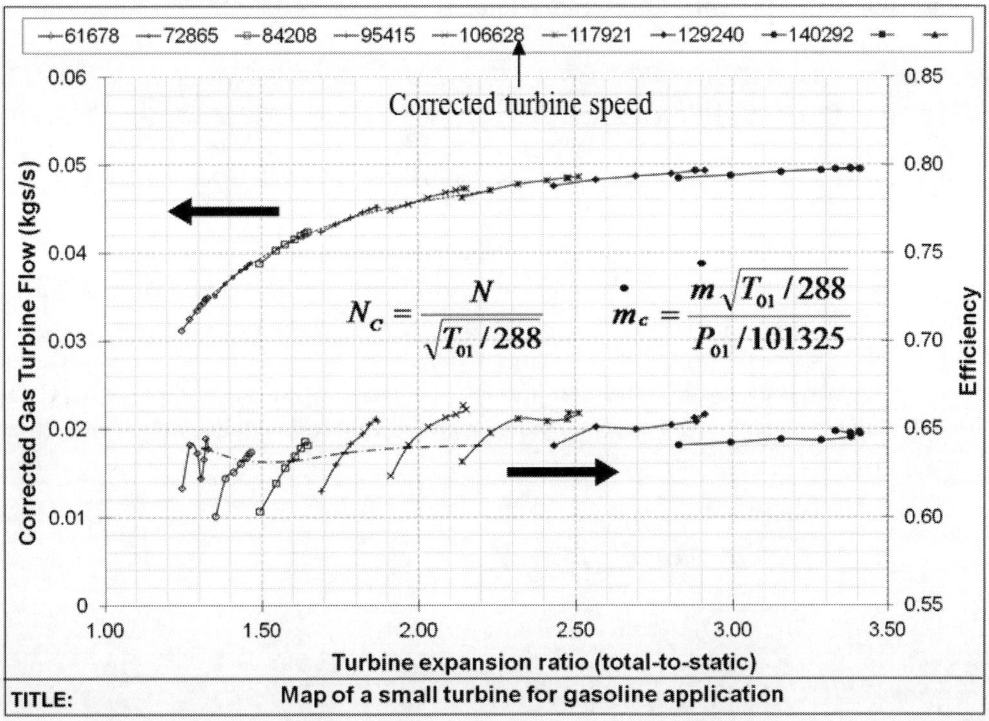

Figure 8. Gas-stand map of a turbocharger turbine.

A compressor is used in a hot gas-stand to load the turbine and the turbine efficiency is calculated from measured compressor power consumption, that is,

$$\eta_{tm,t-s} = \frac{\dot{W}_c}{\dot{W}_{t,is}} \tag{4}$$

where $\eta_{tm,t-s}$ is the turbine's total-to-static, thermal-mechanical efficiency, $\dot{W}_c$ compressor power consumption and $\dot{W}_{t,is}$ turbine isentropic or ideal, total-to-static power. The total power developed by turbine is the power consumed by the compressor $\dot{W}_c$ plus power losses in other parts of the turbocharger such as bearings and seals. This power loss is the turbocharger mechanical loss and can be measured by special test rigs. The compressor power consumption can be calculated from measured compressor inlet and outlet total temperatures:

$$\dot{W}_c = \dot{m}_c[h(T_{c,02}) - h(T_{c,01})] \tag{5}$$

where h is enthalpy, and $T_{c,02}$ and $T_{c,01}$ are compressor outlet and inlet total temperatures, respectively. The turbine isentropic power $\dot{W}_{t,is}$ can be calculated from measured turbine expansion ratio and turbine inlet total temperature:

$$\dot{W}_{t,is} = \dot{m}_t[h(T_{t,01}) - h(T_{t,2is})] = \dot{m}_t\, h(T_{t,01})[1 - \frac{h(T_{t,2is})}{h(T_{t,01})}] \tag{6}$$

where $\dot{m}_t$ is measured turbine mass flow rate, $T_{t,01}$ turbine inlet total temperature, $T_{t,2is}$ turbine outlet static temperature assuming isentropic expansion, and the ratio $h(T_{t,2is})/h(T_{t,01})$ is linked to turbine expansion ratio P_{2s}/P_{01}. Simplifications can be made if assumptions are made that $h = C_pT$, where C_p is the specific heat of the air or gas, and C_p is a constant (The assumptions are fine for purposes such as engine matching calculation, but are unacceptable for precise calculation of turbine and compressor efficiencies. For such a purpose, C_p as a function of temperature must be used). Under such assumptions equations (5) and (6) are simplified into

$$\dot{W}_c = \dot{m}_c\, C_p[T_{c,02} - T_{c,01}] \tag{7}$$

$$\dot{W}_{t,is} = \dot{m}_t\, h(T_{t,01})[1 - \left(\frac{T_2}{T_{t,01}}\right)] = \dot{m}_t\, h(T_{t,01})[1 - \left(\frac{P_2}{P_{t,01}}\right)^{\gamma/(\gamma-1)}] \tag{8}$$

Respectively, where P_2 and $P_{t,01}$ are turbine outlet static pressure and inlet total pressure respectively, and γ is the specific heat ratio of the fluid and is assumed constant.

Knowing the required compressor power and compressor mass flow rate from compressor-engine matching and turbine thermal-mechanical efficiency, the turbine isentropic power can be directly calculated from equation (4), and turbine expansion ratio from equation (8) so the turbine thermal-mechanical efficiency is quite useful in engine-turbocharging matching. However, the compressor power consumption is limited by compressor surge and choke, therefore the hot gas-stand test method usually produces turbine maps with a limited operating range as shown in Figure 8. Turbine dynamometers on the other hand use different means to consume turbine power and usually measure the turbine torque directly to obtain turbine power (power = turbine torque x turbine angular speed). An air turbine [7], eddy current [8] and high speed oil pump/cylinders [9] have all been used. Hot gas-stand usually measures turbine thermal-mechanical efficiency, while turbine dynamometers, depending on their design, may measure directly the total power developed by the turbine and produce turbine thermal or aerodynamic efficiency, or measure the power after bearing loss etc. to produce turbine thermal-mechanical efficiency.

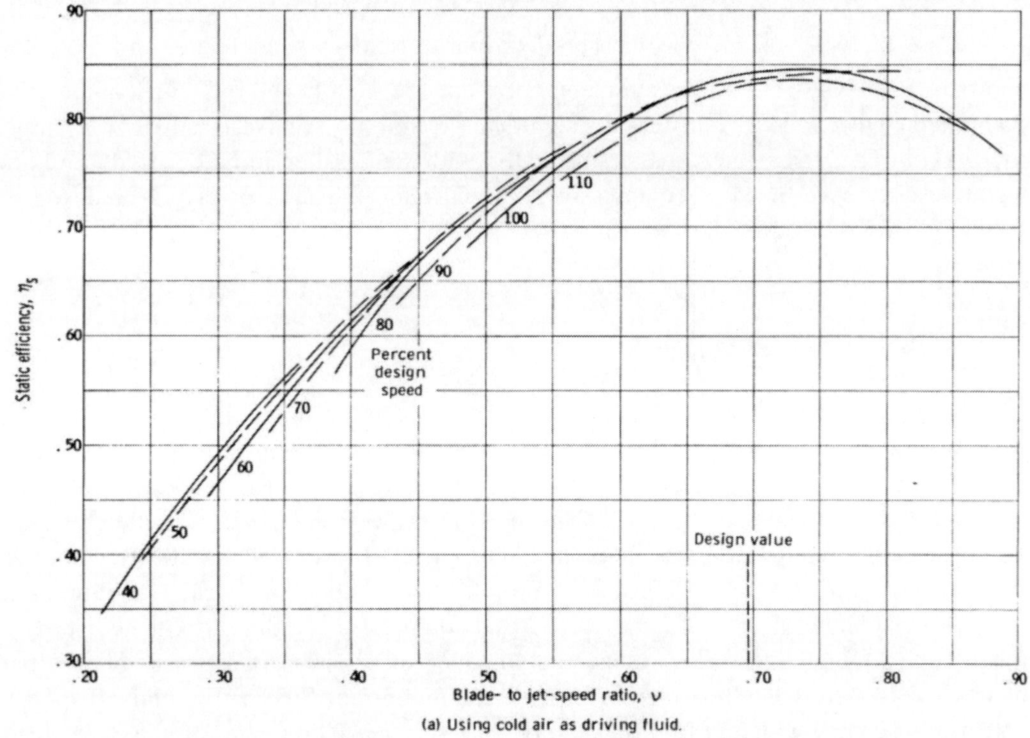

Figure 9. Turbine total-to-static efficiency measured by dynamometer, from [7].

One turbine map from [7] is given here in Figure 9, with turbine efficiency plotted against blade-to-jet speed ratio U_1/C_{is}, where U_1 is the turbine rotor inlet blade speed, and C_{is} the turbine isentropic jet or spout speed and is expressed as:

$$C_{is} = \sqrt{2[h_{01} - h_{2,is})]} \approx \sqrt{2C_p T_{01}[1 - \left(\frac{P_2}{P_{t,01}}\right)^{(\gamma-1)/\gamma}]} \quad (9)$$

The blade-to-jet speed ratio is a useful parameter to plot turbine efficiency against: the efficiency curves at different turbine speeds all fall within a narrow band when plotted against it, and the efficiency curves (turbine thermal efficiency) peak when U_1/C_{is} is at about 0.7. These two features are often used to process turbine test data to extrapolate turbine efficiency curves.

Turbine dynamometer rigs often employ relatively low turbine inlet temperature, 400°K for example, to limit the power generated by the turbine and to reduce the physical speed of the turbine. Recently, new techniques has been developed to measure turbine shaft torque directly under hot gas-stand test conditions, and by using different turbine inlet temperatures, maps with wider operating range can be obtained [10].

Another issue with the compressor loading the turbine on hot gas-stands is that the temperature measurement is affected by the heat transfer from the hot turbine side to the relatively cold compressor side. The heat transfer can raise the compressor outlet temperature and therefore increase $\dot{W}_c$ and the measured turbine efficiency (and reduce the measured compressor efficiency). This measurement error can be significant for small automotive turbochargers where heat transfer is significant, and is relatively small for larger turbochargers. The error increases dramatically with decreasing turbine speed (and the temperature difference between compressor inlet and outlet). The error may be reduced by correcting the test data of this heat transfer effect [11-12].

Compressor Performance Map

A typical compressor map obtained from a hot gas-stand is shown in Figure 10. On the map, compressor total-to-total pressure ratio is plotted against compressor corrected mass flow rate using compressor corrected speed as a parameter. For compressor maps, reference temperature and reference pressure are 298°K and 1bar respectively (SAE). Compressor total-to-total efficiency is plotted as contours on the map. The total-to-total values are used because compressor exit kinetic energy is considered useful. Static-to-total values are sometime used to give a better picture of the static pressure rise in compressor but the exit kinetic energy is then ignored. At high mass flow rates, the impeller or vaned diffuser may choke, and compressor corrected mass flow rate will then remain constant. Just before choke happens, compressor efficiency will start to drop sharply. At lower mass flow rates on the other hand, the same compressor may develop instability. This often shows up as a positive slope of the pressure ratio-mass flow curves (compressor speed lines) on the compressor performance map. When the positive slope is initially developed, the compressor is in a stall condition and will generate pressure fluctuations and acoustic noise, but may still be able to provide boost. When the mass flow is further reduced, the stall will eventually develop into surge with periodic forward and backward flows in the entire compression system, and the compressor will no longer be able to provide boost to engine. In addition, the flow oscillation is violent, and can cause damage to the compressor, so surge should be avoided. The left-most point on a speed line is usually the measurable compressor operating point just before surge; therefore the compressor in general should not operate to the left of this point. However, surge is a system phenomenon and is system dependent, so the mass flow rate of surge under engine operation will differ from that under hot gas-stand operation. Because of more flow restrictions under engine operation, it was found that the compressor often surges at smaller corrected mass flow rates under engine setup than the values shown on a gas-stand compressor map. There is an exception to this, however, since the pulsating nature of single and two cylinder internal combustion engines may lead to earlier compressor surge. The compressor may also choke at smaller mass flows than those shown on the map if the restrictions create partial blockage at compressor inlet, causing the entire compressor map to shift to the left [13].

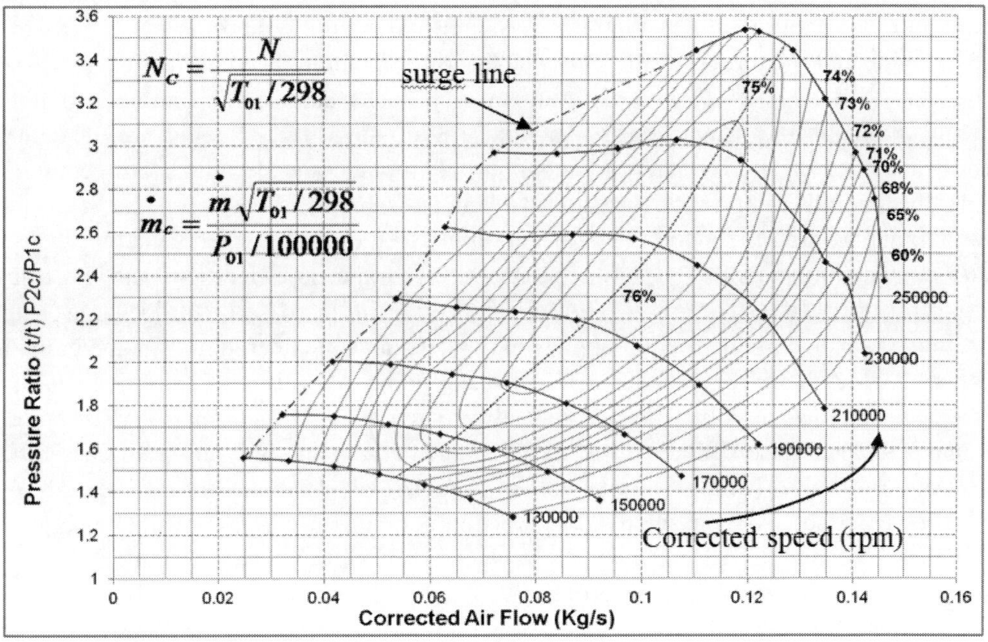

Figure 10. Gas-stand map of a passenger car turbocharger compressor.

The maximum pressure ratio and maximum flow of compressors are limited by the maximum compressor speed. This speed nowadays can reach nearly 600m/s of impeller tip speed for cast aluminium alloy impellers, and more than 600m/s for forged aluminium alloy and cast Titanium impellers. These speeds are at the mechanical limits of the materials.

Compressor efficiency is the ratio of ideal work to real work consumed by the compressor to generate the pressure rise:

$$\eta_{c,t-t} = \frac{\dot{W}_{c,is}}{\dot{W}_c} = \frac{h_{c02,is} - h_{c01}}{h_{c02} - h_{c01}} \tag{10}$$

where $h_{c02,is}$ is the compressor outlet total enthalpy under the condition of isentropic compression, and h_{c01} and h_{c02} are compressor inlet and outlet total enthalpies, respectively. If an approximation is made of constant C_p, then the compressor efficiency can be expressed as a function of compressor pressure ratio and temperature ratio explicitly, as equation (1) shows.

TURBOCHARGER-ENGINE MATCHING

The matching of the turbocharger to the internal combustion engine serves the purpose of selecting an appropriate turbocharger configuration to meet engine requirements for air supply, and engine exhaust back pressure among others. Mismatch of the turbocharger and the engine will result in insufficient air supply to the engine, high engine exhaust back pressure and other consequences that affect engine fuel economy and the safe operation of both the engine and the turbocharger. Only the basics of matching a single stage turbocharger will be described here. Matching of multi-stage turbochargers to the engine will not be discussed. Although the principles of matching a multi-stage turbocharger are the same as those of matching a single-stage turbocharger, additional considerations are required in dividing the overall compressor pressure ratio and turbine work between the stages. For automotive application, the smoothness of air supply to the engine when switching between the different operating modes of turbochargers needs some consideration, and the task of achieving this smoothness becomes more difficult with the increase of turbocharger stages. The method described in this section applies only to steady state operation of both engine and turbocharger. Matching the turbocharger to the engine at transient conditions is more complicated, and may involve the quasi-steady treatment of both turbine and compressor performance maps and sometimes unsteady flow calculations and will not be discussed here. Readers are referred to [2] and the manuals of commercial software such as Ricardo Wave and GT-Power. Also, some engine data or turbocharger data are often unavailable before the matching process is initiated and have to be guessed or extrapolated. The entire matching process also involves a number of iterations.

Matching the Compressor to the Engine

Figure 11 shows a compressor map imposed with the lug line or full load operating line of an automotive gasoline engine. Partial load operation of the engine also needs to be considered if the overall engine fuel economy is to be optimised, but the full load condition is the most demanding of compressor performance and is usually the one initially considered. One requirement of the turbocharger compressor performance is wide compressor flow range or large map width so that the engine lugline can be placed within the compressor map, with good compressor efficiency, and have sufficient surge and altitude margins. Compressor flow range at a given compressor pressure ratio, is expressed as:

$$\text{Compressor flow range} = 1 - (\text{surge flow/choke flow})_{\text{at given pressure ratio}} \qquad (11)$$

where choke flow is defined at the 60% or 65% efficiency contour. To avoid engine knocking, gasoline engine boost pressure is limited and so is turbocharger compressor pressure ratio. Lower (compared with diesel engine application) pressure ratio in gasoline engine application is helpful for the compressor to be able to meet the flow range requirements of the engine because the compressor flow range usually reduces with increasing compressor pressure ratio. Wider engine speed range (compared with diesel engine) on the other hand leads to a larger flow range requirement from gasoline compressors. Compressor flow range is limited by surge and choke as equation (11) shows. Compressor choked flow is dictated by the impeller inducer throat area, and this area can be increased by using a larger inducer blade height or a larger 'trim'. Impeller trim (and that of the turbine rotor) is expressed by:

$$\text{Trim} = (D_{throat} / D_{tip})^2 \times 100 \qquad (12)$$

where D_{throat} is the diameter of the blade shroud where the throat area is calculated (for compressor impellers, it is inlet shroud diameter; for turbine rotors, it is outlet shroud diameter), and D_{tip} is the blade tip diameter (trailing edge diameter of the impellers or leading edge diameter for the rotors). Figure 12 compares compressor maps of the same compressor at two different trims. It can be seen that the surge flow of the compressor is less affected by trim, but compressor choked flow increases with trim. So larger trims can be used to increase compressor flow range. However, as the figure shows, compressor efficiency near surge reduces with the increase of trim so more shaft power is demanded by the larger trim compressor in this map region. Often, turbocharger turbine produces the least power in this region, and this must be taken into consideration in selecting the correct compressor size and trim.

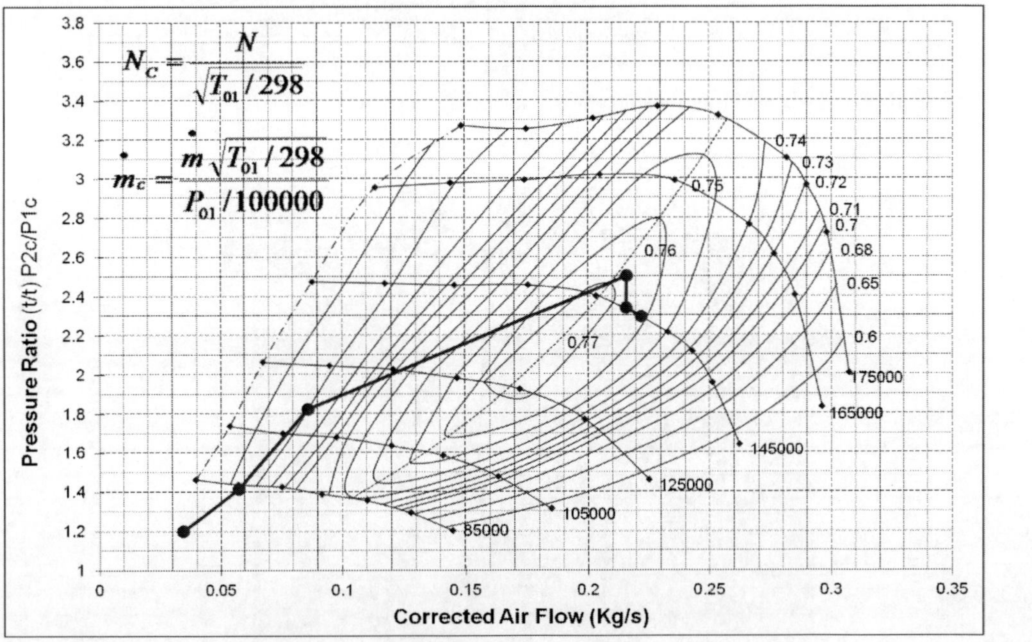

Figure 11. Compressor map with lugline of a gasoline engine.

Surge happens when the compressor pressure ratio increases beyond certain limits. This can occur when the exit or inlet of the compressor is throttled. In gasoline engine application, throttling occurs when the driver of the car releases the accelerator pedal. So to avoid compressor surge when the pedal is released, a relief valve is built in the housing of compressors for gasoline engine application, see Figure 13. When the pressure ratio between compressor inlet and outlet is greater than a preset value, the valve is open to connect compressor inlet and out inlet, compressor pressure ratio is thus reduced and surge avoided.

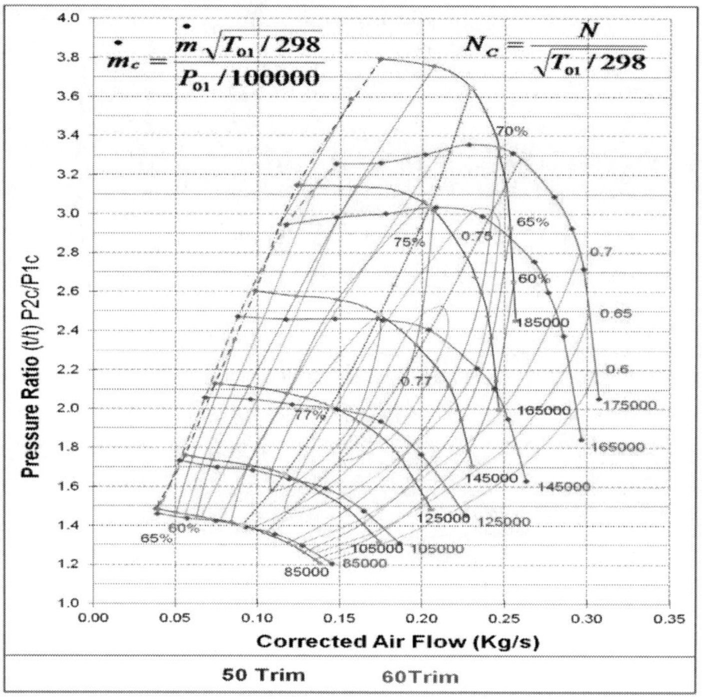

Figure 12. Effects of trim on compressor performance.

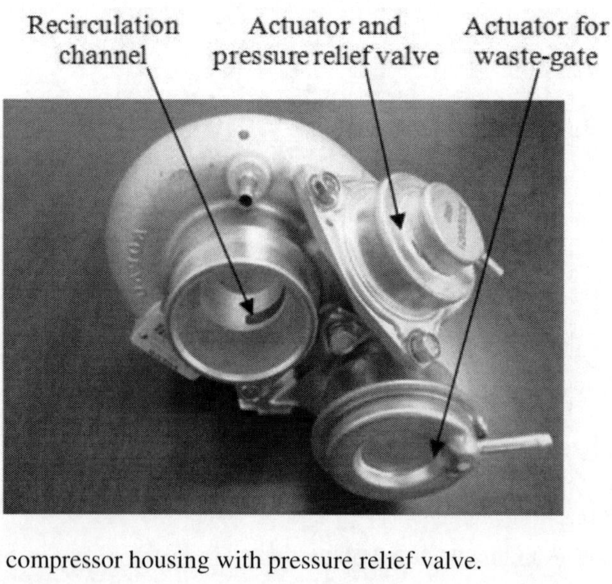

Figure 13. A gasoline compressor housing with pressure relief valve.

As mentioned earlier in the compressor performance map section, on-engine compressors may operate beyond the surge line indicated by the gas-stand map because of the difference in inlet and outlet systems between the gas-stand and the on-engine installation, and in particular the inlet distortion generated by the on-engine installation. This is often the case at low pressure ratios (less than 2). However, if the engine intake process generates strong pulsations of flow, such as in the case of two-cylinder engines, the pulsation may cause early surge of the compressor. If this happens, the lugline must be located inside and away from the surge line of the gas-stand map.

The compression stroke of the diesel engine, on the other hand, compresses only air, so diesel engines can utilise a higher boost pressure than gasoline engine without concern for knocking. Figure 14 shows a compressor map with an imposed lugline of an automotive diesel engine; considerably higher pressure ratio is required than that of the gasoline engine case in Figure 11. As equation (1) shows, compressor outlet temperature will increase with increasing compressor pressure ratio. Ultimately, the maximum pressure ratio achievable by a single stage compressor is limited by the material properties of the impeller at high compressor discharge temperature. In order to limit the compressor outlet temperature and intercooler size, and to control engine inlet air temperature, good compressor efficiency at high pressure ratios is now required. On the surge side of the compressor map, because of the higher pressure ratios, surge is more severe if it occurs, so the engine lugline is usually placed inside the map to the right of the surge line where compressor pressure ratio is greater than 2. Another point for discussion is compressor stall. At high pressure ratio, stall happens with strong pressure fluctuation occurances of compressor discharge and large acoustic noise. On the compressor map, any part of the speedlines with positive slope is associated with compressor stall, and is better avoided.

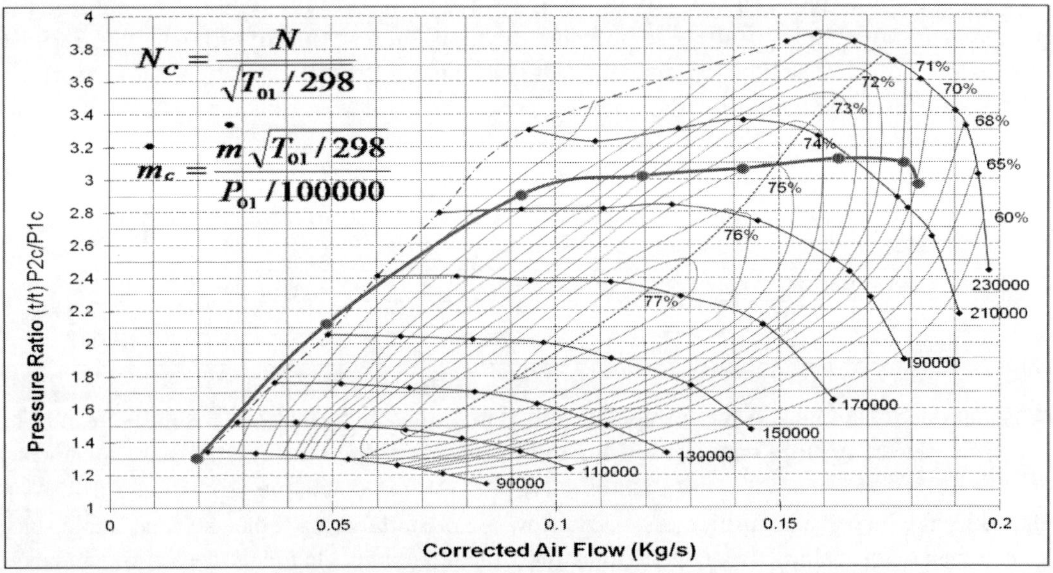

Figure 14. Compressor map with lugline of a diesel engine.

Matching the Turbine to the Compressor

Wih the compressor mass flow, pressure ratio and compressor efficiency, known, the speed and the power consumed by the compressor can be calculated. Since both the compressor and the turbine are on the same shaft, the following matchings condition must be met:

$$N_t = N_c \quad (13)$$

$$\dot{W}_t = \dot{W}_c + \dot{W}_{mech} \quad (14)$$

$$\dot{m}_t = \dot{m}_c + \dot{m}_f = \dot{m}_c\left(1+\frac{\dot{m}_f}{\dot{m}_c}\right) = \dot{m}_c\left(1+\frac{1}{AFR}\right) \quad (15)$$

where N_t and N_c are turbine and compressor rotational speeds, respectively, $\dot{W}_{mech}$ is the mechanical loss of the turbocharger, $\dot{m}_f$ is the fuel mass flow rate, and AFR is the air-fuel ratio. The turbine thermal-mechanical efficiency as obtained from the gas-stand already includes turbocharger mechanical loss and can be used directly in the matching.

The turbine rotor tip diameter D_{tip} is usually smaller than that of the compressor, and this is chosen to provide better turbine efficiency, and to allow the turbine rotor to have lower tip speed than the compressor impeller as the rotor works under a higher temperature environment. The turbine rotor also uses more dense materials than the compressor impeller, so a smaller turbine rotor can also reduce the mass and inertias of the rotating assembly. This benefits the rotordynamics of the assembly and the transient response of turbocharger. But the diameter should not significantly smaller or the turbine efficiency will suffer and the turbine may not be able to utilize all of the engine exhaust flow. The following table provides an approximate guideline only:

Table 1. Compressor impeller-to-turbine rotor tip diameter ratio

Free floating (no bypass)	1.03-1.10
Waste-gate (bypass)	1.05-1.15
Variable geometry (nozzle) turbine	1.10-1.20

Equation (12) applies to the turbine as well. The throat in the turbine casing is located in the rotor exducer. A large trim is used for the turbine to reduce turbine tip diameter and to improve turbine efficiency, as the kinetic energy of turbine discharge, considered a loss, is reduced by the increase of turbine discharge flow area. But large trim means large blade span hence lower blade natural frequencies and high blade root mechanical stress, so a balance needs to be struck between mechanical and aerodynamic considerations.

Unlike the vaneless diffuser compressor where the impeller inducer dictates compressor mass flow, turbine mass flow is less affected by turbine rotor exducer throat area than by the value of turbine housing A/R, where A and R are the cross section area and the centroid radius of the cross section of turbine housing throat, respectively. This is because the turbine housing outflow angle is largely dictated by the value of A/R. The larger the value of A/R is, the smaller the flow angle will be (measured from the radial direction); and the flow at rotor inlet will be more radially inwards. Because of the control A/R has on turbine mass flow, and the difficulty in machining different trims from the same turbine housing casting, turbine housings with different A/R values are often used with the same turbine rotor to meet different flow requirements. However, to achieve the best turbine efficiency, the throat area ratio of turbine housing to turbine rotor should be close to 0.5.

On-engine, the turbine inlet flow conditions are pulsating in nature. This inlet contains more energy than its time averaged total pressure and total temperature indicate. Because time-mean values are used in the matching process, turbine power output can be underestimated despite the turbine efficiency being overestimated in the matching (cycle averaged turbine efficiency is less than that under steady inlet conditions). A simple and effective way to correct this pulsating effect is to multiply the calculated turbine power by a 'pulsating factor' that is ≥ 1. The value of the pulsating factor depends on engine speed: the value should be 1 at high engine speeds (after the engine peak torque point) and should increase when engine speed reduces and may reach 1.6 at low engine speeds. The maximum value of the factor is engine dependent.

Waste-Gated Turbine

In the process of matching the turbocharger to engine, one often finds that it is impossible to meet engine requirements for rated power at high speeds and for boost at low speeds at the same time. Engine rated power requires a large turbine to receive engine flow to reduce engine back pressure, while at low speeds boost requirements dictate a small turbine to increase the turbine expansion ratio and turbine speed. A high trim rotor with a relatively small A/R turbine housing is helpful but will not solve the dilemma which is particularly acute for automotive and on-highway truck applications.

The waste-gated turbine is a technology invented to solve the problem. It uses a small turbine with a built-in flow bypass between turbine inlet and outlet, controlled by a waste-gate valve, see Figure 15. At low engine speeds, the valve is closed and the bypass is not used, so the turbine acts as a small turbine (with small flow area); at higher engine speeds (usually around or after engine peak torque point), the waste-gate valve is opened and approximately 20-30% engine flow then bypasses the turbine. This reduces the engine back pressure and prevents over speeding of the turbocharger. Opening of the waste-gate is controlled by compressor discharge pressure. A hole is drilled on the compressor housing, and a pressure tapping installed and connected to a pneumatic actuator, see Figure 13 (modern turbochargers may also use an electric actuator). When the pressure at the tapping exceeds a preset value, the actuator opens the waste-gate valve. Depending on the location of the pressure tapping at the compressor housing - under the compressor housing tongue or above it - the engine lugline behaves differently after the opening of the valve.

The waste-gate valve and valve seat operation under high temperature are constantly subjected to high speed exhaust gas (which can locally reach the speed of sound). To ensure the reliability of the waste-gate and to reduce the manufacturing cost, a simple structure as shown in Figure 15 has been developed and universally used. Aerodynamics and the actuator mechanism mean that the waste-gate is an on-off device, and this limits the flow control ability of waste-gated turbine.

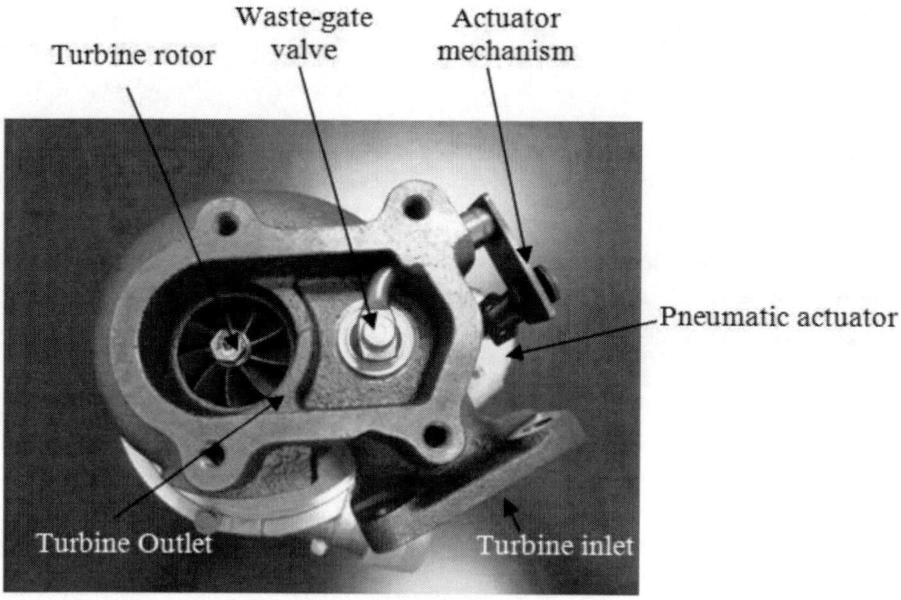

Figure 15. Waste-gate turbine, view from turbine outlet.

Variable Geometry Turbine (VGT)

The variable geometry turbine was introduced to automotive diesel engines in the early1990s to meet increasingly stringent emission regulations and the demand for better turbocharger performance. A nozzle ring, whose geometry is variable, is employed to provide variable guidance of the flow into the turbine rotor, to enlarge the turbine operating range. Two types of variable geometry mechanisms are commonly used: the variable nozzle angle (VNT as is commonly called) and the variable nozzle (axial) opening (VNOP). Figures 16 and 17 show these two types of VGT, respectively. Compared with VNOP, VNT is slightly more complicated hence more expensive to make, but it is more efficient, particularly at small turbine mass flows. Only VNT will be discussed here. Some references on VGT are given in [14-16].

At low engine speeds, the mass flow through the turbine is reduced. For automotive engines, good compressor boost is needed to obtain low speed torque and good transient response, so greater turbine power and speed are required. VNT achieves these by swinging its nozzle ring to a more closed position (nozzle vane angle increased with respect to the radial direction). This action reduces the nozzle ring throat area and increases the flow velocity out of the ring; and it also makes the nozzle outflow more tangential. According to Euler's equation for turbomachinery:

$$\dot{W}_t = \dot{m}_t (U_1 V_{u1} - U_2 V_{u2}) \tag{16}$$

where V_{u1} and V_{u2} are the flow tangential velocity at rotor inlet and outlet, respectively, and U_1 and U_2 are the rotor inlet and outlet blade speeds respectively, V_{u1} is greatly increased and so is the turbine power. At high engine speeds, the nozzle ring of the VNT can be set to a more open position (smaller nozzle vane angle with respect to the radial direction). This increases the nozzle ring throat area and makes the inflow to the rotor more radial, thus reducing the flow restriction provided by the nozzle ring towards the rotor, and allows more flow to pass through the turbine.

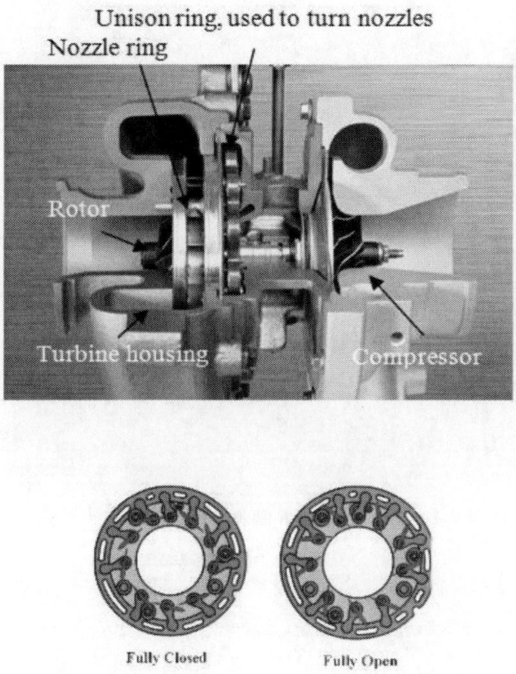

Figure 16. VNT.

Figure 17. VNOP turbine.

Figure 18 shows the gas-stand maps of a VNT at three different nozzle openings. The VNT only achieves good efficiency at the 50% nozzle opening, and its efficiency is poor at the other two extreme nozzle openings. However, the flow range, that is, the ratio of maximum flow to minimum flow at a given turbine expansion ratio (2:1 for example), of the VNT is much larger than that of any fixed geometry turbine such as one shown in Figure 8. In this VNT example, at a turbine expansion ratio 2:1 it has a flow range of 2.6 at which the turbine thermal-mechanical efficiency is no less than 55% (Figure 18D), while the flow range of a free-floating turbine is close to 1 only; this means that at high engine flow conditions, the free-floating turbine will generate higher engine backpressure than the VNT and reduce engine power while at low engine flow conditions, it will not be able to reach 2:1 expansion ratio to provide as much boost as the VNT. The waste-gated turbine has a better flow range than the free-floating turbine, but its flow range is only about 1.3 and this value cannot compete with that achieved by VNT.

VGT (VNT and VNOP) is widely used in modern turbochargers to provide better low speed boost, quicker engine transient response, higher engine power and better engine fuel economy. It has also been used to drive EGR for emissions control purposes. It has even been used for truck engine braking assistance by closing the nozzles of the VNT to increase engine backpressure during engine braking. Such an operation demands the ultimate mechanical integrity from the VNT as well as a good understanding of VNT aerodynamics: when the nozzle is closed and turbine expansion ratio increased, shockwaves can be generated at nozzle exit and this could cause high cycle fatigue of the turbine rotor. Usually turbocharger manufacturers will specify the maximum allowable turbine expansion ratio when the nozzles of the VNT are closed; this expansion ratio must not be exceeded in practice.

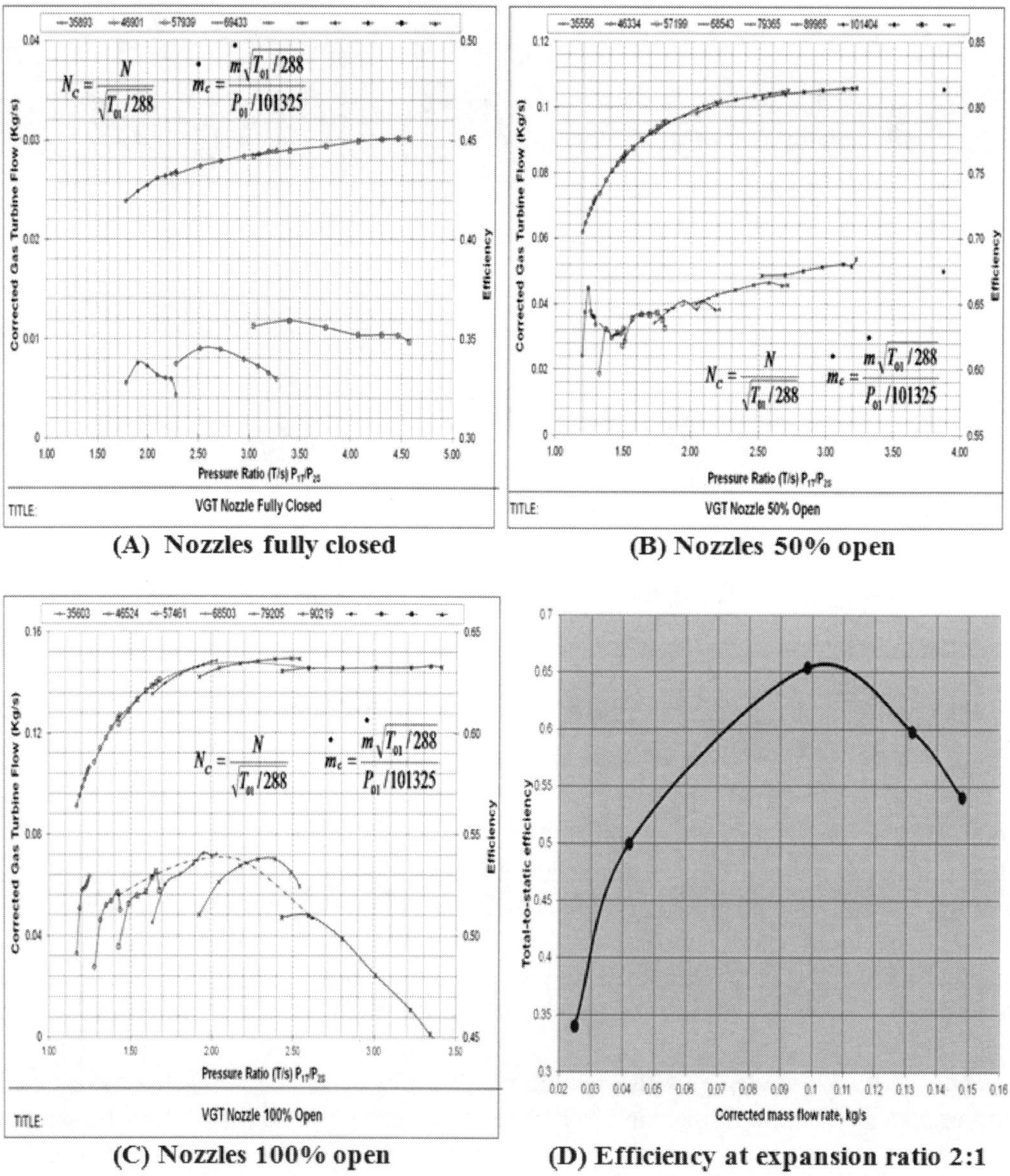

Figure 18. Gas-stand maps of a VNT.

OTHER APPLICATIONS OF TURBOCHARGING FOR WASTE ENERGY RECOVERY

Many methods of using turbocharging and turbomachinery have been developed to increase waste energy recovery and some of them are briefly described in this section.

Two-stage Turbocharging

To further increase waste energy recovery and to improve engine performance, two turbochargers of different size can be connected to form a two-stage turbocharging system, [17-18]. In one of the simplest ways, the two turbochargers are placed in series with bypass control and inter stage cooling, Figure 19. The engine exhaust gas first goes through a relatively small turbine (high pressure or HP turbine) or partially through a bypass valve. After the HP stage, the entire exhaust gas then flows through a relatively large turbine (low pressure or LP turbine). The air is first compressed by a relatively large, compressor (low pressure or LP compressor), which after inter stage cooling, is further compressed by a relatively small compressor (high pressure or HP compressor).

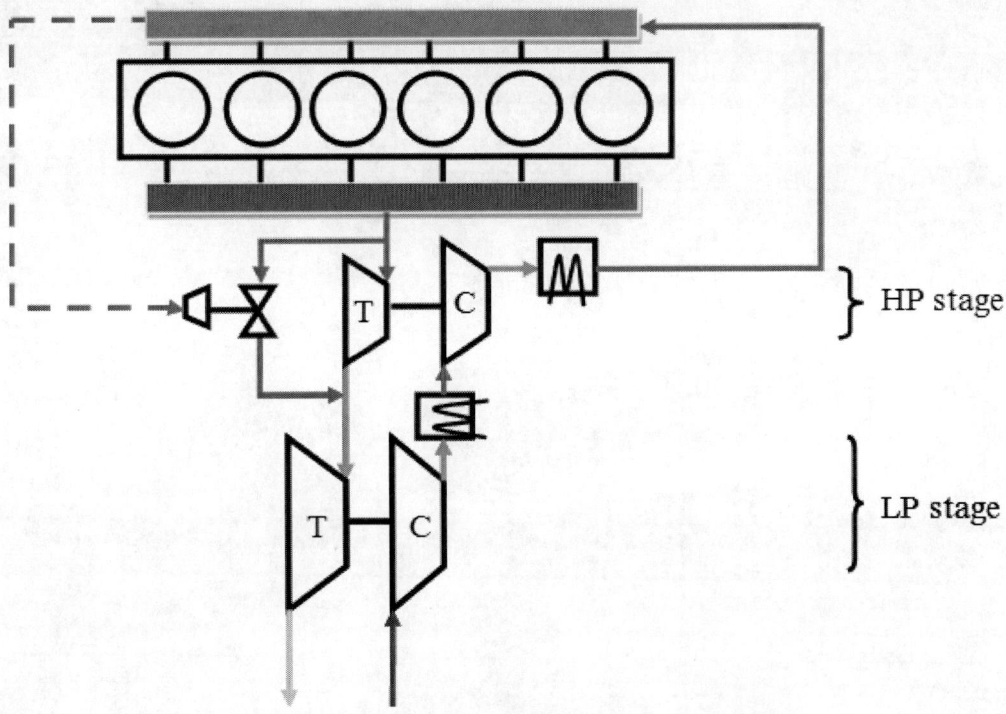

Figure 19. Schematic of a simple, regulated two-stage turbocharging system [17].

At low engine speeds, the bypass valve remains completely closed, and the entire engine exhaust gas goes through the HP turbine and this results in a quick boost pressure rise on the air side; at high engine speeds, the bypass valve opens to reduce engine backpressure, and the exhaust goes through both the HP turbine and the LP turbine to provide high boost for engine power requirement. Many other two-stage configurations are possible by using the two turbochargers differently, for example in parallel rather than in series [19], and each configuration has its own unique characteristics that suit a particular engine architecture and type of application. VGT is often used in the HP stage to enhance low speed performance (Figure 20). Note that the LP compressor often is a higher pressure *ratio* compressor while the HP compressor a lower pressure *ratio* one.

Compared with single stage turbocharging, two-stage turbocharging provides flexibility to meet engine requirements at both low and high speeds. Because of load split, both LP and HP stage can operate at reduced flow and pressure ratio ranges. This enables more efficient turbines and compressors to be specifically designed for two-stage turbocharging. The disadvantages of two-stage turbocharging are complex piping, valve and seal systems and large size of the turbocharger. Control of the turbocharger is more complicated than that of single stage turbocharger to achieve a smooth operation during stage switching. Two-stage systems also have larger flow passage volume and more metal surface than single stage systems, and this can affect the time taken by the turbocharger to warm up from cold start, thus affecting the operation of the downstream catalyst converter and engine cold start emissions.

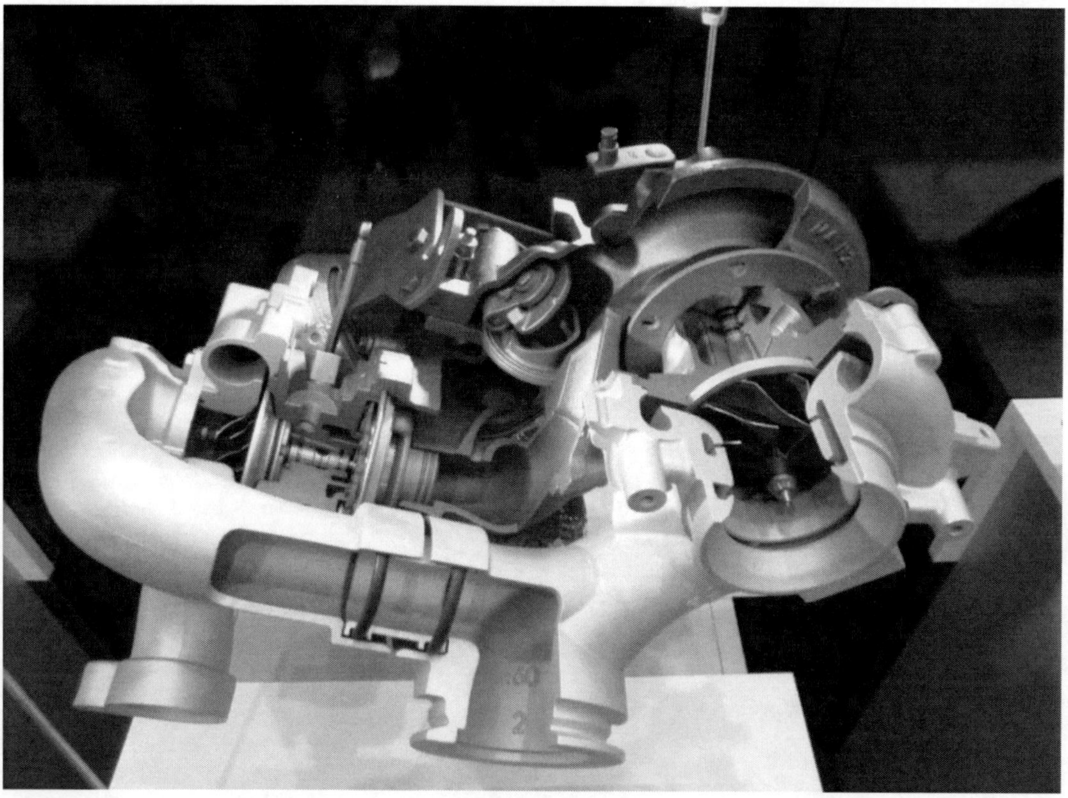

Figure 20. Two-stage turbocharger with VNT as HP turbine, not all components are shown.

Three-stage systems using three turbochargers together for automotive engines have also been developed, see [20].

Turbocompounding

At high engine speeds, turbocharger turbines often have excess power and this power has to be limited, for example by waste-gating, otherwise the turbocharger will over speed. Turbocompound systems [21-22] harvest this redundant turbine power by coupling the turbocharger turbine shaft to the engine drivetrain. The turbocharger turbine rotates at much higher speeds than the engine so a speed reduction gear box is needed to realise mechanical coupling of the two. Modern turbocompounding typically employs a two-stage turbocharging system, and uses the LP stage turbine as a power turbine. Figures 21 and 22 illustrate two such examples. Both radial (Figure 21) and axial (Figure 22) flow turbines can be used as power turbines. For road vehicle operation, however, high engine speed occurs infrequently, so turbocompounding produces relatively little power (about 0.8% fuel saving currently). One option would be the use of a VGT as the power turbine would improve the matching of the power turbine to engine conditions thus increasing the benefit of turbocompounding, but the cost of the turbocompound would also increase.

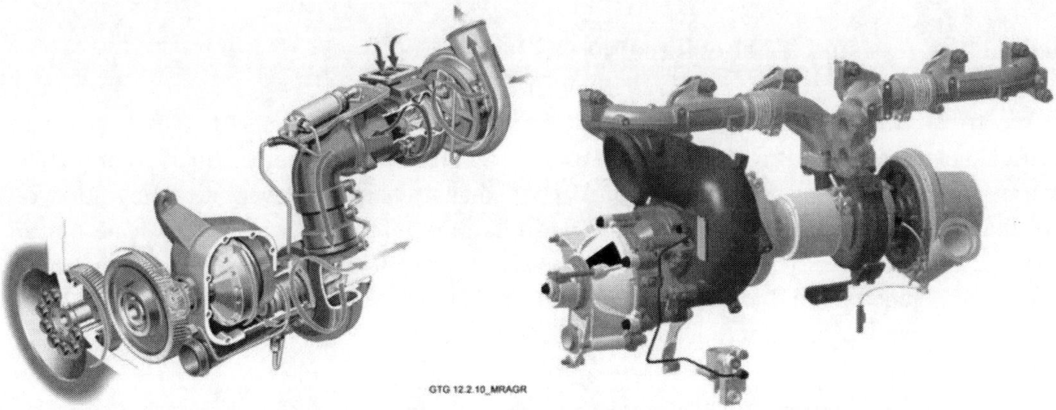

Figure 21. Two turbocompounds, LP turbine as the power turbine (courtesy of HTT).

Unless a variable gear ratio coupling is used, mechanical coupling of turbocompounding fixes the speed ratio of the power turbine to the engine. So the speed of the turbine changes with engine speed and load, and is often away from the speed at which maximum turbine efficiency can be obtained. If, instead of coupling with the engine drivetrain, the power turbine is used to drive an electrical generator, then the turbine speed is less affected by the engine operating condition and better turbine efficiency can be achieved. This electrical turbo compound (ETC), [23-24] is ideally suited for applications where generated electricity can be directly fed into the vehicle's electrical bus (after going through a converter). The cost of a high speed generator and associated electrical devices needs to be balanced with potential fuel savings. With more vehicle electrification, ETC can become more attractive.

Figure 22. Axial flow turbine as the power turbine.

Electrically Assisted Turbocharging (E-Turbo)

Electrical devices have been used to improve turbocharger performance, such as electrical actuators for waste-gated turbine and for VGT, and electrically driven compressors for transient boost. Here discussion is made of exhaust energy recovery by integration of turbocharger and electrical generator [25-27]. Unlike the ETC case where a dedicate power turbine is used for electricity generation purposes, in this integration, an electrical machine (used as both generator and motor) is integrated into the body of main turbocharger, typically inside the central housing of turbocharger in between the turbine and compressor, Figure 23.

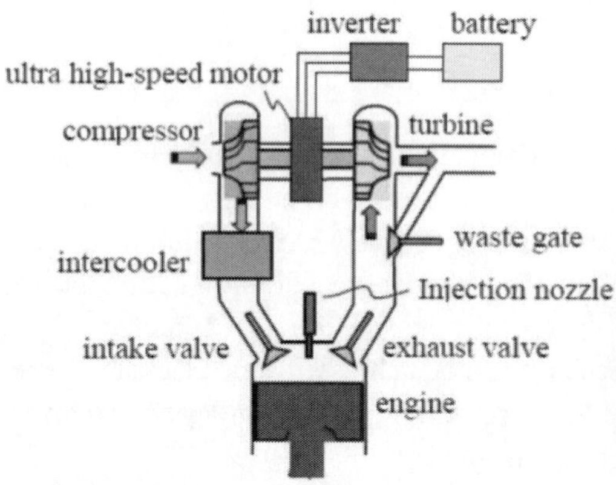

Figure 23. Electrically assisted turbocharger [27].

At low engine speeds or during transient acceleration, the electrical motor can be used to speed up the compressor; while at high engine speeds when there is a surplus of turbine power, the electrical motor turns into a generator, converting this power into electricity that can be used to charge the vehicle battery. Compared with ETC, this integration - called the e-turbo - provides an additional function of accelerating the turbocharger when needed; however, the technical challenge of placing a generator/motor and in between the compressor and the turbine of the turbocharger is much greater. Rotordynamics effects, a high temperature environment and the continuous use of the generator/motor are just a few problems that an e-turbo must overcome. Because of these technical difficulties and their commercial implication, there is no mass produced e-turbo from major turbocharger manufacturers at the time of writing. But efforts are being made to overcome these problems to make the e-turbo successful.

Recovery of Waste Heat after the Turbine

The temperature of exhaust gas after the turbocharger is reduced, but is still significantly higher than ambient. The following table provides the temperature range of various waste heat sources. Of all the waste heat sources, the EGR and the turbocharger turbine exhaust are the most significant sources. The gasoline engine has a higher exhaust gas temperature and smaller expansion ratios at the turbocharger turbine than diesel engines, so the potential of a waste heat recovery system is greater in gasoline engine driven vehicles. However, the gasoline engine is cheaper than the diesel engine, and is only used in passenger cars, so it will be more sensitive to the additional on-cost of any waste heat recovery system.

Table 2. Temperature range of waste heat sources of automotive diesel engines

Component	Temperature range (°C)
Exhaust gas in EGR	350 - 650
Exhaust gas post turbocharger and aftertreatment	150-300
Air in charge air cooler	100-200
Engine coolant	85-100

To further recover the energy in the exhaust gas downstream of the turbocharger turbine and other waste heat sources of automotive engines, various methods have been proposed, but only those involving turbomachinery will be described here. The schemes typically use waste heat to heat up a working medium which is then expanded in a turbine to produce mechanical work. One such scheme employs the Brayton cycle or gas turbine cycle [28], but replaces the combustor with heat exchangers, see Figure 24. In this system, air is used as working medium and the system is simple and of low cost, so is suitable for automotive application, but the ability of the system to recover waste heat energy is limited. Using other gases such as helium can improve the recovery but helium is difficult to seal and a closed cycle is needed. This increases the cost of the system.

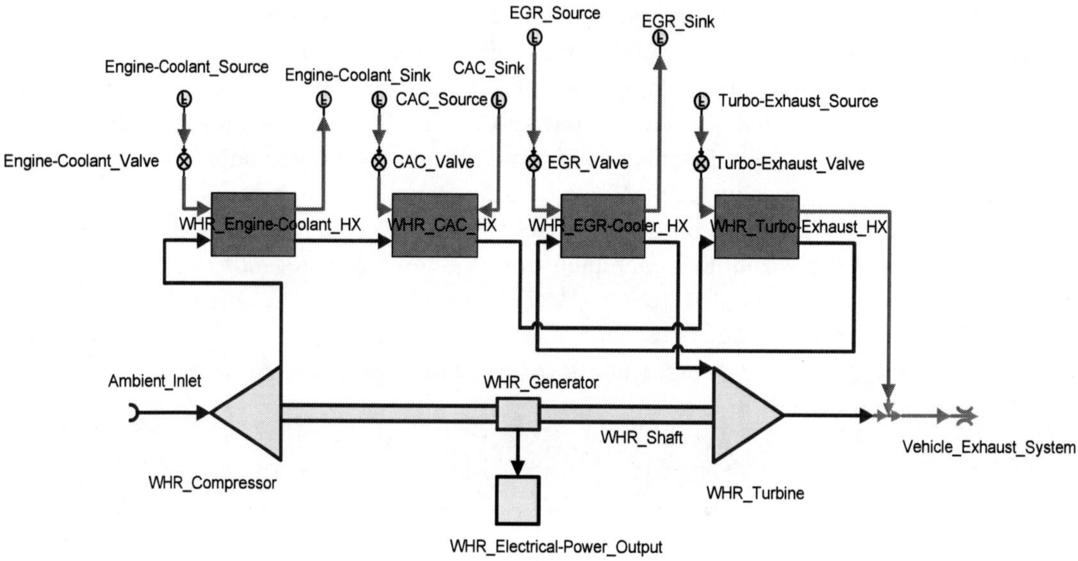

Figure 24. Schematic of a Brayton cycle based waste heat recovery system (WHR: waste heat recovery. Courtesy of HTT).

The Rankine cycle [29-31] or steam turbine cycle uses the waste heat to raise the temperature of the working medium to produce pressurised hot steam, the steam is then expanded in a turbine to produce mechanical work. Water/steam can be used as the work medium, but it is more suitable for high temperature and large power output applications, In Figure 25(a) 'organic' fluids such as R245 refrigerant is illustrated which is better suited for lower temperature applications such as the recovery of waste heat after the turbocharger turbine as it permits dry expansion (expansion without condensation) - Figure 25(b) - enabling a higher energy recovery rate. Studies show that a few percentage points of fuel economy improvement is possible. Figure 26 shows a prototype passenger car fitted with Rankine cycle devices.

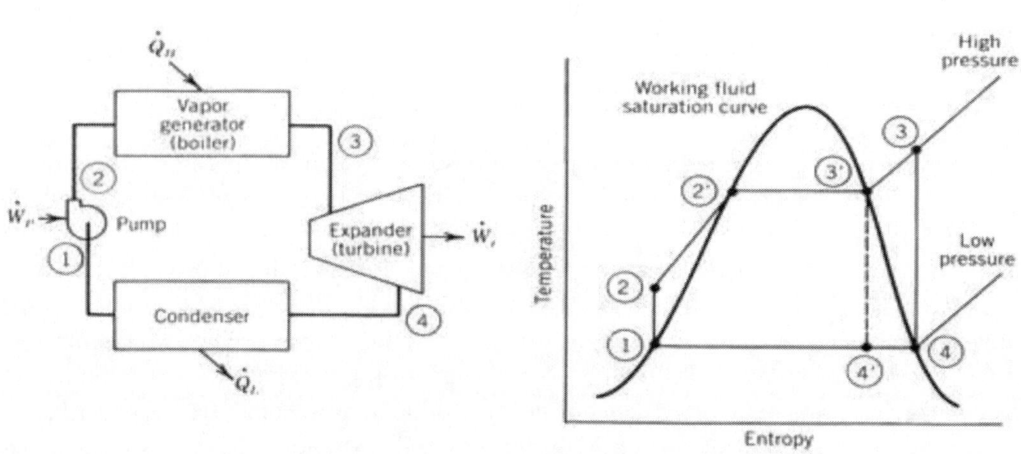

(a) Simple cycle of wet fluid (water)

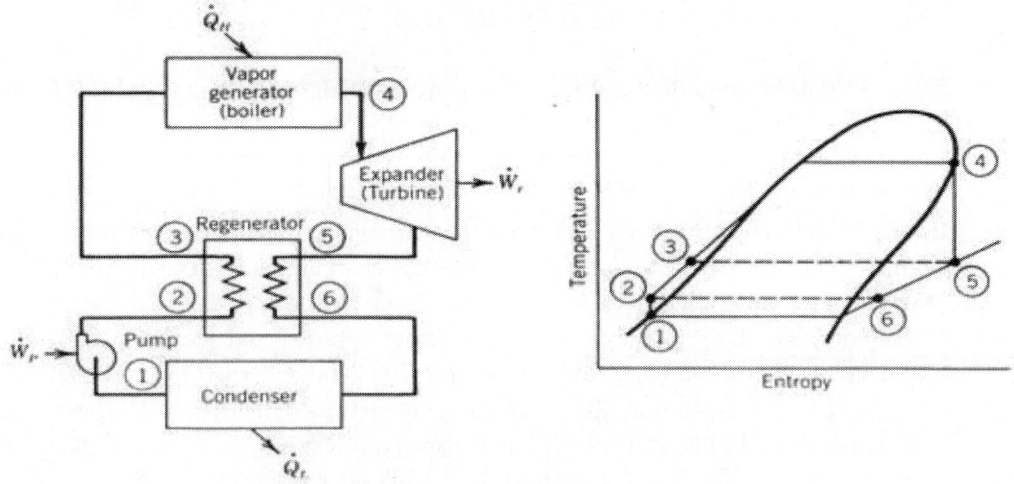

(b) Recuperated cycle of dry fluid (R245)

Figure 25. Rankine cycle for waste heat recovery (courtesy of HTT).

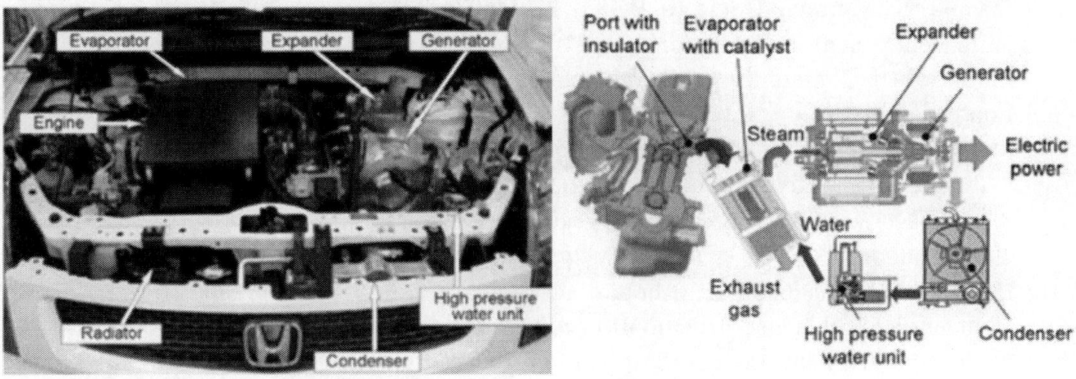

Figure 26. A passenger car fitted with Rankine cycle devices (courtesy of HTT).

Application to Fuel Cells

The turbocharger can also be used in fuel cell powered vehicles to provide waste heat energy recovery. Compressed air from the compressor flows through fuel cells to help generate electricity in the cells; waste heat generated by the full cells can be used to produce hot steam which then expands in a turbine to recover some of the energy in the steam. An electrical motor is integrated, so the system is similar to an e-turbo. However, here the turbine and the compressor work under significantly different conditions from those of turbochargers for internal combustion engines and so are the requirements for such a system. Prototypes of such systems have been studied and built.

REFERENCES

[1] Kindt S, State-of-the-art MAN B&W two-stroke super-long-stroke engines, Paper No. 71, CIMAC Congress 2013, May 2013, Shanghai.

[2] Lilly L C R, (ed.) Diesel engine reference book, Butterworth and Co., 1984. ISBN 0-408-00443-6.

[3] Codan E and Huber T, Application of two stage turbocharging systems on large engines, 10th International Conference on Turbochargers and Turbocharging, IMechE, May 2012, London, UK.

[4] Matsumoto K, Development of high-pressure ratio and high efficiency type turbocharger, Paper No. 69, CIMAC Congress 2013, May 2013, Shanghai.

[5] Schorn N, The evolution of turbocharging, (In German), 17th Supercharging Conference, 13-14 Sept. 2012, Dresden, Germany.

[6] SAE Standards J1826, Turbocharger Gas Stand Test Code, March 1995.

[7] Kofskey M and Holeski D, Cold performance of a 6.02-inch radial inflow turbine designed for a 10-kilowatt shaft output Brayton cycle space power generation system, NASA TN D-2987, Feb. 1966.

[8] Szymko S, Martinez-Botas R, Pullen K, McGlashan N and Chen H, A high-speed, permanent magnet eddy-current dynamometer for turbocharger research, 7th International Conference on Turbochargers and Turbocharging, IMechE, May 2002, London.

[9] McDonnell G, Artt D and Spence S, The design, development and testing of a hydraulic turbine dynamometer, 6th International Conference on Turbocharging and Air management Systems, IMechE Conference Transactions, C554/025/98, November 1998, London.

[10] Luddecke B, Filsinger D, Ehrhard J, Steinacher B, Seene C and Bargende M, Contactless shaft torque detection for wide range performance measurement of exhaust gas turbocharger turbines, ASME Turbo Expo 2013, June 2013, San Antonio, Texas, USA.

[11] Casey M and Fesich T, On the efficiency of compressors with diabatic flow, GT2009-59015, ASME Turbo Expo 2009, June 2009, Orlando, USA.

[12] Luddecke B, Filsinger D and Bargende M, On wide mapping of a mixed flow turbine with regard to compressor heat flows during turbocharger testing, 10th International Conference on Turbochargers and Turbocharging, IMechE, May 2012, London.

[13] Capon G and Morris T, The effect of air inlet system features on automotive turbocharger compressor performance, 9th International Conference on Turbochargers and Turbocharging, IMechE, May 2010, London.

[14] Hawley J, Wallace F, Cox A, Pease A, Bird G and Horrocks R, Use of a VGT to improve the limiting torque characteristics of a D1 automotive diesel engine, 6th International Conference on Turbocharging and Air management Systems, IMechE Conference Transactions, C554/014/98, November 1998, London.

[15] Osako K, Jinnai Y, Samata A, Suzuki H, Ibaraki S and Hayashi N, Development of the high performance and high reliability VG turbocharger for automotive applications, Technical Review, Mitsubishi Heavy Industries, Vol. 43, No. 3, Sept. 2006.

[16] Wilson J, Avila M, Davies P, Theiss N and Zollinger B, Development of a common dual axle VNT for single- and two-stage off-highway applications, 10th International Conference on Turbochargers and Turbocharging, IMechE, May 2012, London.
[17] Pfluger F, Regulated two-stage turbocharging - KKK's new charging system for commercial diesel engines, 6th International Conference on Turbocharging and Air management Systems, IMechE Conference Transactions, C554/035/98, November 1998, London.
[18] Weber O, Christmann R, Gauckler V and Sauerstein R, R2S^{TM} - Modelling and consequences for the boost control, 10th International Conference on Turbochargers and Turbocharging, IMechE, May 2012, London.
[19] HTT developed their unique parallel sequential two-stage in 2006, see HTT's website, http://turbo.honeywell.com/our-technologies/twostage-parallel-turbochargers/
[20] See BW's website, http://www.3k-warner.de/press/newsArticle.aspx?id=395.
[21] Wallace F J and Cox A, The ultimate performance of compounded diesel engines for heavy vehicles, 6th International Conference on Turbocharging and Air management Systems, IMechE Conference Transactions, C554/015/98, November 1998, London.
[22] Kruiswyk R W, The role of turbocompound in the era of emission reduction, 10th International Conference on Turbochargers and Turbocharging, IMechE, May 2012, London.
[23] Thompson I, Spence S W and McCartan C D, Design, validation, and performance results of a turbocharged turbogenerating biogas engine model, 9th International Conference on Turbochargers and Turbocharging, IMechE, May 2010, London.
[24] Ryder O and Sharp N, The impact of future engine and vehicle drivetrains on turbocharging system architecture, 9th International Conference on Turbochargers and Turbocharging, IMechE, May 2010, London.
[25] Fieweger K, Paffrath H and Schorn N, Drivability assessment of an HSDI diesel engine with electrically assisted boosting systems, 7th International Conference on Turbochargers and Turbocharging, IMechE, May 2002, London.
[26] Balis C, Middlemass C and Shahed S M, Design and development of e-turbo for SUV and light truck applications, DEER 2003.
[27] Yamashita Y, Ibaraki S and Ogita H, Development of electrically assisted turbocharger for diesel engine, 8th International Conference on Turbochargers and Turbocharging, IMechE, May 2006, London.
[28] Patterson D J and Kruiswyk R W, An engine system approach to exhaust waste heat recovery, DEER Conference, Aug. 2007.
[29] Regner G, Teng H and Cowland C, A quantum leap for heavy-duty truck engine efficiency - hybrid power system of diesel and WHR-ORC engines, DEER Conference, Aug. 2006, Michigan, USA.
[30] Nelson C R, Exhaust energy recovery, DEER Conference, Aug. 2006, Michigan, USA.
[31] HTiEdition17, See CTT website: http://www.cumminsturbotechnologies.com/ctt/navigationAction.do?url=SiteContent+en+HTML+Downloads+Magazines.

In: Automotive Exhaust Emissions and Energy Recovery
Editor: Apostolos Pesiridis
ISBN: 978-1-63321-493-4

Chapter 8

SMALL, HIGH POWER DENSITY, DIRECTLY INJECTED, TURBOCHARGED ENGINES

Alberto Boretti* and Anthony Tawaf
RMIT, Bundoora, VIC, Australia

ABSTRACT

Downsized, downspeeded, high output, directly injected turbocharged internal combustion engines are receiving more consideration as the preferred thermal engine powering modern vehicles. Internal combustion engines with a small cubic displacement and less cylinders allow for the reduction of weight and the improvement of packaging. Turbocharged downsized engines operate over driving cycles with much larger brake mean effective pressures for improved fuel conversion efficiency while delivering approximately the same peak torque and power of the much larger naturally aspirated engines. This chapter reports on the trends in the turbo gasoline direct injection technology with three way catalytic converter aftertreatment as a measure to recover exhaust energy and reduce exhaust carbon dioxide emissions improving the fuel economy while complying with pollutants emission standards.

Keywords: Direct injection, downsizing, turbocharging, downspeeding, variable valve actuation

INTRODUCTION

Since the beginning of the history of the Internal Combustion Engine (ICE) in the mid-19th century, it has found itself in an unrelenting state of technological progress. However, recent developments of ICEs have mostly been focused towards the reduction of fuel consumption satisfying the emission requirements while retaining previous and increasing peak torque and power levels. Pollutant emissions have been regulated in Europe with

* Email: a.a.boretti@gmail.com.

increased severity starting from the EURO 1 of 1993 up to the present EURO 5/EURO 6 transitional stage [1]. Reduction of fuel consumption has always been a goal of car makers linked to the cost and availability of petrol, but this effort has been strengthened considerably by the claimed effect on climate of the carbon dioxide emission. Following up on a European Commission strategy adopted in 2007, the EU has put in place a comprehensive legal framework to reduce CO_2 emissions through binding emission targets for new car and van fleets [2]. For cars, manufacturers are obliged to ensure that their new car fleet does not emit more than an average of 130 g CO_2/Km by 2015 and 95g by 2020 operating in the cold start New European Driving Cycle (NEDC). This compares with an average of almost 160g in 2007 and 135.7g in 2011. In terms of fuel consumption, the 2015 target is approximately equivalent to 5.6 litres per 100 km (l/100 km) of petrol or 4.9 l/100 km of diesel. The 2020 target equates approximately to 4.1 l/100 km of petrol or 3.6 l/100 km of diesel. Despite these measures, CO_2 rules will be very likely postponed, as the EURO 6 were postponed by 2 years to become effective only in January 2014, however, the path for a cleaner and more efficient transport is clearly set.

The compliance with more stringent, emission standards as EURO 6 [1], the necessity to increase the fuel economy and reduce the carbon dioxide emissions [2] and the strict budgetary limits set up by the financial crisis affecting Europe, the US and Japan suggest the adoption of proven technologies requiring comparatively less research and development effort to progress ICE development in the decade to come, and especially the emissions legislation makes the gasoline stoichiometric engine coupled to the very well developed three way catalytic converter after treatment one of the preferred solutions.

This chapter discusses direct injection, turbocharging and variable valve actuation as the key enablers of more efficient turbo gasoline direct injection (TGDI) engines. While turbocharging and direct injection are very well proven technologies requiring further refinement but not an extensive research and development effort, variable valve actuation is the additional feature that requires further effort to further improve the fuel conversion efficiency especially at part load.

Recovering part of the exhaust heat in the turbocharger turbine to drive an intake compressor increases the mass of air and fuel trapped within the cylinder and therefore engine power density. Variable valve actuation reduces the throttling losses and permits the conditioning of the intake flow to enhance swirl and tumble motions as well as the internal control of the amount of exhaust gas recirculation. Direct injection finally permits fast and precise delivery of the fuel to each cylinder for mostly homogenous stoichiometric or lean stratified low load combustion in addition to improving the knock sensitivity through charge cooling by vaporization.

In TGDI ICEs, the three way catalyst after treatment permits compliance with emission standards, and downsizing and downspeeding through turbocharging reduces the fuel consumption over a driving cycle. The fuel consumption is then further improved through variable valve actuation load, flow and mixture control and direct injection lean stratified combustion.

Not included in this chapter is the recovery of the vehicle braking energy through kinetic energy recovery systems. A traditional powertrain powering one axle of the vehicle coupled to a kinetic energy recovery system (KERS) working on the other axle is the preferred solution for endurance racing and it is certainly a feature passenger cars of the near future would need to drastically improve the fuel economy [3]. KERS, mechanical or electric or electromagnetic, on and off the driveline are enablers of dramatic improvements of the fuel economy [3, 14, 17, 19, 20, 22]. The off driveline configuration offers better costs-to-benefit ratio for passenger cars applications also delivering four-wheel-drive (4WD) performance in otherwise traditional 2WD cars. F1 cars of 2014 will exploit the use of small turbo charged and downsized internal combustion engines with electric KERS [20, 22].

Downsizing and Downspeeding

Improvement of fuel economy follows the reduction of the engine displaced volume thus permitting operation over driving cycles at much higher brake mean effective pressure (BMEP) [4, 5]. Working at higher BMEP in a gasoline engine permits larger brake fuel conversion efficiencies. For the same peak torque and power outputs of a larger naturally aspirated engine, in a TGDI engine the BMEP values for peak torque and power have to increase to compensate for the reduction of displaced volume. Higher BMEP is achieved by boosting the ICE [4, 5, 6, 7 and 8].

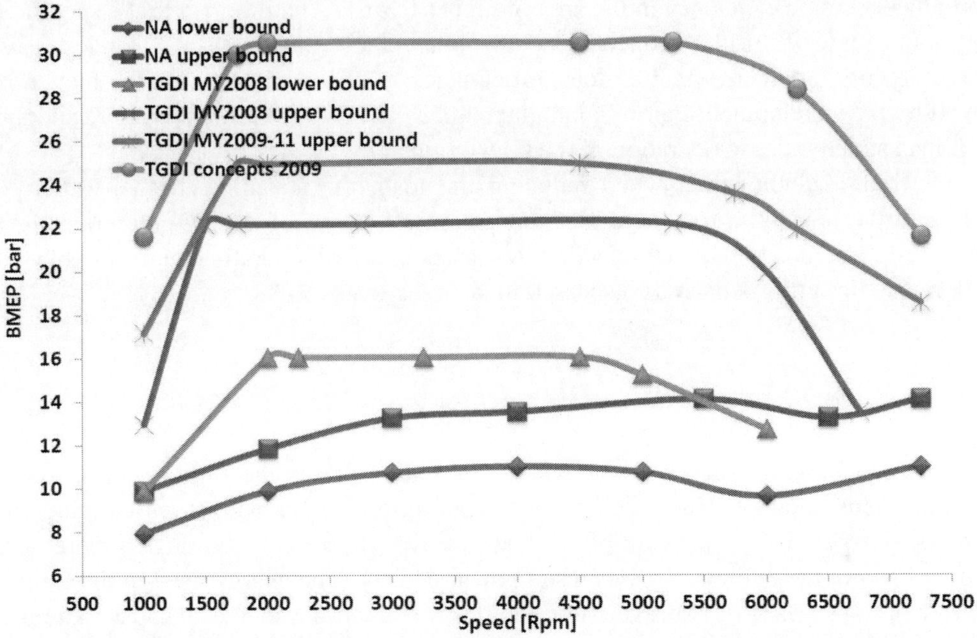

Figure 1. BMEP of naturally aspirated and turbocharged engines (data from [4]).

By introducing within the cylinder more air and fuel, even without any improvement of the indicated fuel conversion efficiency, the indicated mean effective pressure IMEP WILL clearly increase, and despite minimal increments of the friction mean effective pressure FMEP for the small dependence on peak in-cylinder pressure, the BMEP (=IMEP-FMEP) will also increase.

Figure 1 (data from [4]) presents the BMEP values of naturally aspirated and turbocharged engines. While naturally aspirated engines have top BMEP values of 10-12 bar, recent turbocharged engines have top BMEP values of 18–24 bar, and the near future top BMEP values are expected to exceed 30 bar.

In addition, turbocharged engines have much closer values of BMEP for maximum torque and power, and this translates into the opportunity to also downspeed the engine. While the torque, T, is proportional to the product of BMEP by the displaced volume, the power P is proportional to the product of BMEP by the displaced volume by the engine speed. With the ratio $BMEP_{,Pmax}/BMEP_{,Tmax}$ being larger in turbocharged engines, TGDI ICEs can run at lower maximum engine speeds while delivering the same torque and power of their naturally aspirated counterpart.

Downspeeding has some fuel economy advantages through the influence on FMEP in depending more than quadratically on engine speed. Therefore, the next generation turbocharged engines are expected to deliver peak torque with a BMEP of about 30 bar at relatively low speeds of around 1750 rpm, then peak power with BMEP of about 26 to 29 bar at around 5000 rpm i.e., at significantly lower speeds compared to naturally aspirated engines.

The brake specific fuel consumption BSFC, a parameter inversely proportional to the brake fuel conversion efficiency, is presently around 240 g/KW that corresponds to 34% brake fuel conversion efficiency in the area of higher BMEP. The full load brake specific fuel consumption may be further reduced, and the brake fuel conversion efficiencies may be increased accordingly, but the full load efficiencies of the lean burn Diesel may not be achieved by the stoichiometric gasoline, and the 40% mark is a limit difficult to reach during the full load stoichiometric operation of the GDI engine [13, 21].

It is worth including that when reducing the load, the stoichiometric operation with throttling further penalizes the brake fuel conversion efficiency of the GDI engine compared to the Diesel engine, but direct injection for lean stratified operation and variable valve actuation may limit this penalty as discussed in later sections.

BOOSTING

With turbocharging, the energy from exhaust gases drives the turbine that drives the compressor. This allows for increased pressure (but unfortunately also increased temperature despite the charge cooler) upstream of the intake valves ultimately resulting in more air and fuel drawn within the cylinder for increased power density. The energy used in the turbine is waste exhaust gas energy. However, the addition of a turbine in the exhaust system also increases the back pressure, with a small negative impact on the IMEP however, which is more than compensated for by the much larger amount of fuel and air introduced within the cylinder, sharply raising the IMEP.

Since the pressure work in the intake compressor is linked to the expansion work in the turbine, and since the turbocharger has to accelerate to a higher rotational speed to follow an increased load demand, the system suffers from poor response during transients, because a demand for more air and fuel in the cylinder requires more exhaust gas to expand through the turbine and both turbine and compressor to rotate at a higher speed.

In addition, turbines and compressors do not work well in restricted areas of their maps, with reduced efficiencies resulting when moving far from points of nearly optimal operation. Apart from the efficiency islands, surge and choke lines limit the region where a compressor may operate. Figure 2 presents typical turbocharger compressor and turbine maps. For every operating point of the engine in the speed and load map, the mass flow rate through the compressor and turbine changes significantly and changes in the pressure ratio and rotational speed are significant as a result.

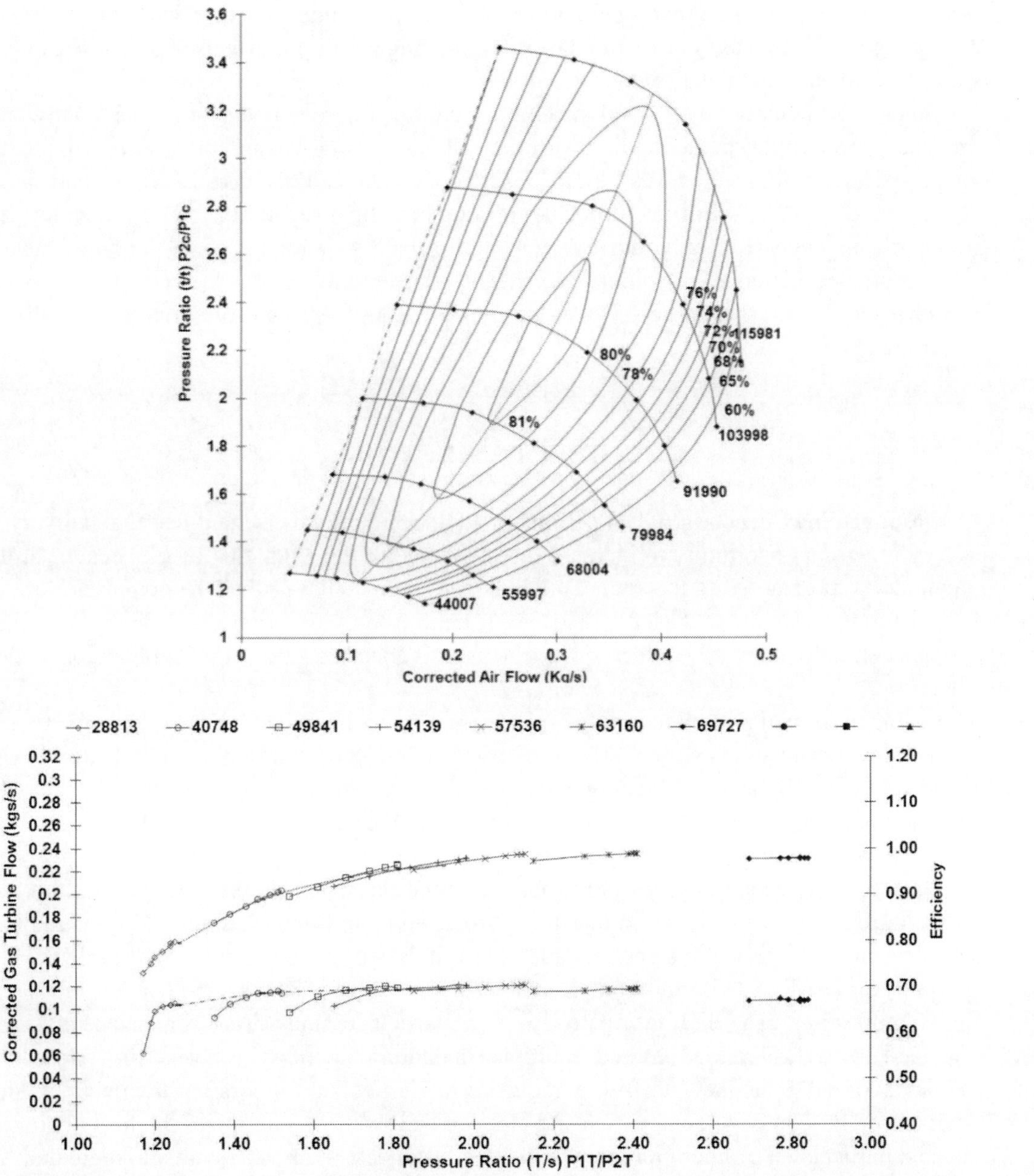

Figure 2. Typical compressor and turbine maps.

It is worth noting that the addition of a turbocharger requires more catalytic converter residence time from the exhaust gas, because the gas initially expands in the turbine, then the expanded gas enters the catalytic converter at a reduced temperature.

To improve steady and transient engine performance, more complex configurations have been proposed in addition to traditional turbocharging with a waste gate on the turbine and a charge cooler after the compressor. Two stage turbochargers, with smaller high pressure and larger low pressure turbines and compressors with a waste gate at the low pressure turbine, and a by-pass valve at the high pressure turbine and compressor are becoming increasingly necessary for downsizing/downspeeding. In addition of a crankshaft or electric motor driven supercharger with by-pass before the turbocharger compressor are proposed in [5]. E-Charger and E-Turbocompound are also considered in [6].

Other turbochargering developments include improvements of the individual turbocharger and turbo compounding with new alloys, composites and processes, twin scroll, compressor recirculation valves, electric actuators, reduced friction bearings and more efficient wheels [10]. Twin flow turbochargers combine flow separation, pulse utilization and low inertia to improve steady and transient performances. Variable nozzle turbines have a wider range of efficient operation and better transient response. In two stage parallel turbochargers, two turbochargers work side by side at high rpm but only one turbo delivers the torque at low rpm.

Direct Injection

Gasoline direct injection (GDI) permits in addition to the precise and fast fuel delivery to each cylinder the opportunity to increase the resistance of the engine to knock because of the in-cylinder vaporization of the fuel. Turbocharged engines have smaller compression ratios than naturally aspirated engines, and GDI engine provide a slightly higher reduced compression ratio alternative which otherwise negatively affects the indicated fuel conversion efficiency.

GDI is beneficial in both naturally aspirated and turbocharged engines [12, 13, 15, 16, 18, 21], even if the charge cooling effects of port fuel injected engines permit better power densities in naturally aspirated racing engines [12]. Port or direct water injection [23] may be a simple but reliable measure to control knock that remains an option should the thermal load increase significantly.

GDI injector technology has not progressed significantly over the last decade, and 200 bar pressure common rail solenoid injectors permitting up to three multiple injections in close sequence are still the leading technologies [11]. It is worth noting that GDI engines need attention to meet EURO 6 emission standards for particulates. The short injection times and rapid combustion can lead to relatively high particulate emissions. Precision metering combined with fast actuation and multiple injection capability reduces the particulate emissions of GDI engines otherwise requiring a particulate filter after treatment. Quick vaporization for complete combustion and reduced wall-wetting minimizes the formulation of soot and particulates. Higher pressures (500 bar GDI injectors, already available more than 10 years ago), further reduce the diameter of nozzle holes and piezoelectric actuation will be likely features of the next generation GDI injectors.

When centrally located in close proximity to the spark plug for high performance, as the latest trend indicates [6, 7], fast actuated, high pressure GDI injectors also enable lean stratified charge operation through multiple injections approaching the top dead center.

Even if combustion is still wall initiated because of the spark plug location, the charge stratification close to the center of the combustion chamber reduces heat losses while keeping the rate of combustion high for optimal pressure build up, without too high a formation of pollutants within the cylinder.

High pressure, fast actuating, direct injection of gasoline plays a significant role in new turbocharged engines. These centrally located direct injectors operate at high pressure and their small nozzles permit large flow rates and fast vaporization. Fast actuation permits exploitation of charge stratification when operating off stoichiometry.

Apart from the fuel vaporization cooling down the charge, direct injection further reduces the knock tendency thereby reducing the time from when the fuel is injected in the combustion chamber, vaporised and mixed with air to when the fuel is burned by the flame before knock can occur. With very high pressures and high flow rate injections of small droplets, the main injection timing may be phased during the early compression stroke so as not to allow enough time for the end gases to knock.

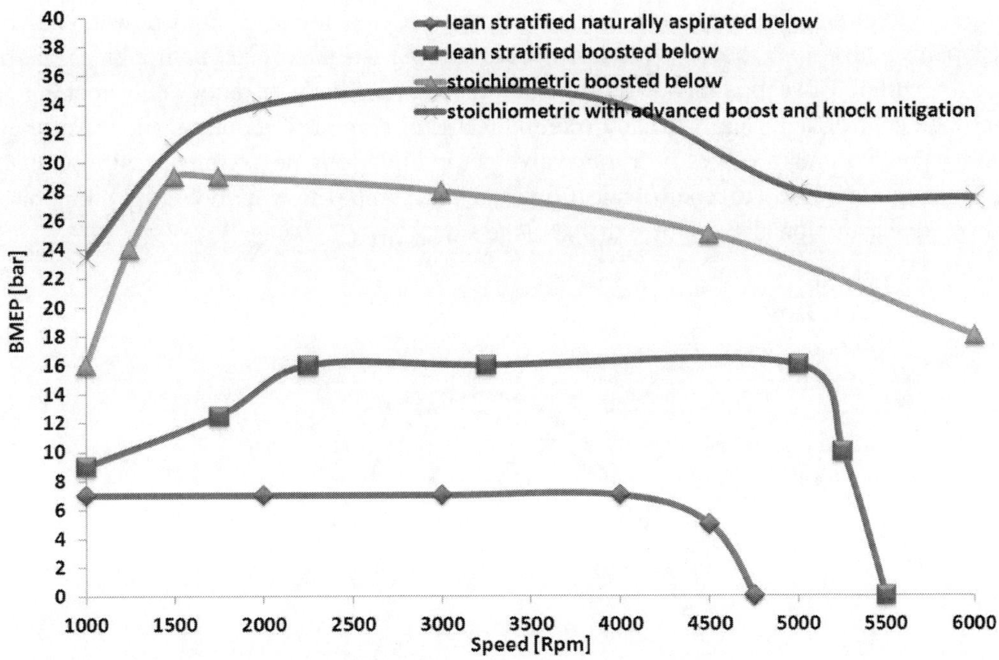

Figure 3. Charge strategies in turbo GDI engines (data from [6]).

Figure 3 (data from [6]) presents charge strategies coupled to advanced boosting and knock mitigation technologies to optimize the engine operation over the full range of loads and speeds. For lower loads, the engine operates more efficiently running at lean stratified operation, with direct injection producing a stratified charge close to the spark plug operating multiple injections close to spark initiation. Very low loads are covered in naturally aspirated operation, and intermediate loads are covered with boost. For higher loads, operation is boosted stoichiometric with a homogeneous mixture resulting from early direct injection. To further increase the BMEP, electrically assisted boosting is used at low speeds below 1500 rpm, and other advanced boosting technologies with knock mitigation are necessary at higher speeds.

While the lean stratified operation is certainly beneficial in indicated fuel conversion efficiency terms, operating at lean conditions may negatively impact the three way catalytic reduction of pollutants.

VARIABLE VALVE ACTUATION

Variable valve Actuation (VVA) permits operation without throttling as well as the conditioning of the flow and the mixture for further improvements of the fuel conversion efficiency over the full range of loads and speeds as well as improvement of transient operation [9]. Multi-Air is a technology developed by Fiat Research in the last two decades now being used in production passenger cars [9]. The Fiat Multi air technology is the first fully variable hydraulic valve actuation system on production vehicles. Figure 4 presents the different strategies for intake valve lift that can be adopted for maximum power with full valve opening, low rpm torque with early valve closing (to maximize air trapping), partial load with partial valve opening for load control, engine idle with late valve opening for further load control, and finally urban cycle multi lift for combustion optimization. Naturally, operating the two intake valves of a four valve per cylinder engine asymmetrically produces even more opportunities to control the flow and the composition as does the adoption of variable valve actuation also on the exhaust valve side.

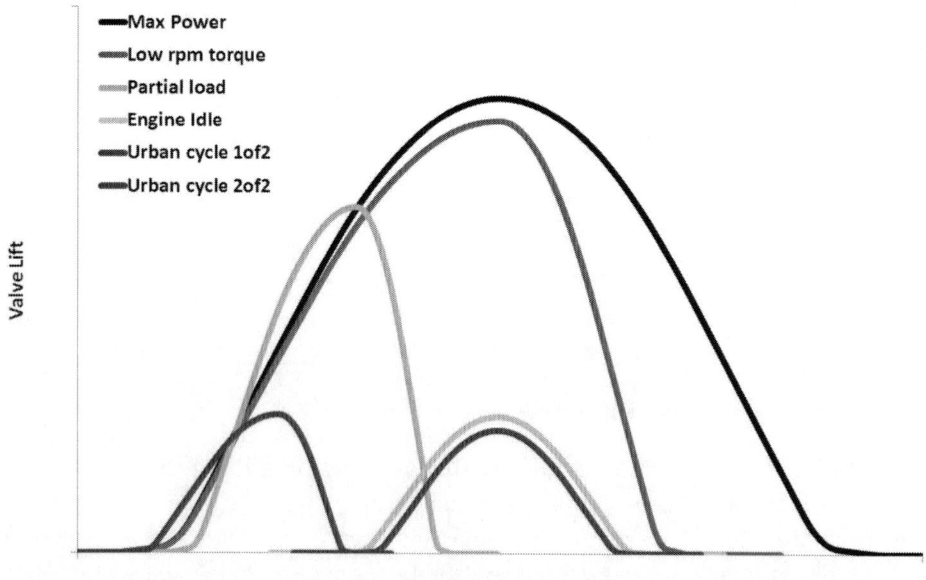

Figure 4. Multi air VVA: a) *Max power* valve lift (high speed full load profile); b) *Low rpm torque* valve lift (low speed full load); c) *Partial load* valve lift (intermediate loads); d) *Engine Idle* valve lift; e) *Urban cycle* valve lift (very low loads).

Essentially, the intake valve lift profile may be shaped free of any restriction if the adapted profile is not within the envelope of maximum power. The system has been so far applied to the intake valves of gasoline engines, but it may be extended to also actuate the exhaust valves and better control the exhaust gas recirculation internally, and to operate the two intake valves independently for swirl and tumble control.

Controlling the solenoid valve opening and closing times produces a wide range of optimum intake valve profiles. For maximum power the solenoid valve is always closed and the valve lift follows the cam profile specifically designed to maximize power at high engine speeds. For low speed torque, the solenoid valve opens near the end of the valve opening event leading to early intake valve closure. Different strategies are available for part load. For medium loads, the solenoid valve opens much earlier causing the partial opening of the valve. For light loads, the solenoid valve may close once the cam event is already started, partially opening the intake valve. The last two actuation modes can be combined in the same intake stroke generating the so-called "Multilift" mode enhancing turbulence and combustion rate at very low loads. All these strategies are effective in controlling the load, reducing the air flow, with the second mode producing much larger velocities of the incoming air thus producing higher in-cylinder turbulence and rate of combustion.

The multi air benefits for gasoline engines are increased maximum power because of the adoption of a power-oriented valve lift profile when power is needed, increase of low speed torque through early intake valve closure, minimizing back flows and reduction of fuel consumption due to the reduction of pumping losses. The multi air benefits also include the option to optimize the warm-up and reduce emissions by introducing internal Exhaust Gas Recirculation by reopening the intake valves during the exhaust stroke and the superior dynamic engine response. The benefits of variable valve actuation are further described in [15, 16, 18].

CONCLUSION

Small, high power density, gasoline direct injection (TGDI) engines are one of the preferred technologies to meet near future emission standards while delivering significant fuel economy advantages.Turbocharging, Direct Injection and Variable Valve Actuation are the main technologies of TGDI.

Traditional turbocharging increases the power density and permits downsizing and down speeding. More complex two-stage turbocharging permits better steady and transient operation but increases cost and complexity.

Downsizing and down speeding permit the a more efficient engine operation while allowing better performances better packaging and reduced weight.

Direct injection is preferably implemented at high pressure and is fast and central. It reduces knock sensitivity and provides a precise and fast delivery of the fuel to individual cylinders while also permitting charge stratification.

Variable valve actuation permits throttle-less engine operation as well as the conditioning of intake flow and the air-fuel mixture for further improvement of the fuel conversion efficiency over the full range of loads and speeds as well as during transient operation.

REFERENCES

[1] delphi.com/pdf/emissions/Delphi-Passenger-Car-Light-Duty-Truck-Emissions
[2] ec.europa.eu/clima/policies/transport/vehicles/index_en.htm
[3] www.williamshybridpower.com/
[4] Whitaker, P, "Turbocharged Spark Ignited Direct Injection – A Fuel Economy Solution for The US", DEER Conference 2009 Directions in Engine-Efficiency and Emissions Research, Dearborn, Michigan, August 3-6, 2009.
[5] Blaxill, H, "The Role Of IC Engine In Future Energy Use", DEER Conference 2009 Directions in Engine-Efficiency and Emissions Research, Detroit, Michigan, October 3-6, 2011.
[6] Boggs, D, King, J, "Application of Synergistic Technologies to Achieve High Level of Gasoline Engine Downsizing", DEER Conference 2009 Directions in Engine-Efficiency and Emissions Research, Detroit, Michigan, October 3-6, 2011.
[7] Yilmaz, H., "DEER 2012 – Bosch Powertrain Technologies". 2012 Directions in Engine-Efficiency and Emissions Research (DEER) Conference, Dearborn, Michigan, October 16-19, 2012.
[8] Tatur M, "Future Directions in Engines and Fuels", 2010 Directions in Engine-Efficiency and Emissions Research (DEER) Conference, Detroit, Michigan, September 27-30, 2010.
[9] www.fptmultiair.com/flash_multiair_eng/home.htm
[10] turbo.honeywell.com/whats-new-in-turbo/
[11] ae-plus.com/technology
[12] Boretti, A.A., Watson, H.C., "Comparison of Pfi and di superbike engines", SAE Technical Papers, 2008.
[13] Boretti, A., "Analysis of design of pure ethanol engines", SAE Technical Papers, 2010.
[14] Boretti, A., "Coupling of a KERS powertrain and a 4 litre gasoline engine for improved fuel economy in a full size car", SAE Technical Papers, 2010.
[15] Boretti, A., "Use of variable valve actuation to control the load in a direct injection, turbocharged, spark-ignition engine", SAE Technical Papers, 2010.
[16] Boretti, A., "Performances of a turbocharged E100 engine with direct injection and variable valve actuation", SAE Technical Papers, 2010.
[17] Boretti, A., "Modeling of engine and vehicle for a compact car with a flywheel based kinetic energy recovery systems and a high efficiency small diesel engine", SAE Technical Papers, 2010.
[18] Boretti, A.A., "On the advantages of E100 over gasoline in down-sized, turbo-charged, direct-injected, variable valve actuated, and stoichiometric S.I. engines", SAE Technical Papers, 2011.
[19] Boretti, A.A., "Improvements of vehicle fuel economy using mechanical regenerative braking", International Journal of Vehicle Design, 2011; 55(1):35-48.
[20] Boretti, A., "KERS braking for 2014 f1 cars", SAE Technical Papers 2012.
[21] Boretti, A., "Towards 40% efficiency with BMEP exceeding 30 bar in directly injected, turbocharged, spark ignition ethanol engines", Energy Conversion and Management, 2012;(57):154-166.

[22] Boretti, A., "F1 2014: Turbocharged and downsized Ice and Kers boost", World Journal of Modelling and Simulation, 2013; 9(2):150-160.
[23] Boretti, A.,"Water injection in directly injected turbocharged spark ignition engines", Applied Thermal Engineering, 2013;52(1):62-68.

In: Automotive Exhaust Emissions and Energy Recovery
Editor: Apostolos Pesiridis
ISBN: 978-1-63321-493-4

Chapter 9

ORGANIC RANKINE CYCLES IN AUTOMOTIVE APPLICATIONS

Antti Uusitalo, Teemu Turunen-Saaresti, Aki Grönman,* Juha Honkatukia and Jari Backman

Laboratory of Fluid Dynamics, Lappeenranta University of Technology, Finland

ABSTRACT

Despite the achieved efficiency improvements of diesel and gasoline engines in the last decades a large portion of the fuel power is still wasted in the engine process in the form of heat. The interest towards utilizing this waste heat in order to improve the energy efficiency of the system has been increasing in the last few years. One of the most promising technical solutions for converting waste heat into usable energy has been identified as the Organic Rankine Cycle (ORC). The principle of the ORC is similar to the conventional steam process. The main idea in most of the ORC-plants is to produce electric power from waste heat. The use of organic fluids provides several benefits compared to conventional steam processes for the utilization of waste heat streams in automotive applications. Typical heat sources for ORC applications are the waste heat from combustion engines, gas turbines, and many industrial processes.

The commercial ORC system efficiencies are varying from 10 to 20 %, depending on the application, working fluid and the power scale of the process. The efficiency is very dependent on process temperatures as well as the selected working fluid and therefore a detailed process of design is always needed. Most of the commercial ORC applications are having a power output in the range of 50 kW to few MW. The energy of automotive exhausts is low compared to industrial ORC installations, which means that the automotive ORC would produce about 1 – 30 kW of electric or mechanical power, depending on the size of the engine. The low power introduces challenges related to process component design, mainly the design of the process expander in small scale systems, and further challenges are related to the restricted available space for the process heat exchangers.

* Corresponding author: Laboratory of Fluid Dynamics, Lappeenranta University of Technology, Finland Email: Aki.Gronman@lut.fi.

Keywords: Waste heat recovery, Organic Rankine cycle, Combined cycle

INTRODUCTION

Despite the efficiency improvements of internal combustion engines in the last decades a large portion of the fuel power is wasted in the engine process in the form of heat. The interest towards utilizing this waste heat of the system has been increasing in the last few years in order to improve the system energy efficiency, fuel economy as well as to meet the emission regulations. One of the most promising technical solutions for converting waste heat into usable energy has been identified as the Organic Rankine Cycle (ORC) [Quoilin and Lemort 2009, Sprouse and Depcik 2013].

The working principle of the ORC is based on the well-known Rankine process but it uses organic working fluids instead of water and steam. A schematic process diagram of a simplified Rankine cycle is presented in Figure 1. The ORC process consists of four main parts: vaporizer, expander, condenser and feed pump. In the cycle, the pressurized liquid is evaporated by introducing heat into the system, and expanded in an expander producing usable mechanical energy or electric power. After the expander the organic vapour is condensed back to a liquid form in a condenser.

The history of ORC technology dates from the 19th century where the first patents were published. However, the first commercial ORC power plants were built as late as in the 1960's. The first commercial ORCs had a power range of 0.2 kW-3 kW and were designed to produce electrical power in remote areas [Bronicki 1988].

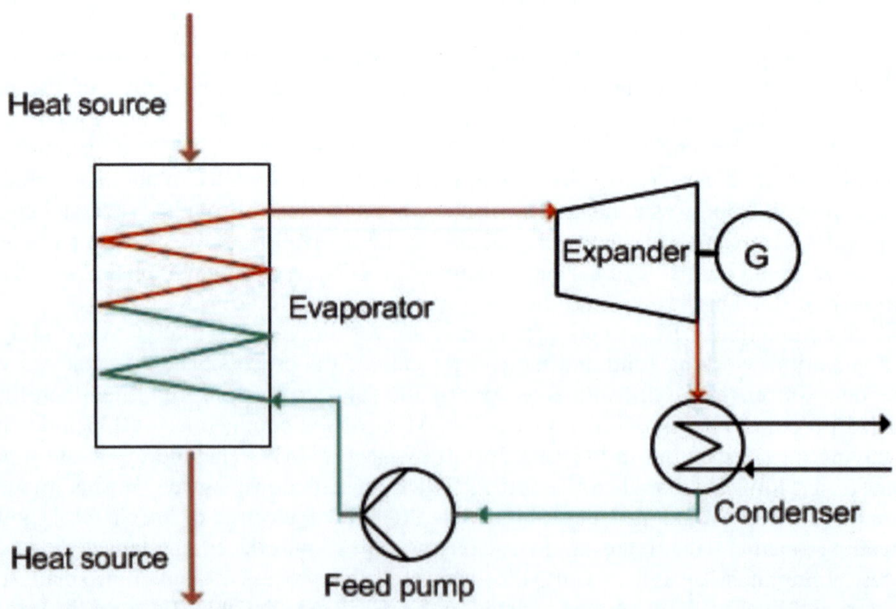

Figure 1. A schematic process diagram of simple Rankine process. Green color indicates liquid working fluid and red color indicates working fluid vapour.

The main idea in ORC technology is to utilize heat sources that have temperature less than approximately 500°C and/or where the amount of heat is relatively low (typically less than 1 MW). Typical ORC applications are for example the waste heat recovery from internal combustion engines, gas turbines or utilizing low grade waste heat from industrial processes. In commercial ORC applications the heat source temperatures are ranging from around 80°C to 400°C [Quoilin and Lemort 2009]. Heat to the process is introduced either directly to the evaporator or by using an intermediate circuit to introduce heat from the heat source into the ORC process. An intermediate circuit is often used in applications where the heat is collected from several heat streams or the heat source temperature is too high for the working fluid to be directly exposed to the heat source.

The reason why ORCs can utilize low heat sources is the lower heat of vaporization of organic fluids compared to water vapour. Another feature that is beneficial when adopting organic fluids is the dry expansion in the expander which enables using processes having little or no superheating in the process without having a risk of liquid droplets damaging the expander. These above-mentioned features allow organic fluid temperature to follow better the heat source temperature when compared to water vapour. The exemplary evaporator temperature diagram demonstrating the difference between using a conventional steam process and an ORC process is presented in Figure 2.

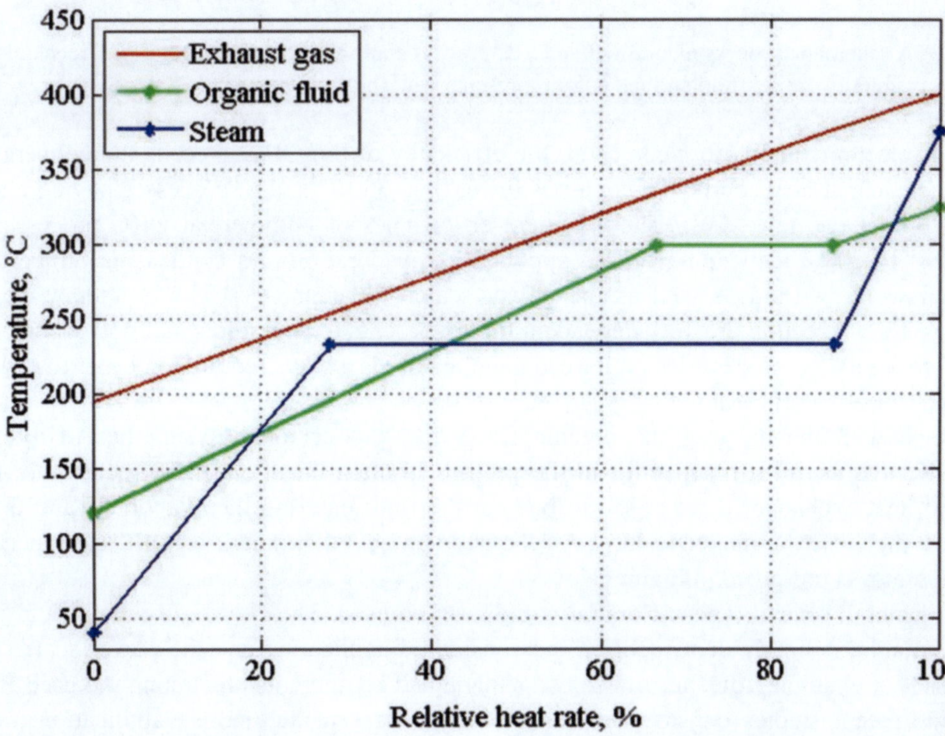

Figure 2. An exemplary evaporator temperature diagram demonstrating the difference between using a conventional steam process and a process adopting organic working fluid.

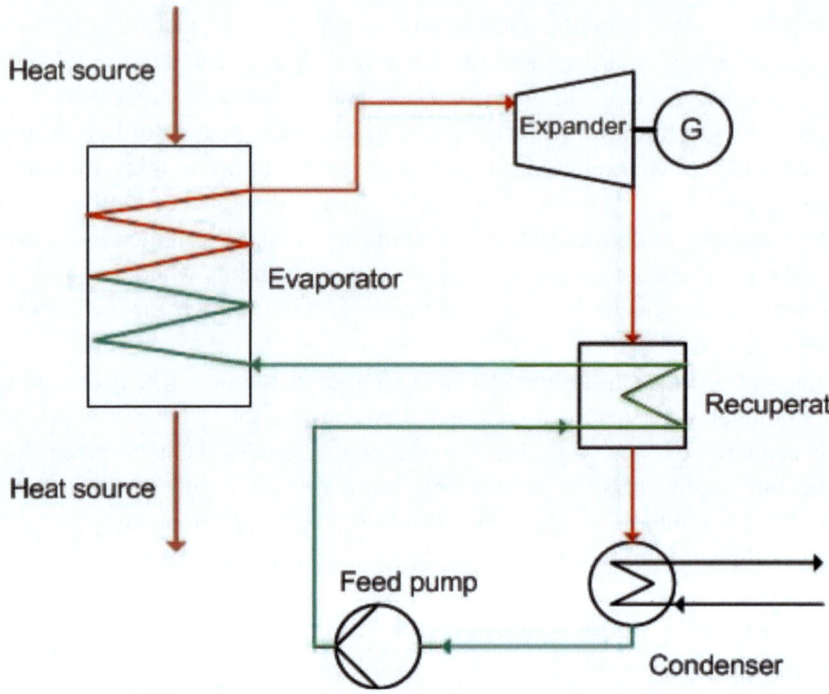

Figure 3. A schematic process diagram of an ORC process equipped with a recuperator. Green color represents liquid working fluid and red color represents working fluid vapour.

By selecting suitable organic fluid, the efficiency of the ORC exceeds the conventional Rankine cycle at heat source temperatures lower than 350°C and in processes having a small power output (< few MW). The electrical efficiencies of commercial ORC processes are typically 10-20 %, depending on the application, evaporation and condensing temperature, selected working fluid as well as the power scale of the process. The efficiency and the working fluid selection are very dependent on the process temperatures and power scale, and therefore a detailed process design for each application is always required in order to evaluate and maximize the efficiency.

Since with the most of the organic fluids the vapour remains superheated after the expander, the use of a recuperator in the process is often beneficial. The recuperator is an internal heat exchanger that is used to preheat the fluid entering the evaporator and thus to increase the cycle thermal efficiency. A schematic process diagram of an ORC equipped with a recuperator is presented in Figure 3.

In general, several types of organic fluids can be used. The choice, of which fluid is the most suitable, is highly dependent on the considered application and the power scale of the process. For example, toluene, siloxanes, pentane and refrigerants are among the used fluids. In some recent studies considering waste heat recovery in automotive applications, water [Seher et al. 2012] or a mixture of water and organic fluid is adopted as working fluid [Daccord et al. 2013, Guillaume et al. 2013].

The first studies on automotive waste heat recovery application by means of Rankine cycles were carried out in the 1970's driven by the concerns over air pollution and escalating fuel prices [Sprouse and Depcik 2013]. Following these studies, few prototypes were designed and built to test the idea of using waste heat recovery in automotive applications showing that 10-15 % improvements on the engine fuel economy was achieved [Sprouse and Depcik 2013]. However, decreasing fuel prices in the 1980's and improved engine technology stalled the implementation of waste heat recovery systems [Sprouse and Depcik 2013].

The number of ORC projects, the number of ORC applications as well as the number of published information concerning the use of ORC technology in different applications has increased significantly during the last 10 years [Quoilin and Lemort 2009]. Despite the fact that there are no commercial ORC systems currently available designed for automotive applications, recently published research papers, sponsored by large engine manufacturers and automotive companies, are clearly indicating the increasing interest towards the waste heat recovery in order to improve engine systems performance and fuel economy [Sprouse and Depcik 2013, Nelson 2008].

PROCESS OF ORC DESIGN

The process of the ORC design is most often based on the heat source type, temperature and power as well as available cooling capacity and its temperature. The most important aspects in the design of the efficient ORC process are the selection of working fluid and the design of the expander. In automotive applications possible heat source types are exhaust gases, charge air, lubrication oil and engine cooling water. The cooling of the ORC can be done either using air radiators or engine cooling water.

Working Fluid Selection

One of the most important aspects in designing ORC systems is the selection of working fluid. Despite the large number of studies concerning working fluid selection for ORC systems e.g. [Hung 2001, Saleh et al. 2007, Wang et al. 2011, Wang et al. 2012], no single working fluid can be identified as the optimal one [Quoilin et al. 2011, Sprouse and Depcik 2013]. Some fluid properties that have to be considered when selecting the working fluid are presented in Table 1.

Table 1. Working fluid selection

GENERAL
• Flammability
• Environmental impacts
• Toxicity
• Price and availability
• Thermal and chemical stability
PROCESS DESIGN
• Efficiency and power output
• Expander type and the size of the heat exchangers
• Evaporation pressure
• Condensing pressure

The most commonly used working fluids can be divided into three main categories: hydrocarbons, refrigerants and siloxanes. High performance can be achieved for cycles adopting heavy hydrocarbons and siloxanes, which are characterized by high expansion ratios over the expander and high critical temperature and pressure. These fluids are most commonly selected as working fluids for high-temperature heat recovery applications including exhaust gas heat recovery or for small biomass power plants. Hydrocarbons which have relatively simple molecular structure and the most of the refrigerants introduce moderate expansion ratios over the expander and high condensing pressures above the atmospheric pressure. On the other hand, these working fluids introduce lower thermodynamic efficiencies for the cycle in high temperature heat recovery applications. These fluids are often selected as working fluids for low- and moderate temperature, approximately 100 – 200 °C, ORC applications. The simplified categorization of the working fluids with respect to waste heat temperature and cycle efficiency is presented in Figure 4.

One drawback of many organic fluids is that they can be identified to be either flammable and/or toxic which have to be taken into account for safety reasons when selecting the working fluid. Also the chemical and thermal stability, which might lead to the decomposition of working fluids, sets limitations on the selection of working fluids, available process maximum temperature as well as on process purity. Some considered working fluids, mainly freons that were used in the past as working fluids for ORC systems, are nowadays banned due to their high ozone depletion potential.

The available space for the process heat exchangers is also restricted in low power applications, especially in case of automotive heat recovery, leading to a demand of using small size process components, which on the other hand is closely related to the selection of working fluid.

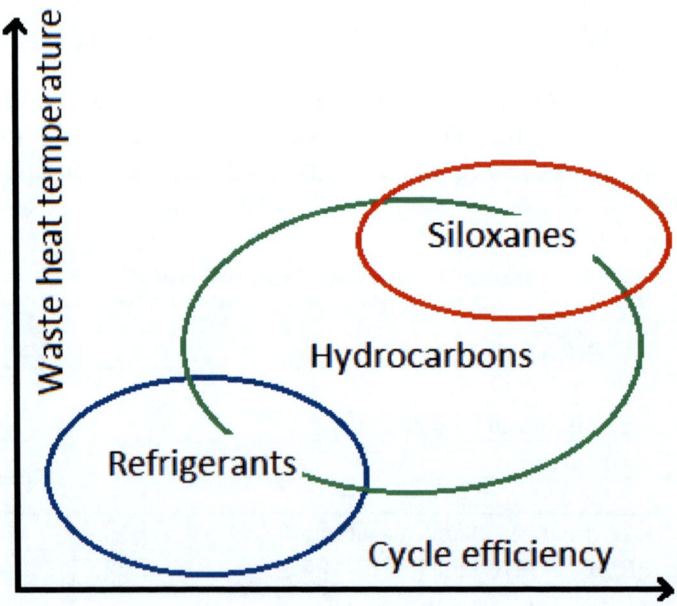

Figure 4. Working fluid categories.

Expander Type

The ORC process expanders can be mainly divided into two categories i) volumetric expanders and ii) turbines. The expander type selection, in addition to the selection of working fluid, is probably one of the most important steps in designing ORC processes and the selection is mainly depending on the scale of the process, process working parameters and the selected working fluid. The most commonly considered volumetric expanders are namely screw [Wang et al. 2011], piston [Seher et al. 2012] and scroll expanders, in which the working principle is based on reversed volumetric compressors that have been widely used in refrigerant cycles and heat pump applications. In case of volumetric expanders lubrication oil system is typically added to the system, which might in some cases have an effect on the working fluid stability. The turbines are mainly axial or radial turbines. Ljungström type multi-stage radial outflow turbine is one of the recently studied turbine types for ORC systems that have potential to decrease the losses related to supersonic flows in ORC turbines [Persico et al. 2013].

In commercial ORC-plants producing electrical power, turbines have been more successful so far. The expanders are mainly based on the conventional turbine technology, especially in large scale ORC's, which includes shaft seals, a reduction gear, an air cooled generator and a lubricating oil system. One technical option is to use a high speed turbo-generator concept in where the organic working fluid is used as a lubricant and as a generator coolant. This concept allows a completely hermetic design, with no need for shaft seals or a separate lubrication system [Larjola 1995, Larjola 2011]. When using hermetic turbo-generators, there is practically no need for service and the oil-free design is also an advantage in ORC systems.

The interest towards small scale ORC heat recovery systems, including the waste heat recovery in automotive applications, has been increasing in the recent years. In systems having power output of few kW, volumetric expanders are often considered following a good off-design performance compared to small scale turbomachinery [Sprouse and Depcik 2013]. The drawback of using volumetric expanders is that only moderate pressure ratios over the expander can be adopted when compared to turbines or alternatively a multistage configuration is needed. Another drawback is that in general the efficiency when using volumetric expanders is lower when compared to turbines, but recent technological developments related to volumetric expanders have narrowed this gap [Sprouse and Depcik 2013].

The energy of automotive exhaust gases is low compared to the industrial ORC installations, which means that the automotive ORC will produce less power compared to their industrial counterparts. The small size introduces also further challenges for the expander design, especially when using turbines, as the turbines tend to be small and fast rotating and the relative losses in turbine are high. Another disadvantage of organic fluids is the low speed of sound which leads often to supersonic turbine flows and, therefore, into increased losses. However, due to the small specific enthalpy drop and large volumetric flow rate of organic fluids, it is possible to design small capacity turbines with relatively simple geometry and relatively good efficiency.

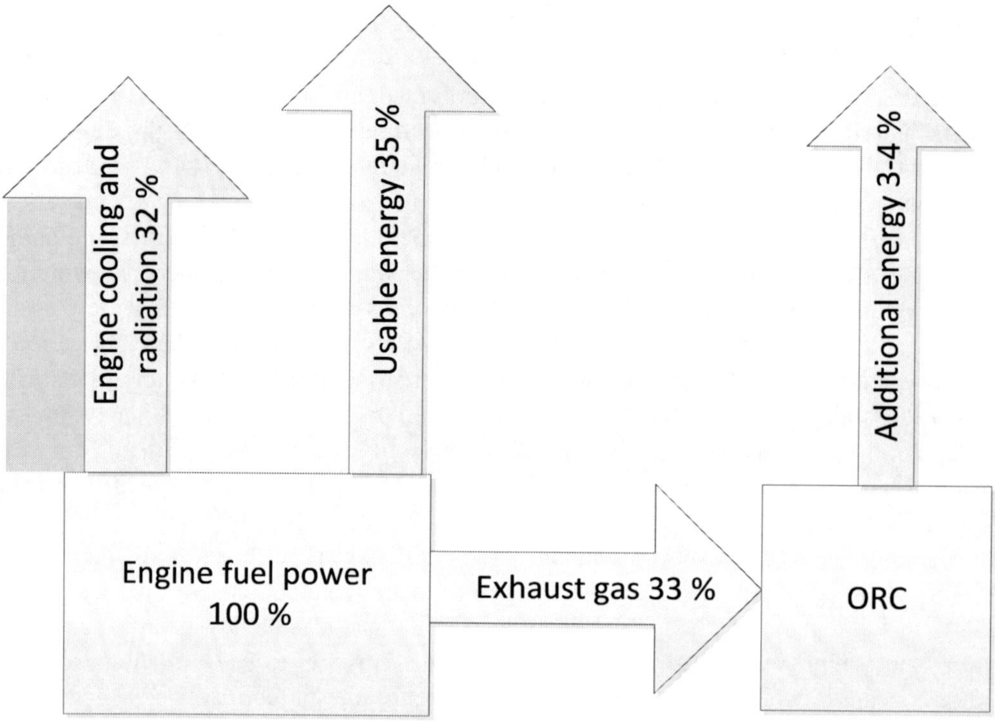

Figure 5. Schematic diagram of diesel engine waste heat streams and exhaust gas recovery.

Utilization of Waste Heat Streams from Vehicle Engines

In internal combustion engines a large amount of heat is not converted into usable mechanical energy but is removed from the system in form of heat. The main waste heat streams can be identified as the exhaust gases, engine jacket cooling system and lubrication oil system as well as the heat of the intercooler in turbocharged engines [Sprouse and Depcik 2013]. In addition a portion of the fuel power is wasted in the system in form of heat radiation into the environment. A Schematic diagram of diesel engine waste heat streams and exhaust gas recovery with ORC is presented in Figure 5.

The largest potential for waste heat recovery is related to exhaust gas utilization because this heat stream exits the engine at high temperature when compared to the other heat streams of the engine. The exhaust gas temperature is dependent on the used engine technology (turbocharger, fuel, exhaust gas after-treatment system, compression ratio of the engine, air-fuel ratio etc.). The process diagram of an ORC process with a recuperator and utilizing exhaust gas heat after the turbocharger turbine is presented in Figure 6. The red color indicates gaseous working fluid and the green color indicates working fluid in liquid form. The use of the recuperator should be evaluated based on the selected working fluid, process parameters and the available space for the process heat exchangers.

The expander power can be converted into electricity by coupling the expander to a generator directly or via a gearbox, or alternatively it can produce mechanical energy.

Another option for recovering engine exhaust heat is to utilize the heat from Exhaust Gas Recirculation (EGR) [Espinosa et al. 2010, Nelson 2008]. The EGR is a commonly used technique in modern diesel and gasoline engines where a portion of the exhaust gases are recirculated back to engine cylinders in order to reduce NOx emissions.

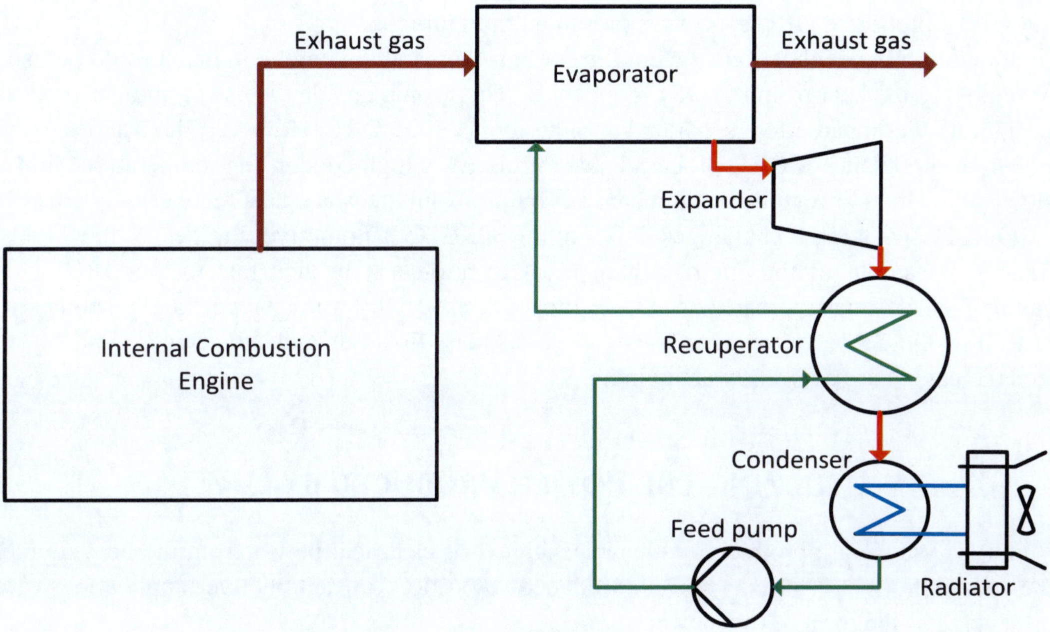

Figure 6. Process diagram of ORC process with recuperator.

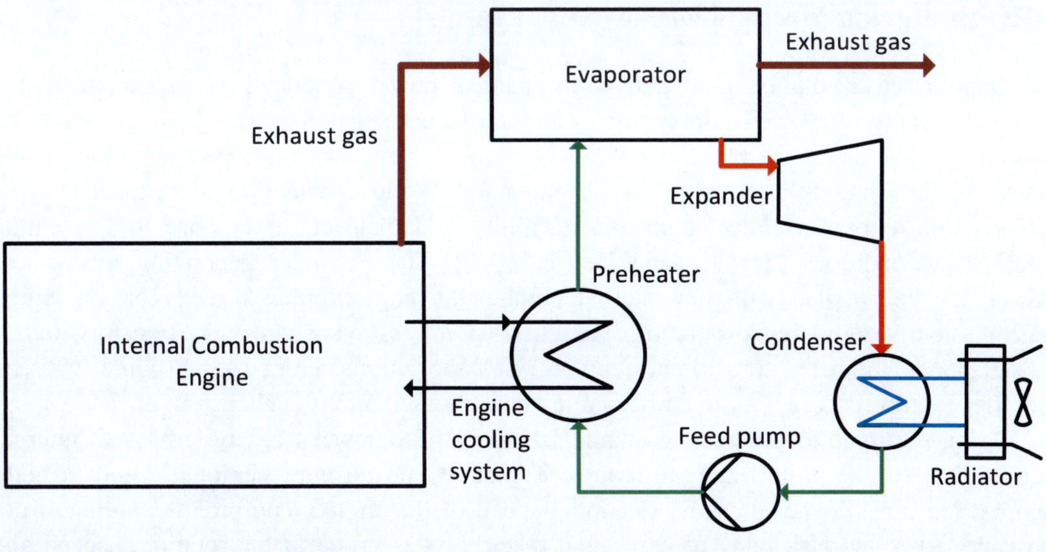

Figure 7. Process diagram of ORC process with preheating.

The temperatures of engine cooling circuit and lubrication oil systems are typically low (ca. 80-100 °C). Thus the potential for using engine cooling system and lubrication oil system in recovering waste heat is restricted. A portion of the lower temperature heat streams, for example charge air heat in turbocharged engines or engine cooling circuit, could be used for preheating the liquid working fluid entering the evaporator and thus replacing the recuperator in the process in a case if low condensing temperature for the ORC can be adopted. An example of preheated ORC process is presented in Figure 7.

In automotive applications the engine cooling loop and the engine radiator could be used for removing the heat from the ORC condenser. This would enable the use of simpler process configuration compared to a separated cooling loop for the ORC. However, the use of engine coolant as a coolant for ORC process leads to relatively high condensing temperatures in the ORC cycle which lowers the available power output from the waste heat recovery system.

One of the largest challenges in coupling ORC to automotive engines is the highly dynamic behaviour of the engine which leads to non-constant heat rate to the ORC. This requires special design care on ORC process, especially to achieve fast response in controlling the ORC system when the exhaust gas mass flow rate and exhaust gas temperature from the engine are constantly changing.

UTILIZING THE POWER PRODUCED BY ORC

ORC systems are producing either mechanical or electrical power from the heat source. The possibilities to utilize power produced by ORC in automotive applications are influencing to the form of this power.

ORC Producing Mechanical Power

One attractive solution is to deliver mechanical power generated by an expander of a Rankine cycle to the power train system of an internal combustion engine. In such a case, the bottoming cycle can be an ORC or a steam Rankine cycle where e.g. a piston engine can be considered as suitable alternative for an expander. While considering the expander type, reciprocating expanders have good controllability with respect to response to fluctuating waste heat conditions [Sprouse and Depcik 2013]. The expander generating mechanical power can be coupled with the internal combustion engine either directly or via a belt transmission. From the oscillation reduction point of view, a belt transmission is advantageous compared to a direct coupling with the internal combustion engine whereby significant effort is necessary to dampen unfavourable oscillations [Seher et al. 2012].

Using a turbine to generate mechanical power to the power train system of an internal combustion engine is not straightforward. Typically, the turbine rotational speed is high, approx. ten times higher than the rotational speed of the engine, and direct coupling is not possible. Also, the efficiency of expansion in turbines is prone to the rotational speed and matching the rotational speed of the engine and the ORC turbine is very demanding even if a gearbox is used.

ORC Producing Electrical Power

ORC producing electrical power is using a generator coupled to the same shaft as an expander or can be connected via a gearbox. Power electronics has developed greatly over the last two decades and the electric power generated by ORC can be easily transformed into suitable form to be utilized in automotive electrical systems. There are standard topologies available for many converter circuits simply as modules from manufacturers [Rashid 2008].

The rapid development has been seen in the various equipment coming to the vehicles and the list is getting wider every year: electric tools, route navigators, driver management, music amplifiers etc., which all consume electricity. Therefore, different options are welcome for the additional electricity generation.

Vehicles are equipped with alternators, which convert part of the mechanical power of the engine into electricity. As ORC can produce up to 10 percent of the engine power, the simplest way to utilize it is to feed the power to the alternator. The ORC producing electrical power could also replace a portion of the electricity produced by the alternator.

In vehicles, hybrid systems have been developed to improve the energy efficiency. Thus, hybrid systems contain at least two power sources and they enable to combine the best properties of an internal combustion engine and an electric motor for various situations. Batteries are usually used as energy storages and they can be charged by generators included in the hybrid system. In plug-in hybrids, it is possible to charge the batteries from the electric grid as well. An electric machine can act as a motor to drive a vehicle or as a generator to charge batteries, and the latter function enables to convert braking energy into electricity. Because there is a need of electric power in hybrid systems, an ORC producing electric power provides an interesting opportunity to recover waste heat of the internal combustion engine. Thus, an ORC generator acts as an additional energy source in the hybrid system which improves fuel economy [El Chammas et al. 2005].

ECONOMICS

When evaluating the economics of coupling waste heat recovery systems into automotive applications different point of views can be identified. One point of view is related to increasing fuel prices, and thus the possibility to improve the fuel economy by increasing the power output of the system by utilizing waste heat would be beneficial. In this case the justification for investing to the waste heat recovery equipment can be evaluated based on the payback time of the investment. Another point of view that can justify the use of waste heat recovery is related to emission regulations and specific emissions of the system. [Sprouse and Depcik 2013]

The current estimation of the specific investment prices of commercial ORC power plants vary from 1000 €/kW$_e$ to 4000 €/kW$_e$. In the case of automotive exhaust recovery the specific price and operational costs of the system would be highly dependent on the selected technological solutions and the size of the production series and modularity.

Despite the large number of the research papers and the development of the ORC components, developments related to maintenance-free processes, reduced size of the heat exchangers, technological maturity and operational reliability of small scale waste heat recovery systems will still play the major role in the future. These developments are still needed in order to widely adopt ORC systems to utilize waste heat streams in automotive applications.

CONCLUSION

Waste heat recovery technologies can be identified as potential future technologies in improving the fuel economy of diesel and gasoline engines excluding the improvements in in-cylinder or turbocharging techniques. One of the most promising and most studied waste heat recovery technologies is related to ORC processes that are based on Rankine processes using organic substances as working fluids. The use of ORC process enables to increase the system power output by converting a portion of the waste heat into usable energy.

Several different technological paths and different working fluids have been studied and can be identified to be potential in realizing small ORC systems. The major differences are related to the choice to produce either mechanical or electrical power with the ORC system. Another significant difference can be found to be related on the selection of the expander type, mainly the choice of using volumetric expanders or systems based on turbine technology.

The automotive applications set further challenges for the ORC process components size and weight since the space is restricted in vehicles which sets further challenges especially when designing the heat exchangers. In addition, automotive applications have a highly dynamic behaviour which requires good off-design performance for the ORC and control strategies with a quick response.

Despite the large number of the research papers and development work related to the ORC components, the improvements related to maintenance-free processes, reduced-size heat exchangers, technological maturity as well as the operational reliability of small scale waste heat recovery systems, are still required and will play the major role in the future development work related to ORC technology.

REFERENCES

Bronicki, L. (1988). Experience with high speed organic Rankine cycle turbomachinery. In *Proceedings of Conference on High Speed Technology*, 21–24 August 1988, Lappeenranta, Finland. Lappeenranta University of Technology, Department of Energy Technology, Publ. No ENTE D-15, 47–61.

Daccord, R., Kientz, T. & Melis, J. (2013). Automotive Heat Recovery with Piston Expanders and Wet Fluids. ASME ORC 2013 *2nd International Seminar on ORC Power Systems*. October 7-8, 2013, Rotterdam, The Netherlands

El Chammas, R. & Clodic, D. (2005). Combined Cycle for Hybrid Vehicles. *SAE Int. Publication*, 2005-01-1171.

Espinosa, N., Tilman, L., Lemort, V., Quoilin, S. & Lombard, B. (2010). Rankine cycle for waste heat recovery on commercial trucks: approach, constraints and modelling, Presented at the *Diesel International Conference and Exhibition*, France, 2010.

Guillaume, L., Legros, A., Quoilin, S., Declaye, S. & Lemort, V. (2013). Sizing Models and Performance Analysis of Waste Heat Recovery Organic Rankine Cycles for Heavy Duty Trucks. *ASME ORC 2013 2nd International Seminar on ORC Power Systems*. October 7-8, 2013, Rotterdam, The Netherlands

Hung, T. C. (2001). Waste heat recovery of organic Rankine cycle using dry fluids. *Energy Conversion and Management*, *42*(5), 539-553.

Larjola, J. (1995). Electricity from industrial waste heat using high-speed organic Rankine cycle (ORC). *International journal of production economics*, *41*(1), 227-235.

Larjola, J. (2011). Organic Rankine Cycle (ORC) Based Waste Heat/Waste Fuel Recovery Systems for Small Combined Heat and Power (CHP) Applications. In: *Small and micro combined heat and power (CHP) systems, Advanced Design, Performance, Materials and Applications*, Editor Robert Beith. Woodhead Publishing, 206-232. ISBN 978-1-84569-795-2.

Nelson, C. (2008). Exhaust Energy Recovery. In: *2008 DEER Conference*, August 3rd, 2008.

Persico, G., Pini, M., Dossena, V. & Gaetani, P. (2013). Aerodynamic Design and Analysis of Centrifugal Turbine Cascades. *In Proceedings of ASME Turbo Expo*, San Antonio, Texas, USA, 2013, GT2013-95770.

Quoilin, S. & Lemort, V. (2009). Technological and Economical Survey of Organic Rankine Cycle Systems. *In proceedings of 5th European Conference on, Economics and Management of Energy in Industry*, April 14-17 2009, Villamoura, Portugal.

Quoilin, S., Declaye, S., Tchanche, B. F. & Lemort, V. (2011). Thermo-economic optimization of waste heat recovery Organic Rankine Cycles. *Applied Thermal Engineering*, *31*(14), 2885-2893.

Rashid M. (2008). Power Electronics For Alternative Energy Sources. *Power Electronics Congress, CIEP*.

Saleh, B., Koglbauer, G., Wendland, M. & Fischer, J. (2007). Working fluids for low-temperature organic Rankine cycles. *Energy*, *32*(7), 1210-1221.

Seher, D., Lengenfelder T., Gerhardt J., Eisenmenger, N., Hackner, M. & Krinn, I. (2012). Waste Heat Recovery for Commercial Vehicles with a Rankine Process. *21st Aachen Colloquium Automobile and Engine Technology*.

Sprouse, C. & Depcik, C. (2013). Review of organic Rankine cycles for internal combustion engine exhaust waste heat recovery, *Applied thermal engineering*, *51*, 711-722

Wang, E. H., Zhang, H. G., Fan, B. Y., Ouyang, M. G., Zhao, Y. & Mu, Q. H. (2011). Study of working fluid selection of organic Rankine cycle (ORC) for engine waste heat recovery. *Energy*, *36*(5), 3406-3418.

Wang, Z. Q., Zhou, N. J., Guo, J. & Wang, X. Y. (2012). Fluid selection and parametric optimization of organic Rankine cycle using low temperature waste heat. *Energy*, *40*(1), 107-115.

In: Automotive Exhaust Emissions and Energy Recovery ISBN: 978-1-63321-493-4
Editor: Apostolos Pesiridis

Chapter 10

AUTOMOTIVE EXHAUST POWER AND WASTE HEAT RECOVERY TECHNOLOGIES

Srithar Rajoo[1]*, Alessandro Romagnoli[2], Ricardo Martinez-Botas[3], Apostolos Pesiridis[4], Colin Copeland[5] and A. M. I. Bin Mamat[6]

[1]Transport Research Alliance, Universiti Teknologi, Malaysia
[2]Energy Research Institute at Nanyang Technology University (ERI@N), Singapore
[3]Department of Mechanical Engineering, Imperial College London, UK
[4]College of Engineering, Design and Physical Sciences, Brunel University London, UK
[5]Department of Mechanical Engineering, University of Bath, UK
[6]Faculty of Mechanical Engineering, University of Technology MARA, Malaysia

ABSTRACT

Internal combustion engines (ICE) almost entirely responsible to power the land and sea transportation needs for the past many decades. Amid the increasing awareness on environmental impact and volatility of fossil fuel dependence, internal combustion engines are envisioned to dominate way into the future. Looking ahead to the year 2020, many new technologies which encompass alternative engines and fuels may have increasing share of the transportation and industrial markets, however major proportion of prime movers will be powered by advanced internal combustion engines. The United States Department of Energy (DOE) stated that improving the efficiency of internal combustion engines is the most promising and cost-effective approach to increasing fuel economy of vehicle over the next 30 years. They put forward the targeted efficiency improvement of 25-40% and 20% for passenger and commercial vehicles respectively. Tackling the waste heat of exhaust gas will be one of the most significant technological efforts in achieving the required tall order efficiency improvements. Approximately 20 – 45% of the fuel energy into an internal combustion engine is wasted in the exhaust, which could be enhanced for system level efficiency improvement. DOE has

* Transport Research Alliance, Universiti Teknologi Malaysia. Email: srithar@fkm.utm.my.

demonstrated via a 2nd law estimate for exhaust gas recovery systems, that, peak brake thermal efficiency of a 1.9L diesel engine can be pushed from 41% to 44.5% - 53.6%, an improvement of approximately 8.5% - 30.7%. Thus, exhaust waste heat recovery holds great potential of improvement in engine brake thermal efficiency. There are some key technologies which have been tested and proven to recover exhaust heat and turn it into useful energy such as electricity. These are, among others, Thermoelectric Generator (TEG), Bottoming cycles and Turbocompounding (TC). The main objective of this chapter is to critically review available waste heat recovery technologies based on latest developments, research trends and economic feasibility. In addition, this chapter will also discuss and provide scientific-cum-economic assessment on groundbreaking and disruptive technologies to recover exhaust waste heat energy.

Keywords: Internal combustion engine, turbocharging and turbocharger, waste heat recovery

INTRODUCTION

Since the inception of modern mass transport, the Internal Combustion Engine (ICE) has been largely responsible in powering the world land and sea transportation needs. In addition, despite the increasing awareness of the environmental impact and the volatility of fossil fuels, industry experts predict the combustion engine will maintain its dominance in the near future. Looking ahead to 2020, many new technologies which are currently classed as 'alternative' may have an increasing share of the transportation market. According to Dr. Thomas Weber of Mercedes-Benz, at least 90% of vehicles will be powered by advanced technology internal combustion engines by 2020 [1]. Intensified research and development is thus essential to improve the efficiency of the internal combustion engine in order to meet the increasingly stringent demands that are to be placed on future vehicles. This is clearly demonstrated in the significant investment of most major engine and vehicle manufacturers in measures to improve fuel economy. The United States Department of Energy, through its Vehicle Technologies Program, stated that improving the efficiency of the internal combustion engine is the most promising and cost-effective approach to increasing fuel economy of road vehicles over the next 30 years. They put forward the targeted efficiency improvement of 25-40% for passenger vehicles and 20% for commercial vehicles by 2014 [2]. This chapter gives an outlook on the research and development activities that aim to deliver significant efficiency improvements to vehicle internal combustion engines using waste heat recovery technologies.

The internal combustion engine converts a relatively low percentage of fuel energy into useable work. Figure 1 shows the energy distribution from a typical internal combustion engine with a range of peak energy conversion efficiencies between 25% for gasoline, spark ignition engines and 40% for large diesel engines. It is therefore clear that approximately two-third of the fuel energy is typically lost as heat to the environment. If some of this 'wasted' heat could be recovered and converted into mechanical or electrical energy, this could supplement the vehicles' energy demand and improve overall system efficiency.

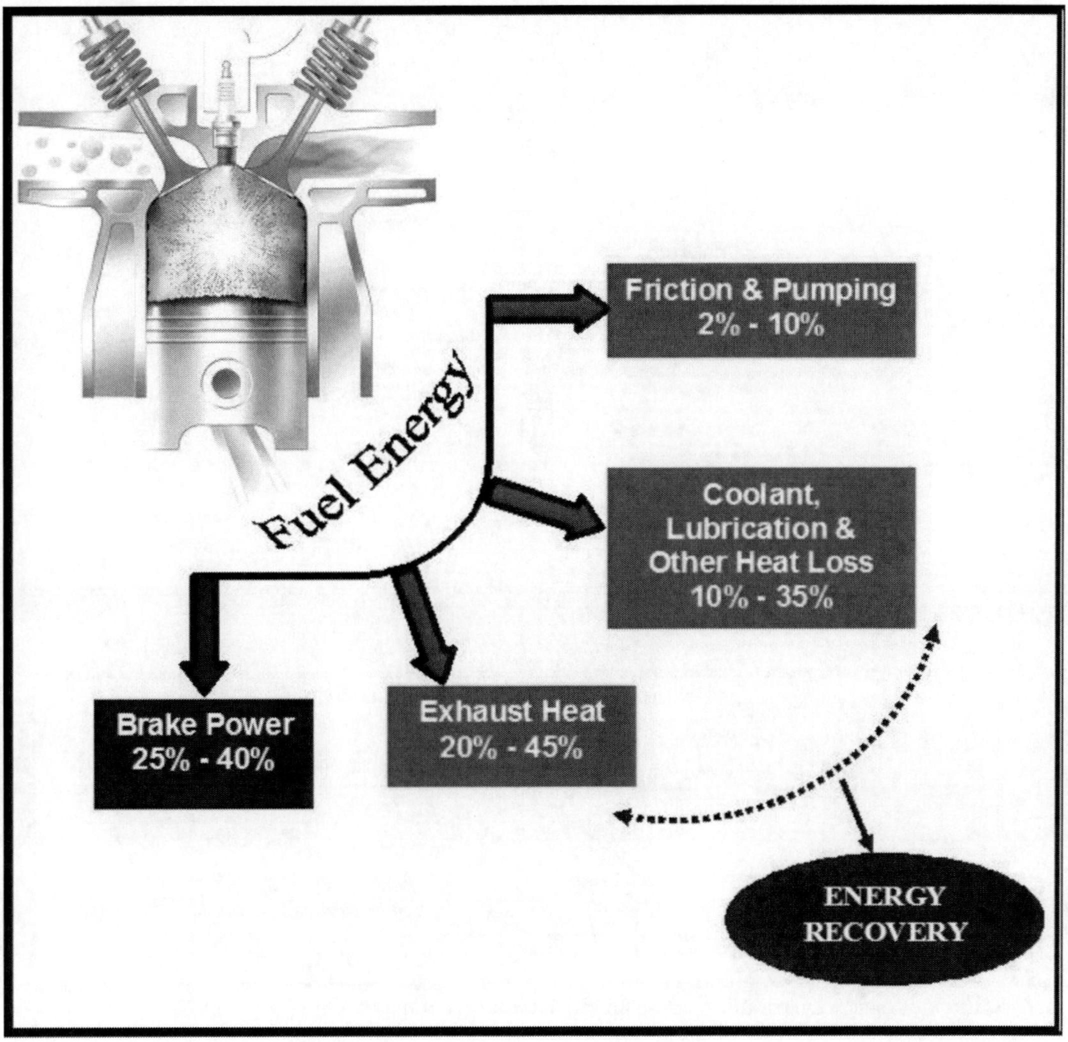

Figure 1.Typical Energy Distribution in an Internal Combustion Engine.

Recovering energy from the heat that is rejected from an internal combustion engine is one viable solution to improve its efficiency. A portion of the fuel energy is also lost to friction in transmitting the mechanical energy from the engine to the vehicle wheels. As a result, Figure 2 shows that according to United States Department of Energy, only 15% of fuel energy is used for positive work to run a vehicle and its auxiliary components [3]. Thus, apart from thermal energy, some forms of mechanical energy can also be recovered in order to raise the fuel efficiency of the vehicle. One such example is the recent development of the Kinetic Energy Recovery System (KERS) that aims to reclaim the kinetic energy of the vehicle that would otherwise be lost to friction. Automotive KERS systems are an example of regenerative braking, where the energy used to slow a vehicle's speed is stored rather than dissipated at the wheels. Regenerative braking has proven ideal for hybrid and electric vehicles since the electrical machine can be switched to generating mode to slow the vehicle and store the reclaimed potential energy within the available battery pack.

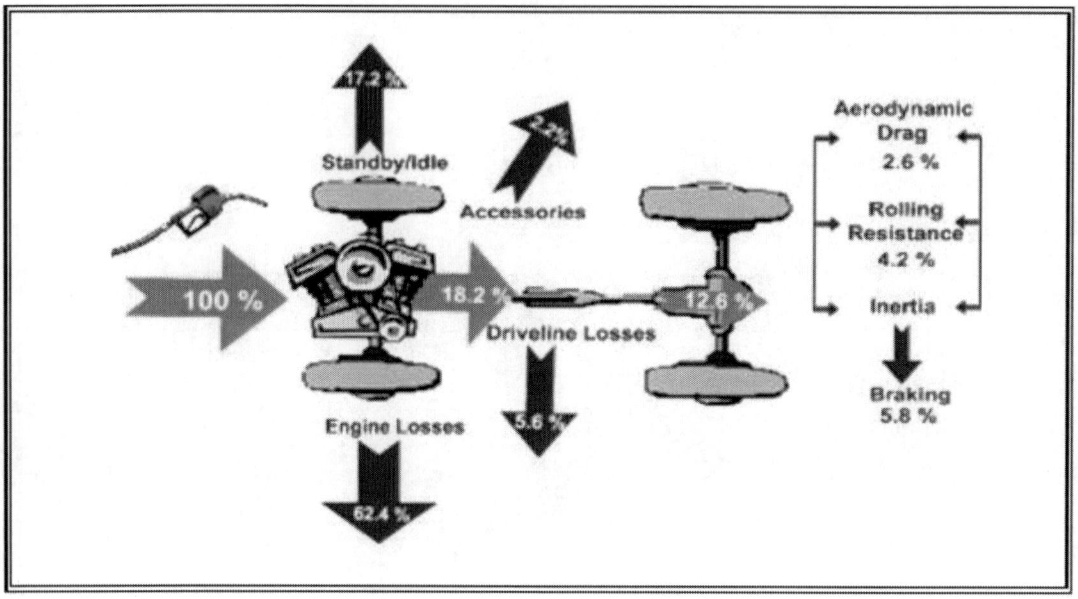

Figure 2. Fuel Energy Losses in a Vehicle [3].

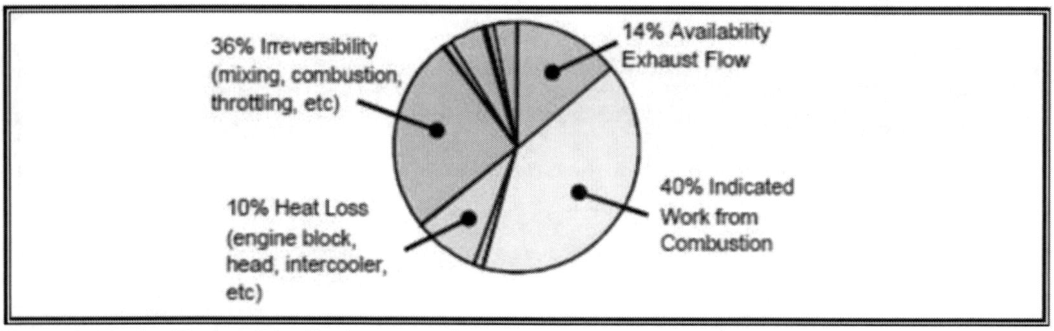

Figure 3. Engine Energy Distribution according to Thermodynamic 2nd Law [4].

Referring to Figure 2, another large source of loss is the energy wasted during standby or idle operation. To address this, a number of engine manufacturers employ engine Stop/Start to limit this wasted energy. In this strategy, the engine will stop during the period where the vehicle is stationary and start when the driver removes the braking signal. The challenge is to develop a strategy which can start the engine as quickly and as unobtrusively as possible.

Thermal energy recovery systems for internal combustion engines have not yet been employed widely in the transportation sector. The main exception is the adoption of different forms of heat recovery in large diesel engines which power ships and power plants where the added weight and expense of such interventions can be justified. Thus, its application for passenger and commercial sectors is yet to become common-place. There is therefore ample opportunity to use heat recovery technology much more widely in order to increase engine efficiency and consequently lower fuel consumption. However, further research is needed in order to ensure the additional technology meets the competing requirements of cost, engine durability and emissions.

TECHNOLOGY OVERVIEW

Most thermal energy recovery technologies seek to utilize high temperature combustion exhaust gases which carry approximately half of the rejected heat energy from an engine. The other half of the heat is rejected through liquid coolant and lubrication oil as well as through charge-air heat exchangers. It is also useful to note here that although it has been stated that approximately two-thirds of the fuel energy is converted into 'waste' heat, much of this rejected thermal energy is the result of the limitations of the thermodynamic cycle of a heat engine. Thus, it is impossible to reclaim all of this heat energy for useful work without contravening the laws of thermodynamics.

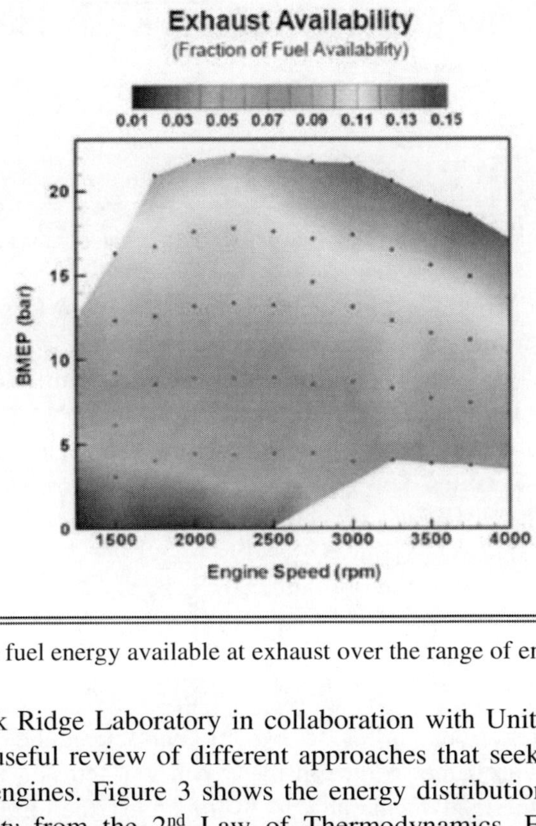

Figure 4. Fraction of the fuel energy available at exhaust over the range of engine operations [4].

The National Oak Ridge Laboratory in collaboration with United States Department of Energy conducted a useful review of different approaches that seek to recover energy from internal combustion engines. Figure 3 shows the energy distribution in an engine using the concept of *availability* from the 2nd Law of Thermodynamics. From this diagram, it is possible to recover approximately 14% of the exhaust gas energy for useful work. However, this value is likely associated with the peak or full-load operation. Thus, Figure 4 is provided to demonstrate the fraction of fuel energy available for heat recovery over the range of engine loads as indicated by the Brake Mean Effective Pressure (BMEP). This demonstrates that there are real opportunities to apply energy recovery technologies with the ultimate aim to significantly raise the efficiency of the internal combustion engine.

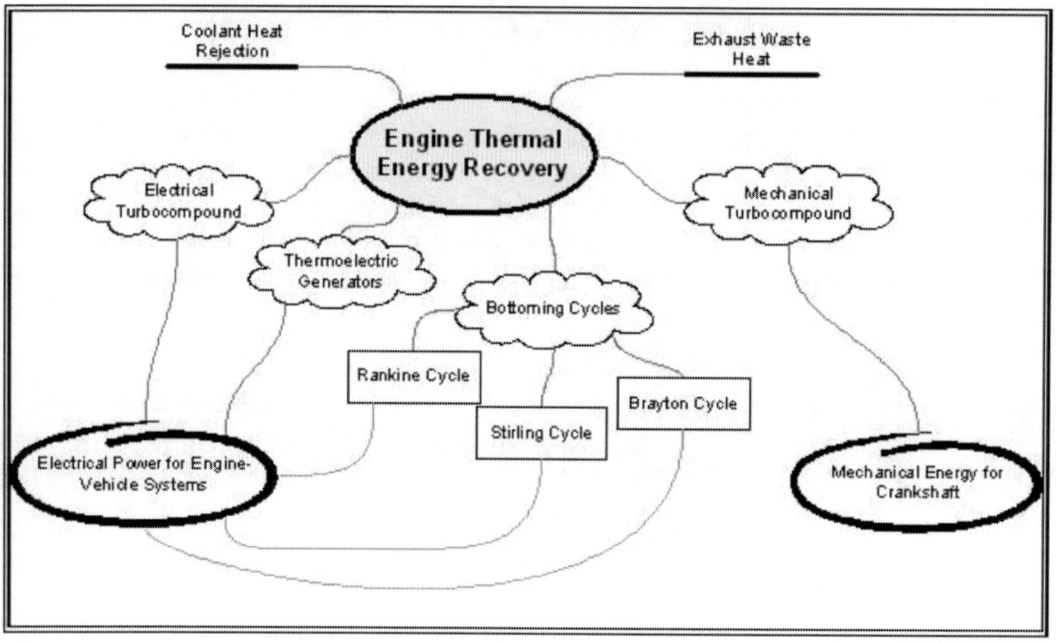

Figure 5. Various Methods of Thermal Energy Recovery from Internal Combustion Engines [5].

Ricardo PLC. UK prepared a report for the United Kingdom Department of Transport that reviews low carbon technologies for heavy goods vehicles. In this report, waste heat recovery is emphasised as one of the approaches that is able to significantly reduce the carbon impact of future vehicles. Figure 5 shows a useful map of the approaches that are currently available that aim to recover heat energy from an engine. These technologies are not only meant for heavy goods vehicles but are also applicable across a wide range of the transportation industry. The different approaches and technologies discussed in this chapter are largely based on those shown in Figure 5.

Mechanical Turbocompound

Turbo-compounding relies on a turbine to extracts energy from the hot exhaust gases that exit an internal combustion engine. Although the expansion across a turbine does create a temperature drop, unlike thermodynamic heat recovery cycles, the turbine relies on pressure remaining from the combustion process in order to operate. Thus, it could be argued that a turbine does not truly recover heat in the same way that a bottoming cycle does. The turbine can be mechanically linked to the engine crankshaft in order to re-use the extracted power directly. For an engine already fitted with a turbocharger, the turbo-compounding turbine is typically placed downstream to make use of the remaining exhaust gas enthalpy. Since the technique is based on mature turbomachinery technology and thus nearer to market, it can be more cost effective than other approaches. Research has also shown this approach can have a fuel economy benefit of between 3-5%. However, the complexity of mechanically coupling a high-speed turbine to the engine crankshaft has meant that that this technique has traditionally only been applied to large diesel engines where slower speed turbomachinery and the cost-to-benefit equation work in favour of larger engine displacements.

Electric Turbocompounding

As the name implies, electric turbo-compounding is similar to the mechanical system except that the extracted energy is converted into electrical power by coupling the turbine to a high speed electrical generator. If the turbine-generator unit is made separate from the turbocharger, it is referred to as a power turbine. In some cases however, the electrical generator is built as an integral part of the turbocharger, in which case it is referred to as a turbo-generator. Turbo-generators are challenging devices to engineer since the design must address heat transfer and rotor-dynamic issues while still producing electrical power efficiently. Whatever approach is taken, the electrical power produced from the exhaust gas flow can be used directly to power vehicle components or stored in a battery pack. In the case of a turbo-generator, the electric machine can also act as motor in order to return power to the system during engine transient events. This technology has demonstrated fuel economy benefits of between 3% and 10% depending on the application. This system should also be simpler to apply to small-to-medium sized automotive engines since is not limited by a mechanical connection to the drivetrain. It also potentially matches well to the current trend towards increasing electrification in the automotive market.

Bottoming Cycles

This technology involves utilizing thermodynamic cycles to extract heat energy from the exhaust gases. These cycles centre on a heat exchanger which exchanges waste energy from the internal combustion engine to a working fluid that is used to drive a power turbine. Rankine and Brayton cycles form the basis for most approaches although the Stirling cycle has also been used in some marine engine applications. The Rankine cycle is widely used in powerplants around the world wherever electricity is produced from a steam turbine coupled to electric generator. The Brayton cycle is most commonly used in jet engines, where a coupled turbine and compressor system extracts energy from hot, pressurized gases produced within a combustor. Research has demonstrated that bottoming cycles can produce an improvement in fuel economy of between 3-6% when applied to an internal combustion engine. However, it is widely acknowledged that packaging all the necessary components of a bottoming cycle into a vehicle can be a significant challenge. Accepting this as a limitation, this technology is currently most suitable for heavy duty applications such as marine power-plant where packaging and weight are less of a challenge.

The Stirling cycle operates using an external temperature gradient and potentially delivers a higher thermal efficiency than an internal combustion engine. The cyclic expansion and contraction of gas within the hot and cold sources of the cycle produce shaft power thus making it suitable for energy recovery purposes. However, the Stirling cycle suffers from a relatively low power density which makes it harder to package and justify the added weight within a vehicle. Considering this, the Stirling cycle has primarily been used for heat energy recovery in the large marine and stationary power plant applications. Interestingly, there are also examples where the Stirling cycle is used as the 'prime mover' in place of an internal combustion engine [6]. As it is an external combustion engine, any type of fuel whether it is a solid, liquid or gas can be used to produce the heat that drives the cycle and does so with a better thermal efficiency compared to an internal combustion engine.

Thermoelectric Generators

Thermoelectric generators convert exhaust heat directly to electricity using the *Seeback effect* whereby electricity is generated in between two metals or semiconductors due to a temperature difference. Applying the thermoelectric effect to large scale heat recovery is still largely in the research sphere and hence can be more costly more than all the other approaches. This can be attributed to the novel technology and materials that are employed to produce electricity efficiently. Research has demonstrated that the technology could yield an improvement in fuel efficiency of approximately 2%.

SYSTEM SPECIFICS

Three of the most widely accepted technologies for exhaust energy recovery are Organic Rankine Cycles (ORCs), Thermoelectric Generators and Turbo-compounding. Some of these techniques are already well established outside of the automotive sector and thus the increasing demand for greater fuel efficiency has encouraged a transfer of heat recovery technologies to automotive internal combustion engines.

Organic Rankine Cycles (ORC): One of the most important benefits of the Rankine cycle for automobile applications is that it offers an increased thermal efficiency without an increased engine exhaust back-pressure. Since an ORC does not require an expansion of exhaust gases to generate power, it does not obstruct the flow through the engine and therefore does not increase pumping loss. In fact, the working principle of Rankine cycles in an automotive application is essentially identical to a steam driven power plant. The exhaust gases leaving the engine are passed through a heat exchanger (evaporator) which absorbs the thermal energy which drives the cycle. As a result of the thermal exchange between the organic fluid and the exhaust gases, the working fluid which enters the evaporator in liquid state leaves as superheated vapour. The high energy vapour is then expanded through a turbine that is physically connected to an electric generator thereby producing energy for electric ancillaries. After expanding through the turbine, the fluid is then directed through another heat exchanger (condenser) where the working fluid rejects any remaining heat in order to return to a fully condensed state ready to start a new cycle.

Hence the application of these three technologies to on-road and off-road vehicles is currently undergoing intensive research in order to find solutions to the significant challenges particular to each approach. Table 1 provides a comparison between the three technologies with an outline of the working principle as well as their advantages and disadvantages. While not exhaustive, this table serves as a quick reference for the current exhaust energy recovery systems with more detail provided in the forthcoming paragraphs.

Table 1. Comparative study of different exhausts energy recovery system

EXHAUSTS ENERGY RECOVERY SYSTEMS	WORKING PRINCIPLE	ADVANTAGE	DISADVANTAGE
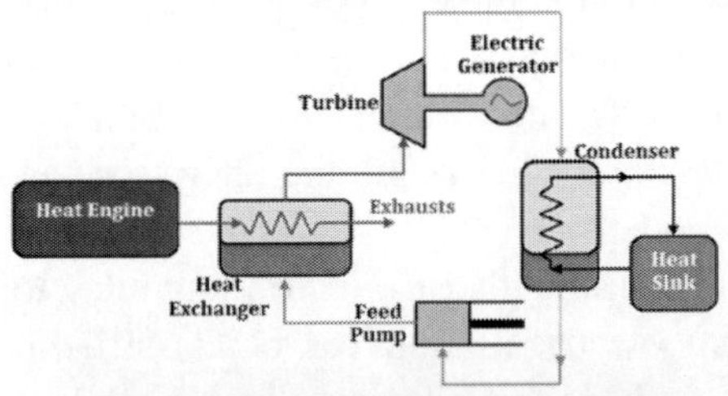	• In the heat exchanger, a highly compressed organic fluid absorbs exhausts energy • The organic fluid exits as superheated vapour and expands in the turbine • The remaining energy discharges to the surrounding in the condenser before it is compressed by a feed pump	• No increase in pumping work • Large benefit in fuel consumption	• Installation problem • Hazardous liquid • Cost ineffective
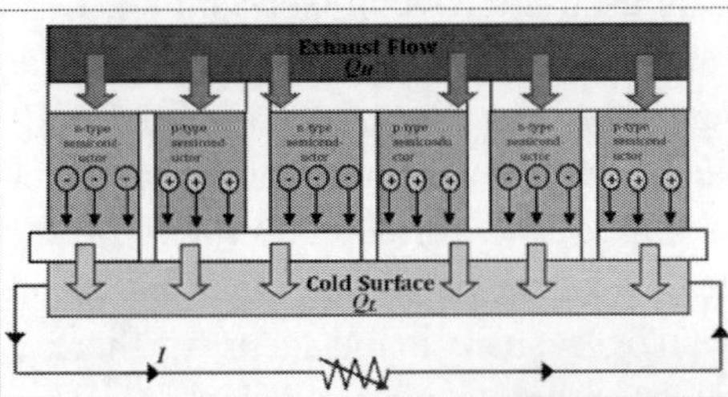	• Electricity is generated from the Seebeck effect • Highly doped n-type (excess electron) and p-type (defiance electrons) semiconductor place between the hot and cold surface • Temperature difference between the surface of exhausts pipe and thermoelectric material produces electric current	• Lightweight • Large variety of application areas • No increase in pumping work • High reduction of *BSFC*	• Large temperature difference between surfaces (extra load on the cooling system) • Expensive • Poor conversion efficiency
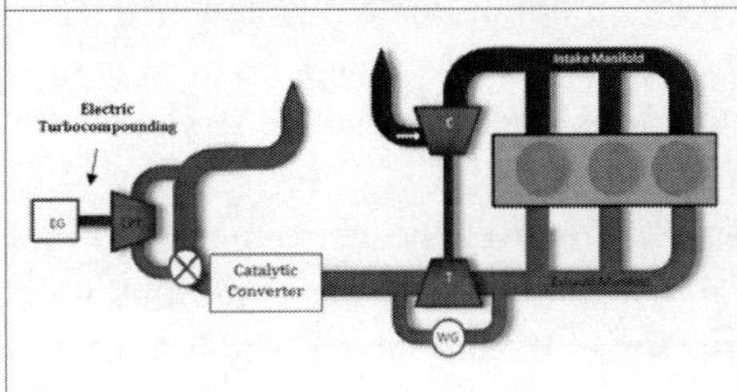	• An additional turbine is installed downstream of the main turbine • At the exit of the main turbine, the exhaust pressure is low • Secondary turbine has to recover energy at low pressure • Recovered energy is used to power electric auxiliaries or to mechanically add energy to the engine	• Bolt-on unit which can be added to the powertrain • Low mass flow capacity	• Increase in pumping work • Low energy available • Electric generator limitation

Research that applies the Rankine cycle to an internal combustion engine has been ongoing for more than thirty years [7-18]. There is also an exponential growth in investment in the area both financially and with respect to research and intellectual property generation. Close to one-thousand patent applications have been filed to date in this area including a number of notable patents granted of late. While the fundamental theoretical understanding has been present for some time, there remains very few examples of an ORC being applied within the on-road or off-road sectors. The most active industrial groups currently investing in research and development in the area are Cummins Engines, General Electric, BMW and United Technologies Corporation. The Cummins Supertruck programme, for example, co-funded by the U.S. Department of Energy (DOE) aimed to develop the next generation of tractor-trailer that included a waste heat recovery system that delivers a 6% fuel efficiency improvement. Another notable research program, the BMW Turbosteamer project, was based on a compact version of a two-stage (dual-cycle) system typically used in larger powerplants. The primary element is a high-temperature circuit which employs a heat exchanger to recover energy from the engine exhaust gases while the secondary circuit collects heat from the engine cooling system. Preliminary data using this system on a four-cylinder gasoline engine has demonstrated an efficiency improvement of 15%. On the other end of the size spectrum, General Electric and United Technologies Corporation main area of focus is the application of the organic Rankine cycles to stationary 'gen-set' engines.

Thermoelectric Generation: Similar to an organic Rankine cycle, thermoelectric generation does not increase exhaust back-pressure while recovering energy from the exhaust gases. The working principle and associated layout of a thermoelectric generator element is given in Table 1. The principle behind thermoelectric generation relies on the Peltier-Seebeck effect which describes the mechanism whereby thermal energy is converted into electrical energy. The electricity is produced from a temperature difference which causes a heat flow through the device thereby driving a diffusion of charge carriers in the thermoelectric semiconductors and creating an electric current.

Thermoelectric generators consist of three elements: the thermoelectric module, the support structure and heat exchangers/sinks. The thermoelectric module can be classified by its material, shape and size. The most common materials used are Silicon Germanium, Lead Telluride, Bismuth Telluride and new oxide materials such as $Na_xCo_2O_4$. The selection of the materials strongly depends on the range of operational temperatures it is designed for. In relation to the shape and size of the module, two basic configurations are currently considered viable: the traditional square and linearly shaped modules. The former is suited for flat surfaces whereas the latter is most suitable for circular exhaust pipes. The structure of the thermoelectric heat exchanger also plays a fundamental role since it must ensure there is maximum energy extraction over a short length and deliver an optimum conversion efficiency over a large range of exhaust gas temperatures. The use of a secondary heat exchange surfaces such as fins is potentially one of the best solutions but needs to be designed for minimal pressure drop to limit the exhaust backpressure [19].

The implementation of thermoelectric technology to recover heat from an internal combustion engine is shown to reduce fuel consumption without increasing emissions [20]. Nissan, BMW, Porsche and Jaguar Land Rover are some of the manufacturers in the automotive sector assessing thermoelectric generation as a possible route to improved system efficiency [21-25]. Despite the lightweight nature of the thermoelectric generator and the variety of applications in which they could be applied, the technology must address a number of issues such as costs, low efficiency (5% to 10%) and the large temperature difference required for adequate energy conversion. In particular, the latter challenge requires a cold junction temperature that is as low as possible. Thus, this typically involves a cooling mechanism that aims to create the maximum temperature gradient without consuming heat useful for the thermoelectric process [26].

Turbo-compounding: By adding a turbine downstream of the main turbocharger turbine, any available pressure can be used to recover some of the exhaust gas heat energy (refer to Table 1). This is a simple and well established technology which is easier to engineer compared to a bottoming cycle and typically less expensive than current thermoelectric generators. Despite an increase in engine pumping work due to presence of an additional turbine, an overall decrease in Brake Specific Fuel Consumption (BSFC) has been demonstrated with the addition of the technology [27-38].

Turbo-compounding can be divided into two main categories, mechanical and electrical. The former feeds the energy recovered from the exhaust gases directly into the engine crankshaft whereas the latter uses the recovered energy in order generate electricity which can be used for the vehicle electric ancillaries. A variation on electric turbo-compounding is the Turbo-generator or Electrically Assisted Turbocharger. This consists of an electric machine directly coupled to the shaft of the main turbocharger which can either be used as an electric generator to recover exhaust gas energy or as an electric motor to add energy to the turbocharger compressor during a transient manoeuvre.

Figure 6 provides a more detailed description of current turbo-compounding technology development. It is clear from this figure that most of the turbo-compounding units (Iveco, Scania, Caterpillar, Volvo, Detroit Diesel, John Deere and Bowman Power) have been developed to recover exhaust gas energy from large diesel engines. Controlled Power Technologies is one exception which has developed an electric turbo-compounding unit for small capacity internal combustion engines. This therefore suggests that there is significant potential in a viable turbo-compounding solution for wider adoption within the small to medium car segment.

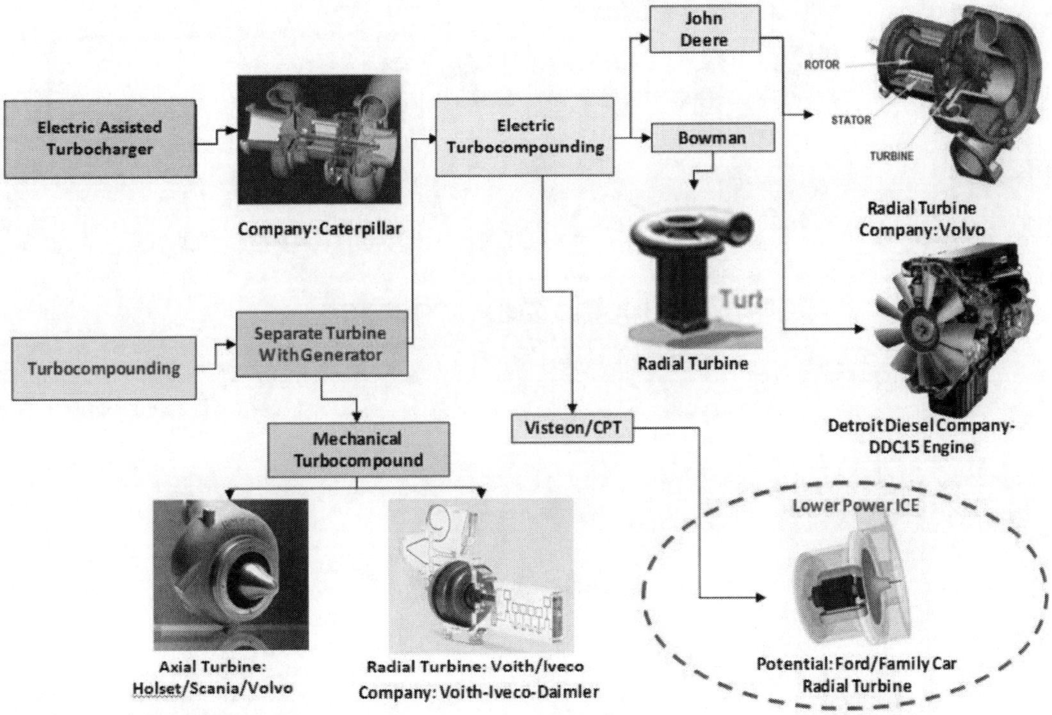

Figure 6. Family of turbocompounding.

One notable example of the application of turbo-compounding in the automotive sector is a recent development for C-segment passenger vehicles, namely, the HyBoost project (Hybridised Boosted Optimised System with Turbo-compounding). This ambitious research and development project aimed to deliver a 1.0 litre, four-cylinder, downsized gasoline engine rated at 116kW, 240Nm and offering a 35% potential reduction in fuel consumption and CO_2 emissions. That is, reducing the carbon emissions from 169.0 g/km to 99.7 g/km over the New European Driving Cycle while still matching the performance of an equivalent 2.0L engine (50% engine downsizing). The project focussed on technologies and systems available in the market and hence capable of practical implementation in the near term. This included micro-hybrid functionality with stop/start, a 200 Farad ultra-capacitor pack, an electric supercharger for transient lag mitigation, an electric turbo-compounding unit, an efficient liquid charge air cooler and the addition of advanced knock mitigation technologies. These technologies were synergistically integrated to achieve the required CO_2 emission and performance targets. In addition, further changes in the cam profile enabled a degree of Atkinson cycle operation and the engine compression ratio was increased to improved part load efficiency. This resulted in a combined 35% reduction in fuel consumption from engine downsizing alone. The HyBoost engine layout is shown in Figure 7 [39].

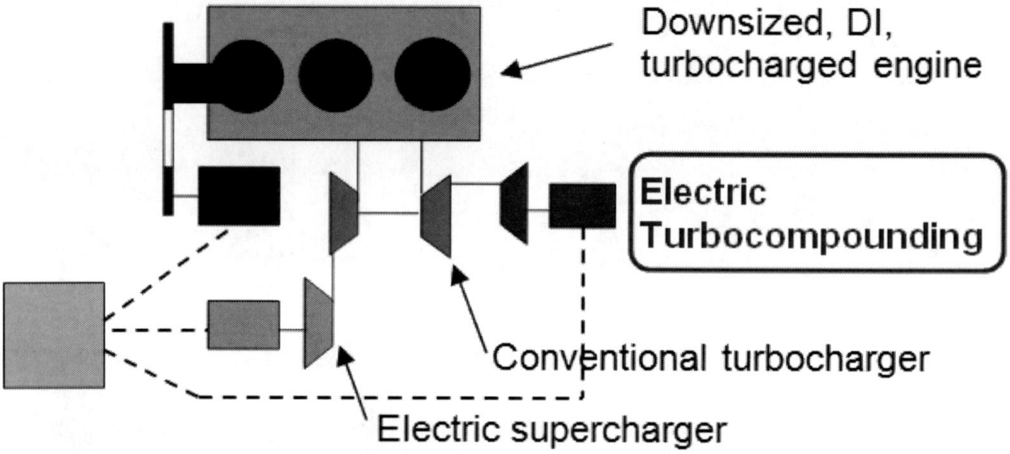

Figure 7. HyBoost engine layout.

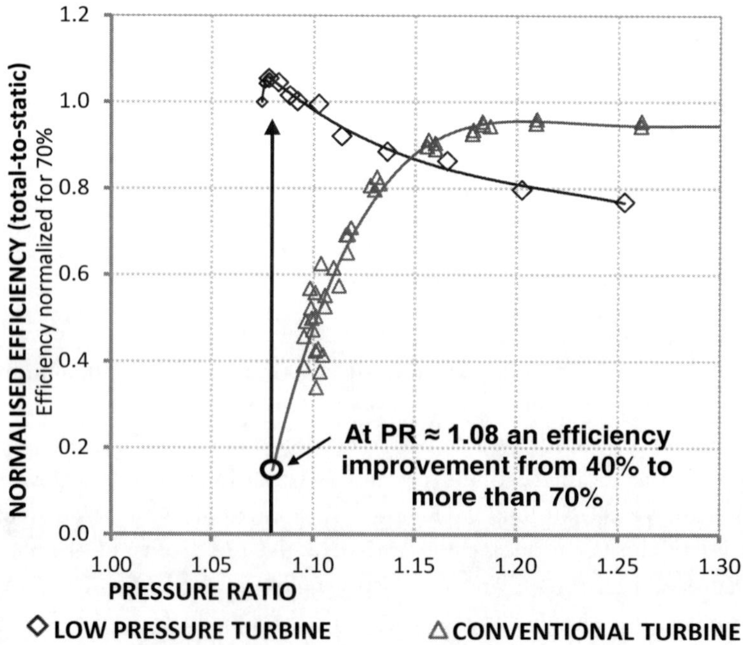

Figure 8. Comparison of normalized efficiency between LPT and conventional off-the-shelf turbines.

The development of a bespoke turbine for the electric turbo-compounding unit was one of the challenges that the HyBoost consortium needed to tackle. The application of turbo-compounding technology in small-to-medium sized passenger vehicles has always been perceived unfavourably from the automotive industry due to the increased back-pressure that an additional turbine may impose on the engine. Hence, the goal for the new turbine design was to design a highly efficient Low Pressure Turbine (LPT) that presents minimal flow restriction. The design specifications for the LPT were obtained from a validated 1-D engine model which showed that the pressure and temperature of the exhaust gases leaving the main turbocharger turbine spanned between 1.05bar and 1.2bar at 990K and 1170K respectively. Thus, the design point of the turbine was taken to be PR_{LPT}=1.1 and T_{LPT}=1100K. In addition, the limitation associated with the electric generator constrained the rotational speed and turbine power output at N_{LPT}=50000 rpm and W_{LPT}=1.0 kW respectively. The former limitation was considered a good compromise between bearing life and implementation costs whereas the latter was set to achieve a continuous power output during the drive cycle. The design process was accomplished by means of successive steps from mean-line analysis to 3-D Computer Aided Design (CAD) and Computational Fluid Dynamic (CFD) modelling. This resulted in a mixed-flow turbine design with 9 blades, a leading edge RMS wheel radius of 24.2 mm and blade length of 33.5 mm. A detailed description of the design process is given in references [40-41].

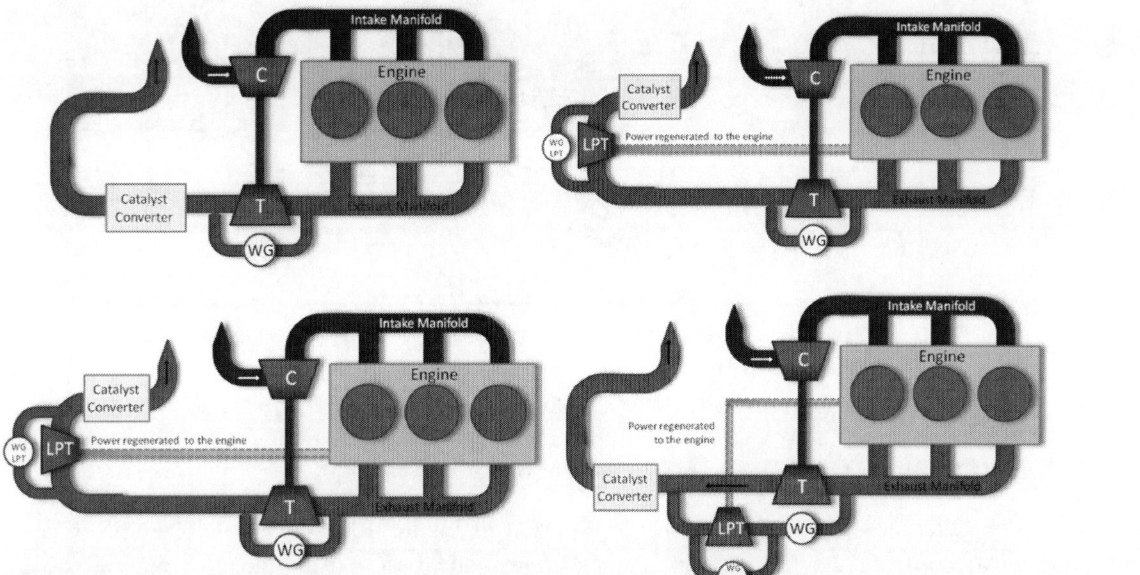

Figure 9. Positions of the Low Pressure Turbine (LPT) as compared with the baseline engine.

A prototype of the Low Pressure Turbine was developed and tested at Imperial College London. The results of testing are summarized in Figure 8 as normalized data where the performance of the LPT is compared with an equivalent conventional radial turbine designed to operate at higher pressure ratios. The figure clearly shows that the LPT operates with a very low expansion ratio between 1.07 and 1.3 with much greater efficiency compared to a conventionally designed turbine. Operating with a similar expansion ratio, a conventional turbine could only deliver an efficiency of approximately 40% compared to 70% isentropic efficiency from the LPT.

In order to assess the impact of the LPT on engine performance, a validated 1-D engine simulation was run under full load conditions for three different positions of the electric turbo-compounding. Figure 9 provides a schematic of these positions which include post-catalyst, pre-catalyst and within the wastegate path of the main turbocharger turbine. The turbo-compounding unit was modelled in 1-D software using the LPT experimental maps and by limiting the rotational speed and power output to 50,000 rpm and 1.0 kW respectively. Since no mathematical model was available to replicate the characteristics of the electric generator, the additional power generated by the turbo-compounding turbine was directly supplied to the engine crankshaft with the assumption of 100% mechanical efficiency.

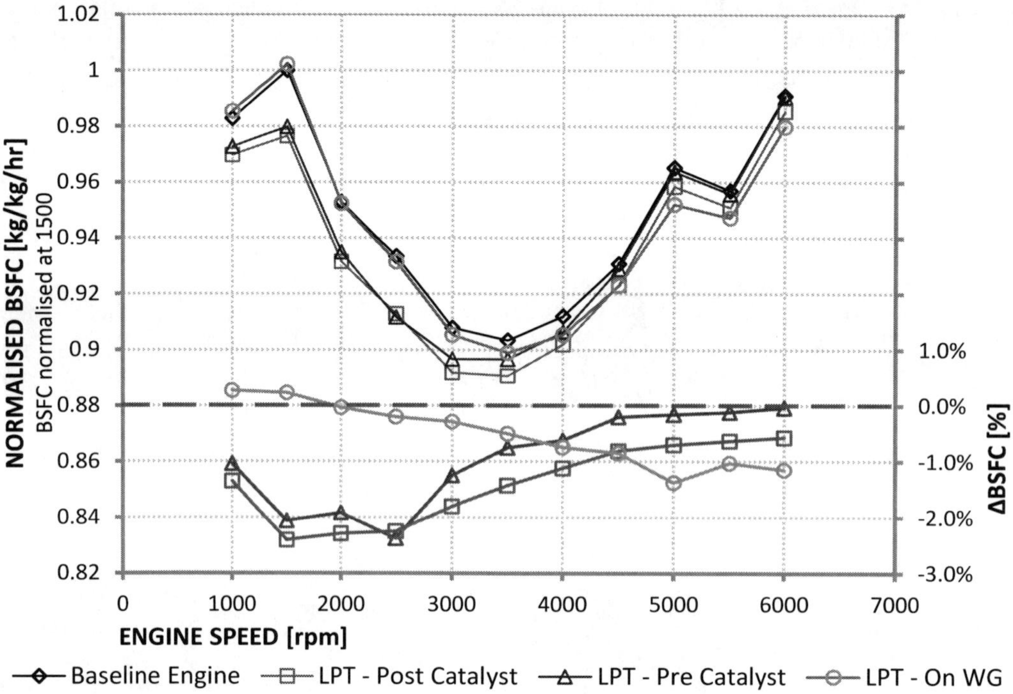

Figure 10. LPT impact on BSFC for full load engine operating conditions.

The outcomes of the simulation are reported in Figure 10 using normalised BSFC as a metric to compare the three arrangements against the baseline engine. For engine speeds lower than 2000 rpm, the pre and post-catalyst position exhibit similar impact on engine BSFC with a predicted 2% fuel economy benefit (ΔBSFC=-2.41% at 1500rpm). Nevertheless, as the engine speed increases the benefits associated with the turbo-compounding deteriorates at a faster pace in the pre-catalyst position compared to the post-catalyst position. This can be attributed to a greater back-pressure due to the position of the turbo-compounding unit which acts as a small but notable restriction to the flow through the engine. However, when the LPT is positioned in the wastegate path of the turbocharger, a 1% BSFC reduction is predicted at high engine rpm. This is almost twice as much as that provided by the turbo-compounding unit placed in the pre and post-catalyst positions. It is likely that this is related to the difference in exhaust backpressure imposed on the engine resulting from the positioning of the Low Pressure Turbine.

REFERENCES

[1] Excerpt from Automotive Engineering International, June 2009.

[2] Excerpt from Vehicle Technologies Program report, DOE/GO-102009-2762, March 2009

[3] http://www.fueleconomy.gov/FEG/atv.shtml ; last accessed 19th September 2009.

[4] Report on 'Achieving and Demonstrating Vehicle Technologies Engine Fuel Efficiency Milestones (ACE 16)', 2009 DOE Hydrogen Program and Vehicle Technologies Annual Merit Review.

[5] Review of low carbon technologies for heavy goods vehicles, RD.09/182601.6, June 2009.

[6] K. Hirata and M. Kawada, 'Discussion of Marine Stirling Engine Systems', Proceedings of the 7th International Symposium on Marine Engineering, Tokyo, October 24th to 28th, 2005.

[7] Leising, C. J., Purohit, G. P., DeGrey, S. P., and Finegold, J. G., "Waste Heat Recovery In Truck Engines," *West Coast Meeting*, Society of Automotive Engineers (SAE), Paper No: 780686, Warrendale, Pensylvania, USA, 1978.

[8] Miller, E. W., Hendricks, T. J., Wang, H., and Peterson, R. B., "Integrated dual-cycle energy recovery using thermoelectric conversion and an organic Rankine bottoming cycle," *Proceedings of the Institution of Mechanical Engineers, Part A: Journal of Power and Energy*, Vol. 225, No. 1, 2011, pp. 33-43.

[9] Larjola, J., "Electricity from industrial waste heat using high-speed organic Rankine cycle (ORC)," *International Journal of Production Economics*, Vol. 41, No. 1GÇô3, 1995, pp. 227-235.

[10] Yamada, N., Minami, T., and Anuar Mohamad, M. N., "Fundamental experiment of pumpless Rankine-type cycle for low-temperature heat recovery," *Energy*, Vol. 36, No. 2, 2011, pp. 1010-1017.

[11] Srinivasan, K. K., Mago, P. J., and Krishnan, S. R., "Analysis of exhaust waste heat recovery from a dual fuel low temperature combustion engine using an Organic Rankine Cycle," *Energy*, Vol. 35, No. 6, 2010, pp. 2387-2399.

[12] Peterson, R. B., Wang, H., and Herron, T., "Performance of a small-scale regenerative Rankine power cycle employing a scroll expander," *Proceedings of the Institution of Mechanical Engineers, Part A: Journal of Power and Energy*, Vol. 222, No. 3, 2008, pp. 271-282.

[13] Wang, H., Peterson, R. B., and Herron, T., "Experimental performance of a compliant scroll expander for an organic Rankine cycle," *Proceedings of the Institution of Mechanical Engineers, Part A: Journal of Power and Energy*, Vol. 223, No. 7, 2009, pp. 863-872.

[14] Wang, H., Peterson, R., Harada, K., Miller, E., Ingram-Goble, R., Fisher, L., Yih, J., and Ward, C., "Performance of a combined organic Rankine cycle and vapor

compression cycle for heat activated cooling," *Energy*, Vol. 36, No. 1, 2011, pp. 447-458.

[15] Harinck, J., Turunen-Saaresti, T., Colonna, P., Rebay, S., and Buiktenen, J. V., "Computational Study of a High-Expansion Ratio Radial Organic Rankine Cycle Turbine Stator," *Journal of Engineering for Gas Turbines and Power*, Vol. 132, No. 5, 2010.

[16] Miller, E. W., Pettersson, R., and Hendricks, T. J., "Modeling Energy Recovery Using Thermoelectric Conversion Integrated with an Organic Rankine Bottoming Cycle," *Journal of Electronic Materials*, Vol. 38, No. 7, 2009.

[17] Facao, J., Palmero-Marrero, A., and Oliveira, A. C., "Analysis of a solar assisted micro-cogeneration ORC system," *International Journal of Low-Carbon Technologies*, Vol. 3, No. 4, 2008, pp. 254-264.

[18] Schimpf, S., Uitz, K., and Span, R., "Simulation of a solar assisted combined heat pump-Organic Rankine Cycle-system," *World Renewable Energy Congress 2011*, Linköping University Electronic Press, Sweden, 2011.

[19] S. R. Jumade, V. W. Khond, "Survey on Waste Heat Recovery from Internal Combustion Engine Using Thermoelectric Technology", International Journal of Engineering Research & Technology (IJERT), Vol. 1 Issue 10, December- 2012, ISSN: 2278-0181

[20] Fairbanks JW. The 60 percent efficient diesel engine; probable, possible, or just a fantasy. 2005 Diesel Engine Emissions Reduction (DEER) Conference Presentations, U.S Department of Energy, Chicago, Illinois, USA, 2005.

[21] Ikoma, K., et al."Thermoelectric Module and Generator for Gasoline Engine Vehicle". Proc. 17th International Conference on Thermoelectrics. 1998, Nagoya, Japan

[22] Shinohara, K., et al."Application of Thermoelectric Generator for Automobile". Journal of the Japan Society of Powder and Powder Metallurgy

[23] Ikoma, K., et al."Thermoelectric Generator for Gasoline Engine Vehicles Using Bi2Te3 Modules". J. Japan Inst. Metals. Special Issue on Thermoelectric Energy.

[24] Serksnis, A.W. "Thermoelectric Generator for Automotive Charging System". Proc. 11th Intersociety Conversion Engineering Conference. 1976, New York, USA, pp. 1614-1618

[25] http://www.powerdriver.info/

[26] Converting Waste Heat from Automobiles to Electrical Energy, Feasibilty and potential of thermoelectric conversion, *Megha Tak, Gandhinagar, Mohit Setia,* 2012 IEEE 7th International Power Electronics and Motion Control Conference - ECCE Asia June 2-5, 2012, Harbin, China

[27] Kapich D. Turbo-hydraulic engine exhaust power recovery system. SAE Powertrain & Fluid Systems Conference & Exhibition, San Diego, USA, 2002.

[28] Ryder O, Sharp N. The impact of future engine and vehicle drivetrains on turbocharging system architacture. 9th International Conference on Turbochargers and Turbocharging, Institution fo Mechanical Engineers, London, UK, 2010, pp. 1-10.

[29] Bin Mamat AMI, Padzillah MH, Romagnoli A, Martinez-Botas, R. A High performance low pressure ratio for engine electric turbocompounding. ASME Turbo Expo 2011: Power for Land, Sea and Air, ASME, Vancouver, Canada, 2011.

[30] Bin Mamat AMI, Romagnoli A, Martinez-Botas, R. Design and development of Low pressure turbine for turbocompounding applications. International Gas Turbine Congress, Gas Turbine Society of Japan, Osaka, Japan, 2011.

[31] Ono Y, Kimura M, Shiraishi K, Teshima T. Approaches to energy conservation for the environment through met turbochargers. 9th International Conference on Turbochargers and Turbocharging, Institution of Mechanical Engineers, London, UK, 2010, pp. 23-31.

[32] Thompson IGM, Spence SW, McCartan CD, Talbot-Weiss JMH. Design, validation, and performance result of a turbocharged turbogenerating biogas engine model. 9th International Conference on Turbochargers and Turbocharging, Charlesworth Group, Wakerfeild, London, UK, 2010, pp. 11-22.

[33] Millo F, Mallamo F, Pautasso E, Mego GG. The potential of electric exhaust gas turbocharging for hd diesel engines. SAE 2006 World Congress & Exhibition, SAE International, Warrendale, USA, 2006.

[34] Hopmann U, Algrain MC. Diesel engine electric turbo compound technology. Future Transportation Technology Conference & Exhibition, SAE International, Warrendale, USA, 2003.

[35] Hountalas DT, Katsanos CO, Lamaris VT. Recovering energy from the diesel engine exhaust using mechanical and electrical turbocompounding. SAE Technical Papers-Australasia, SAE INTERNATIONAL, 2007, pp. 65-81.

[36] Darlington TE, Frank AA. Exhaust gas driven generator with altitude compensation for battery dominant hybrid electric vehicles. Fall Technical Conference of the ASME International Combustion Engine Division, October 24, 2004 - October 27, 2004, American Society of Mechanical Engineers, Long Beach, CA, United states, 2004, pp. 1-10.

[37] Chris B. Application of a variable drive to supercharger & turbo compounder applications. SAE TECHNICAL PAPER SERIES, 2009.

[38] Brands MC, Wener JR, Hoehne JL, Kramer S. Vehicle testing of cummins turbocompound diesel engine. SAE International Congress and Exposition, Society of Automotive Engineers (SAE), Detroit, Michigan, 1981.

[39] King. J., Heaney. M., Bower J., Jackson. N., Owen. N., Saward. J., Fraser. F., Morris. G., Blore. P., Chang. T., Borges-Alejo. Criddle. M., *"HyBoost – An intelligently electrified optimised downsized gasoline engine concept"*, Proceedings of the Institution of Mechanical Engineers: 10th International Conference of Turbochargers and Turbocharging, pp. 3-14, 2012.

[40] Bin Mamat AMI, Padzillah MH, Romagnoli A, Martinez-Botas, R. A High performance low pressure ratio for engine electric turbocompounding. ASME Turbo Expo 2011: Power for Land, Sea and Air, ASME, Vancouver, Canada, 2011.

[41] Bin Mamat AMI, Romagnoli A, Martinez-Botas, R. Design and development of Low pressure turbine for turbocompounding applications. International Gas Turbine Congress, Gas Turbine Society of Japan, Osaka, Japan, 2011.

INDEX

C

E

F

J

K

L

M

N

O

P

S

T

U

V

W

Y

Z